Physical Chemistry and its Applications

Physical Chemistry and its Applications

Edited by
Jamie Langdon

C WILLFORD PRESS

www.willfordpress.com

Published by Willford Press,
118-35 Queens Blvd., Suite 400,
Forest Hills, NY 11375, USA

ISBN: 978-1-68285-791-5

Cataloging-in-Publication Data

Physical chemistry and its applications / edited by Jamie Langdon.
    p. cm.
Includes bibliographical references and index.
ISBN 978-1-68285-791-5
1. Chemistry, Physical and theoretical. 2. Chemistry. I. Langdon, Jamie.
QD453.3 .P49 2020
541--dc23

For information on all Willford Press publications
visit our website at www.willfordpress.com

WILLFORD PRESS

Contents

Preface

Physical Chemistry is the study of macroscopic, atomic, particulate and sub-atomic phenomena in chemical systems. It is studied in terms of principles, practices, and concepts of physics. Energy, thermodynamics, motion, statistical mechanics and force are some common concepts of physics applied to this field. Quantum chemistry and spectroscopy are the two primary sub-fields of physical chemistry. The application of quantum mechanics to solve chemical problems falls under quantum chemistry. It provides tools to determine the shape of bonds, strength of the bonds, movement of nuclei and how light can be absorbed or emitted by a chemical compound. Whereas, spectroscopy is concerned with the interaction of electromagnetic radiation with matter. The objective of this book is to give a general view of the different areas of physical chemistry and its applications. Most of the topics introduced herein cover new techniques and applications of physical chemistry. This book will serve as a valuable source of reference for graduate and post graduate students.

The information contained in this book is the result of intensive hard work done by researchers in this field. All due efforts have been made to make this book serve as a complete guiding source for students and researchers. The topics in this book have been comprehensively explained to help readers understand the growing trends in the field.

I would like to thank the entire group of writers who made sincere efforts in this book and my family who supported me in my efforts of working on this book. I take this opportunity to thank all those who have been a guiding force throughout my life.

Editor

Application of Food-Grade Proteolytic Enzyme for the Hydrolysis of Regenerated Silk Fibroin from *Bombyx mori*

Ji An Joung⃝, **Mi Na Park**⃝, **Ji Young You**⃝, **Bong Joon Song**⃝, and **Joon Ho Choi**⃝

Department of Food Science and Biotechnology, Wonkwang University, Iksan 54538, Republic of Korea

Correspondence should be addressed to Joon Ho Choi; jhchoi1124@wku.ac.kr

Guest Editor: Sang-Hyun Park

In vitro biodegradation of *Bombyx mori* silk fibroin (SF) was studied using food-grade proteolytic enzymes to replace acid hydrolysis. Based on the residual protein quantity and yield of amino acids (AAs) after enzymatic hydrolysis, we evaluated the proteolytic enzyme process of SF. FoodPro and Alcalase that are classified as alkaline proteases are selected as two of the best candidate enzymes for hydrolysis of SF. The activity of these enzymes exhibits a broad range of pH (6.5 to 9.0) and temperature (50°C to 65°C). The single enzyme treatment of SF using FoodPro exhibited a hydrolytic efficiency of 20–25%, and >2 g/L AAs were released after reaction for 3 h. A 2-stage enzymatic treatment using a combination of FoodPro and Flavourzyme in a sequence for a reaction time of 6 h was developed to enhance the efficiency of the proteolytic process. The yield of AAs and residual protein quantity in the enzymatic hydrolysates obtained from FoodPro-treated regenerated SF in the 1st step was 2,040 ± 23.7 mg/L and 70.6%, respectively. The yield of AAs was > two-fold (4,519 ± 42.1 mg/L), whereas the residual protein quantity decreased to 55.1% after the Flavourzyme treatment (2nd step) compared to those of the single FoodPro treatment. In the mixed treatment by simultaneously using FoodPro and Flavourzyme, approximately 45% of SF was degraded and 4.5 g/L of AAs were released within 3 h of reaction time. The regenerated SF and its enzymatic hydrolysates were characterized by performing UV-visible spectra, gel electrophoresis, and size-exclusion chromatography analyses. In the 2-stage treatment using FoodPro initially and subsequently Flavourzyme, the aggregates and high molecular weight proteins of SF were dissociated and degraded into the low molecular weight proteins/peptides (10–15 kDa and 27 kDa). SF hydrolysates as functional food might be enzymatically produced using the commercial food-grade proteolytic enzymes.

1. Introduction

The silk fibroin (SF) from *Bombyx mori* (silkworm) cocoons present as a double-stranded fibroin fiber that is coated with adhesive sericin proteins. The raw silk fiber is composed of 20–30% sericin and 70–80% SF with trace amounts of waxes and carbohydrates [1]. Pure SF is separated from sericin by the degumming procedure [2], and a variety of aqueous or organic solvent processing methods are used to generate silk biomaterials for a wide range of applications [3–6]. The predominant protein of silk, i.e., fibroin, is a hydrophobic structural protein that is insoluble in hot water and consists of heavy and light chains with 390 kDa and 25 kDa molecular weights (MW), respectively, linked by a disulfide bond [7, 8]. The SF heavy chain is rich in hydrophobic β-sheets that form

blocks linked by small hydrophilic linkers or spacers. The predominant regions of SF primarily consist of approximately 76% of amino acids (AAs) with nonpolar side chains. The main AAs of SF are glycine (43.7%) and alanine (28.8%) that form the dominant crystalline β-sheet regions that act as reinforcements and contribute to the strength and stiffness of silk [9, 10]. Although silk is defined as a nondegradable material by the United States Pharmacopeia owing to a negligible loss in tensile strength *in vivo*, the enzymatic degradation behaviors of SF as biomaterials have been reviewed for the medical application *in vivo* [11, 12].

The *in vitro* biodegradation of *B. mori* SF was studied using proteolytic enzymes (collagenase F, α-chymotrypsin I-S, and protease type XXI) to degrade SF fibers and films [13]. The enzymatic hydrolysis of SF was performed using

α-chymotrypsin, collagenase IA, and protease XIV in SF yarns [14], protease XIV in porous SF sheets [15–17], collagenase IA and protease XXIII in 3D-scaffolds SF [18–20], proteinase K, protease XIV, α-chymotrypsin, collagenase, and matrix metalloproteinase in SF hydrogels and films [21], and actinase in SF solution [22]. Due to the unique chemical, mechanical, and biocompatibility properties, *B. mori* SF was investigated as a biomaterial and functional food source for years. Fibroin-derived bioactive proteins/peptides and hydrolysates were developed as the functional components of foods, cosmetics, pharmaceutical preparations, and alternative sources of food additives to improve human health [23]. The applications of SF in processed foods were proposed, and currently, the three forms of SF (solution, gel, and powder) are commercially available [24]. The SF hydrolysates that are prepared by performing acid hydrolysis or enzymatic hydrolysis consist of a mixture of AAs and oligopeptides that are known to exhibit beneficial effects on animal models. The bioactive peptides derived from SF are reported to reduce blood cholesterol level, increase antigenotoxicity, inhibit angiotensin-converting enzyme activity, enhance insulin sensitivity and glucose metabolism, improve alcohol metabolism, and stimulate osteoblastic differentiation [23, 25–28].

Fibroinase is a native silk digestion enzyme characterized as a cysteine proteinase and digests fibroin and sericin [29, 30]. Alcalase [25–27], actinase [22], and alkaline protease [31] were used to prepare the bioactive peptides and enzymatic hydrolysates (EHs) from SF/silkworm powder. The fibroin solubilization conditions affected the MW distribution of EHs produced using Alcalase supplied from Novozymes [32]. Additionally, actinase was introduced to hydrolyze regenerated silk fibroin (RSF) at 37°C for 12 h [22]. The silkworm powder was modified by enzymatic treatment using alkaline protease (FoodPro), and the MW of silk protein was decreased [31]. To prepare SF hydrolysates, enzymatic hydrolysis is more efficient than acid hydrolysis with aspect to the recovery yield and quality of products. However, very few previous studies reported the production of SF hydrolysates using commercial food-grade enzymes. Therefore, we performed the enzymatic hydrolysis of SF using commercial food-grade enzymes and characterized these EHs.

2. Materials and Methods

2.1. Materials.
Degummed SF fiber acquired from a Korean sericulture farm (Kimjae, South Korea) was provided by SSBIO PHARM Co., Ltd. (Cheonan, South Korea). Commercial food-grade proteolytic enzymes, FoodPro® and Alphalase® (DuPont Industrial Biosciences, Brabrand, Denmark), Collupulin™ MG (DSM Food Specialties, Alexander Fleminglaan, Netherlands), Alcalase®, Neutrase®, Protamex®, and Flavourzyme® (Novozymes A/S, Bagsvaerd, Denmark), Promod™ 192P, 279MDP, and 278MDP (Biocatalysts Ltd., Wales, UK), and Bromelain (PT Bromelain Enzyme, Lampung, Indonesia) were purchased and stored at 4°C until their use in experimental reactions. Anhydrous citrate and Na_2HPO_4 to prepare citrate-phosphate buffer

(CPB) [33], $CaCl_2$ and 95% ethanol to prepare SF solubilization reagent, and other chemicals were of reagent-grade and purchased from Samchun Pure Chemical Co., Ltd. (Gyeonggi-do, South Korea). Standard protein, i.e., albumin from bovine serum (BSA) and Folin and Ciocalteu's phenol reagent to perform quantification of protein concentration and ninhydrin reagent and glycine to analyze the AAs were purchased from Sigma-Aldrich (St. Louis, MO, USA). The Precision Plus Protein™ standards and Silver Stain Plus™ kit to perform SDS-PAGE and silver staining, respectively, were purchased from BIO-RAD Laboratories (Hercules, CA, USA).

2.2. Preparation of RSF Solution.
Degummed SF fiber was dissolved in Ajisawa's reagent [34] composed of $CaCl_2$: ethanol : distilled water in $1:2:8$ molar ratio at 121°C for 15 min, thus yielding a 20% (W/V) solution. This solution was dialyzed against deionized water using a cellulose membrane dialysis tube (MW cut-off 14,000 kDa, Sigma-Aldrich) for >3 days to remove excess salt and ethanol. The RSF solution was centrifuged to remove insoluble materials, and final concentration was determined as 130–180 g/L using the Lowry method [35] with 73–76% recovery yield. RSF solutions at various pH levels were prepared using CPB. The pH-conditioned RSF solutions were stored at room temperature for 48 h to achieve homogeneity. The premature precipitates in RSF solutions were removed by centrifugation at 13,000 rpm for 10 min (Smart-R17, Hanil Scientific Inc., Gimpo, South Korea).

2.3. Estimation of Specific Activities of Proteolytic Enzymes.
The proteolytic activity of enzymes was determined using casein as a substrate by performing Sigma's nonspecific protease activity assay [36] with a minor modification. Casein solution (5 mL; 0.65% (W/V)) was mixed with the enzyme solutions (1 mL) of various concentrations. After 10 min incubation at 37°C, 0.4 M trichloroacetic acid (5 mL) was added to terminate the reaction. The reaction mixture was incubated at 37°C for 30 min and centrifuged to remove insoluble precipitates. The supernatant (1 mL) was mixed with 0.55 M sodium carbonate (5 mL) and subsequently 0.5 mM Folin and Ciocalteu's phenol reagent (1 mL). After 30 min incubation at 37°C, the absorbance at 660 nm was measured using a spectrophotometer (Optizen POP, Mecasys Co., Daejeon, South Korea). Based on the standard curve obtained using L-tyrosine, the proteolytic activity was determined in terms of units (U), that is, the quantity of tyrosine equivalents in micromoles released from casein per min. The specific pH of each enzymatic reaction was regulated using 0.25 M CPB at corresponding pH. The protein concentrations of proteolytic enzyme solutions were determined using the Lowry method. The proteolytic activity of enzymes was expressed as specific activity in terms of U/mg-protein.

2.4. Proteolytic Hydrolysis of RSF.
Enzymatic reaction was performed using RSF solution along with food-grade

enzymes at specific conditions. Each sample acquired from the reaction mixture was boiled at 100°C for 10 min to inactivate the enzyme and centrifuged at 13,000 rpm for 10 min to remove the insoluble precipitate. The supernatant of each sample was used to measure the protein and AAs concentrations by performing the Lowry method and ninhydrin reaction, respectively [37]. The protein and AAs concentrations were expressed in terms of BSA and glycine equivalent, respectively. The protein concentration of SF solution used for enzymatic reaction was approximately 30 g/L. The efficiency of protein degradation by enzymatic hydrolysis was expressed as the percentage of remaining protein using the following equation: residual protein (%) = $(P_f/P_i) \times 100$ where P_i is the initial concentration of the sample and P_f is the protein concentration of reaction mixture at a specific condition. To screen the proteolytic enzymes to perform SF hydrolysis, each enzyme (6 U) was incubated with 330 mg of RSF for 3 h. The effects of temperature and pH on the reactions involving the candidate enzymes (FoodPro and Alcalase) were determined by performing a combination reaction using enzyme (5 U) and SF (400 mg) for 5 h. The effect of substrate-to-enzyme ratio was determined using the initially selected enzyme, i.e., FoodPro, and an enzyme concentration of 5 U was used during the subsequent enzymatic treatment experiments. To enhance the efficiency of enzymatic hydrolysis by performing the 2-stage treatment, SF was initially allowed to react with FoodPro and then with any other selected proteolytic enzyme.

2.5. Gel Electrophoresis. The MW of SF and its EHs were determined by performing sodium dodecyl sulfate polyacrylamide gel electrophoresis (SDS-PAGE) using 4–20% Mini-PROTEAN® TGX™ precast protein gel (BIO-RAD). The total protein samples of RSF, initial EHs obtained using FoodPro, and final EHs obtained using Flavourzyme were loaded at protein concentration values of 28–84, 19–57, and 15–45 μg/well, respectively. The samples were resolved in SDS-PAGE with Precision Plus Protein standards as MW markers and stained using a Silver Stain Plus kit (BIO-RAD).

2.6. Size-Exclusion Chromatography (SEC). The high-performance liquid chromatography system combined with a photodiode array detector (Atlus™, PerkinElmer, Waltham, MA, USA) and 5Diol-300-II size-exclusion column (COSMOSIL, 7.5 mm ID × 300 mm, 300 Å pore size, Nacalai Tesque, Inc., Kyoto, Japan) was used to perform SEC. The protein quantities used during analyses were 700 μg RSF, 250 μg EHs obtained after 1st step using FoodPro, and 200 μg EHs obtained after 2nd step using Flavourzyme. 0.25 M CPB (pH 6.0) was used as the mobile phase at a flow rate of 1 mL/min, and the eluted protein concentration was measured at 280 nm at 30°C.

2.7. Statistical Analysis. All the experiments were performed in triplicates, and results were represented as mean value ± standard deviation. Analysis of variance was conducted, and

the mean variations were analyzed using Duncan's multiple range test ($p < 0.05$). The statistical analyses were conducted using the Statistical Package for the Social Sciences (SPSS) for Windows 12.0 software (SPSS Inc., Chicago, IL, USA).

3. Results and Discussion

3.1. pH Dependency of RSF Solubility in Aqueous Solution. pH affects the charges of carboxyl as well as amino groups in RSF, enables electrostatic interactions, and promotes intramolecular or intermolecular interactions such as hydrophobic interactions and hydrogen bonding. The effect of pH on the RSF solubility in aqueous solution was determined using 0.25 M CPB solutions with pH range 3.0–9.0. The protein concentration of each sample equilibrated at a specific pH for 48 h was determined before and after centrifugation (Figure 1). The protein concentration in RSF solution prior to centrifugation was relatively stable (26.8 ± 0.81 g/L) at pH between 5.0 and 9.0. After centrifugation, the stable protein concentration (26.7 ± 0.14 g/L) was obtained at a narrow pH range 5.0–6.5. Some of the soluble RSF was coagulated in alkaline condition; therefore, the soluble protein concentration after centrifugation decreased at pH > 7. In pH < 5 condition, the soluble protein concentration after centrifugation decreased owing to colloidal precipitation. The stable protein concentration was maintained at pH between 5.0 and 6.5 that might be associated with the spinning process (decrease from 6.9 to 4.8 along the path) in the silk glands [38]. A similar effect of pH on RSF solubility was reported in a study that indicated pH dependency between the shear sensitivity of SF and β-sheet crystallization [39].

3.2. Screening of Candidate Proteolytic Enzyme to Hydrolyze RSF. The specific activities of food-grade proteolytic enzymes and results of enzymatic hydrolysis at particular pH and temperature conditions are summarized in Table 1. The reaction conditions such as pH and temperature of each enzyme were selected based on the manufacturer's recommendations. FoodPro and Alcalase are classified as endotype and alkaline proteases that exhibit higher specific activity values than other proteolytic enzymes. An equal unit of each enzyme was used to screen the candidate proteolytic enzyme to perform RSF hydrolysis. FoodPro, Alcalase, Protamex, and/or Alphalase were potential candidates to efficiently hydrolyze SF owing to the concentration of AA in the EHs which was >2,000 mg/L. Flavourzyme is classified as mixed endotype/exotype and was not selected as the SF started to precipitate after 1 h reaction time. With an increase in the AA concentrations in EHs, the residual protein (%) decreased to approximately 80% in Alcalase-, FoodPro-, Protamex-, and Alphalase-treated SF. The AA concentrations in EHs represent not only free AAs but also peptides because the ninhydrin reaction is used to determine free amino and amine groups in AAs, peptides, and proteins [40]. Alcalase and FoodPro were selected as candidate proteases to hydrolyze SF. These two enzymes are produced by *Bacillus* strains, known as serine proteases, and classified

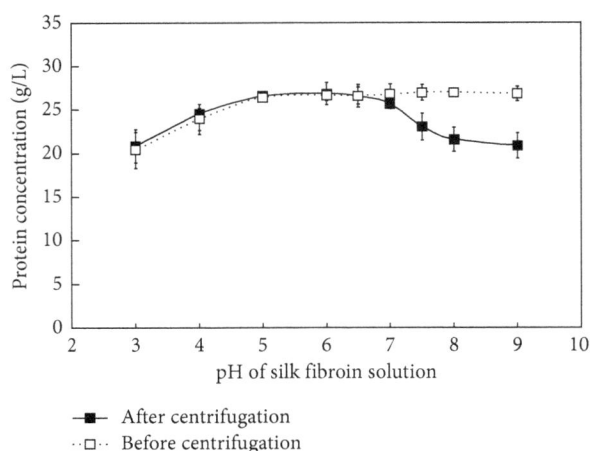

FIGURE 1: Effect of pH on the solubility of regenerated silk fibroin in 0.25 M citrate-phosphate buffer.

as alkaline proteases. Although fibroinase is the native silk-digesting enzyme, it was identified as a cysteine protease [29, 30]; Collupulin and Bromelain are classified as cysteine proteases, and they exhibited low activities in enzymatic hydrolysis of SF in our study.

3.3. Effects of Temperature, pH, and Substrate-to-Enzyme Ratio on RSF Hydrolysis.

The effect of temperature on the candidate enzymes was determined by performing the reactions at range 50°C–65°C and pH 8.0 (Table 2). The proteolytic activities of Alcalase and FoodPro were efficient at a broad temperature range between 50°C and 60°C and drastically decreased at >65°C. The optimal temperature to conduct RSF hydrolysis was determined as 60°C. The effect of pH on the efficiency of enzymatic hydrolysis of SF was determined at pH range 6.5–9.0 and 60°C (Table 3). The proteolytic activities of Alcalase and FoodPro were effective at a broad pH range 6.5–9.0, and the optimal pH was determined as 8.0. Finally, FoodPro was selected as the proteolytic enzyme to perform the 1st step of 2-stage RSF hydrolysis. To determine the efficiency of enzymatic hydrolysis, the substrate-to-enzyme ratio was estimated using a fixed quantity of RSF (400 mg) and FoodPro (1, 5, 10, or 25 U) at pH 8.0 and 60°C (Figure 2). A decrease in residual protein (%) and an increase in AAs of EHs were proportional to an increase in FoodPro concentration ranging from 1 to 10 U per RSF (400 mg).

3.4. Enhancement of Efficiency in Enzymatic Hydrolysis.

The enzymatic hydrolysis using FoodPro was almost saturated after 3 h of reaction and did not extend during a prolonged reaction time. To improve the efficiency of enzymatic hydrolysis in RSF, a 2nd enzyme (5 U) was introduced after 3 h of reaction initiation using FoodPro. The 2nd enzymatic treatment was conducted for additional 3 h of reaction time at specific conditions, and the results are summarized in Table 4. The AAs in the EHs obtained from 1st step FoodPro-treated RSF was 2,040 ± 23.7 mg/L, and it increased > two-fold (4,519 ± 42.1 mg/L) during the

subsequent Flavourzyme treatment (2nd step). The drastic increase in AA concentration in EHs might be associated to the characteristics of Flavourzyme that exhibits mixed activities of endotype and exotype enzymes. The residual protein was 70.6 ± 5.39% after FoodPro treatment (1st step), and it further decreased to 55.1 ± 5.28% during Flavourzyme treatment (2nd step). Therefore, RSF was efficiently hydrolyzed by the 2-stage enzymatic treatment by sequentially using FoodPro and Flavourzyme at the 1st and 2nd step, respectively, within 6 h of reaction time. To reduce the enzymatic reaction time, the AAs and protein yields of mixed proteolytic enzyme treatment by simultaneously using the combination of FoodPro and Flavourzyme were compared to those of the 2-stage enzymatic treatment (Figure 3). In the mixed proteolytic enzyme treatment, the enzymatic hydrolysis was conducted at 60°C and pH 7.0. The residual protein and AAs concentration in EHs during the mixed proteolytic enzyme treatment for 3 h were 54.4 ± 2.35% and 4,663 ± 23.0 mg/L, respectively. Additionally, enzymatic hydrolysis in the mixed proteolytic enzyme treatment was not extended at prolonged reaction time up to 6 h. Therefore, the RSF in aqueous solution was hydrolyzed by mixed proteolytic enzyme treatment which reduced the reaction time compared to that by the 2-stage enzymatic treatment by sequentially using FoodPro and Flavourzyme.

3.5. Characterization of RSF and Its EHs Using UV-Visible Spectra.

To characterize RSF and its EHs, UV-visible spectra analysis was performed using the same concentrations of samples, and results are presented in Figure 4. All the samples with 2.0 g/L concentration exhibited a characteristic protein absorption band in UV-visible spectra. The absorption intensities at 280 nm increased after enzymatic treatments, indicating that the hydrophobic and self-assembled regimes in RSF protein were initially relaxed, and subsequently the degradable protein/peptide content increased compared to those of untreated RSF. Moreover, similar changes were reported using alkaline protease-treated silkworm powder in UV-visible spectra analysis [31].

3.6. MW Determination of RSF and Its EHs Using SDS-PAGE.

The MWs of RSF and its EHs were analyzed by performing SDS-PAGE using 4–20% gradient gel (Figure 5). The RSF exhibited a broad MW distribution ranged from 60 to 150 kDa, and the protein bands were visible at approximately 10–22 kDa. The broad protein bands of RSF at approximately 100 kDa might be related to the intermolecular interaction between the SF molecules, and similar results were reported in previous reports [2, 31]. After enzymatic treatment using FoodPro in the 1st step, the broad protein bands of RSF disappeared and distinct protein bands at 350–420 kDa and 60–70 kDa MW were observed. In the EHs obtained from FoodPro-treated RSF, a novel protein band at 27 kDa was detected and the intensities of several bands at approximately 10–12 and 15 kDa were enhanced. In enzymatic treatment using FoodPro at the 1st step, the RSF aggregates were dissociated and gradually degraded into low MW proteins at

TABLE 1: The specific enzyme activities of food-grade enzymes and the results of enzymatic hydrolysis of regenerated silk fibroin during 3 h reaction.

Food-grade enzyme	Specific activity (U/mg)	Conditions		Enzymatic hydrolysis	
		pH	Temperature (°C)	Residual protein (%)	Amino acid (mg/L)
Alcalase® AF 2.4 L	14.0 ± 0.66	8.0	60	71.8 ± 1.07[b]	2,209 ± 37.7[ab]
Alphalase® NP	4.86 ± 0.18	7.0	60	84.4 ± 0.53[d]	2,074 ± 25.8[c]
Bromelain BR1200	0.10 ± 0.00	7.0	50	98.2 ± 2.39[h]	580 ± 87.4[f]
Collupulin™ MG	2.09 ± 0.02	6.0	60	92.4 ± 0.71[f]	1,228 ± 40.8[e]
Flavourzyme® 500 MG	2.38 ± 0.02	7.0	50	[(1)]54.0 ± 1.67[a]	2,011 ± 90.7[c]
FoodPro® alkaline protease	15.0 ± 0.80	8.0	60	78.0 ± 0.19[c]	2,271 ± 51.8[a]
Neutrase™ 0.8 L	1.72 ± 0.02	7.0	50	97.1 ± 0.16[g]	489 ± 69.6[f]
Promod™ 192P	[(2)]0.59 ± 0.01	5.0	50	NA	NA
Promod™ 278MDP	0.88 ± 0.01	7.0	60	90.8 ± 0.40[f]	1,461 ± 12.4[d]
Promod™ 279MDP	[(2)]2.62 ± 0.05	5.0	60	NA	NA
Protamex®	3.11 ± 0.03	7.0	50	81.8 ± 2.51[e]	2,118 ± 89.6[bc]

[(1)]The silk fibroin started to precipitate during the enzymatic reaction with Flavourzyme. [(2)]Enzymatic reaction was performed at pH 7 instead of pH 5 as the substrate precipitated at acidic pH 5. NA: enzymatic reaction was not observed as these enzymes are classified as exotype proteases. a–g: means in the same column followed by different letters differ significantly ($p < 0.05$).

TABLE 2: Effects of temperature on the enzymatic hydrolysis of regenerated silk fibroin using FoodPro and Alcalase at pH 8.0.

Enzyme		FoodPro		Alcalase	
Temperature (°C)	Time (h)	Residual protein (%)	Amino acid (mg/L)	Residual protein (%)	Amino acid (mg/L)
50	0	100.0 ± 1.16[d]	62 ± 1.5[h]	100.0 ± 1.14[f]	62 ± 1.5[h]
	3	86.1 ± 3.65[c]	2,456 ± 4.5[e]	86.9 ± 1.67[e]	2,417 ± 15.7[e]
	5	86.3 ± 0.60[c]	2,671 ± 19.5[a]	87.3 ± 2.97[e]	2,662 ± 26.2[c]
55	0	100.0 ± 1.16[d]	64 ± 1.6[h]	100.0 ± 1.14[f]	64 ± 1.6[h]
	3	80.5 ± 1.29[b]	2,441 ± 13.1[e]	84.5 ± 3.92[d]	2,426 ± 19.3[e]
	5	80.9 ± 1.09[b]	2,602 ± 30.6[c]	83.1 ± 3.40[c]	2,536 ± 38.7[d]
60	0	100.0 ± 1.13[d]	62 ± 1.5[h]	100.0 ± 1.10[f]	62 ± 1.5[h]
	3	77.2 ± 2.17[a]	2,537 ± 18.9[d]	75.8 ± 0.72[b]	2,769 ± 17.4[b]
	5	75.9 ± 2.41[a]	2,645 ± 4.5[b]	74.2 ± 0.37[a]	2,891 ± 23.4[a]
65	0	100.0 ± 1.14[d]	67 ± 1.6[h]	100.0 ± 1.13[f]	67 ± 1.6[h]
	3	98.0 ± 1.50[d]	1,843 ± 10.7[g]	100.5 ± 3.24[f]	1,787 ± 23.0[f]
	5	98.2 ± 1.12[d]	1,876 ± 5.4[f]	99.2 ± 2.93[f]	1,750 ± 4.7[g]

a–h: means in the same column followed by different letters differ significantly ($p < 0.05$).

TABLE 3: Effects of pH on the enzymatic hydrolysis of regenerated silk fibroin using FoodPro and Alcalase at 60 °C.

Enzyme		FoodPro		Alcalase	
pH	Time (h)	Residual protein (%)	Amino acids (mg/L)	Residual protein (%)	Amino acids (mg/L)
6.5	0	100.0 ± 0.92[c]	27 ± 0.7[i]	100.0 ± 2.55[c]	27 ± 0.7[f]
	3	84.9 ± 0.88[b]	1,857 ± 10.0[h]	86.4 ± 2.38[b]	1,826 ± 42.8[e]
	5	83.4 ± 4.87[b]	2,058 ± 13.2[g]	87.2 ± 2.53[b]	1,895 ± 2.4[d]
7.0	0	100.0 ± 0.49[c]	27 ± 0.7[i]	100.0 ± 1.62[c]	27 ± 0.7[f]
	3	80.6 ± 5.74[ab]	2,223 ± 2.4[f]	81.3 ± 0.69[a]	2,072 ± 16.4[c]
	5	80.7 ± 2.30[ab]	2,254 ± 19.0[e]	82.0 ± 2.45[a]	2,156 ± 55.7[c]
8.0	0	100.0 ± 1.92[c]	27 ± 0.7[i]	100.0 ± 1.83[c]	27 ± 0.7[f]
	3	79.7 ± 1.55[ab]	2,355 ± 10.0[d]	79.2 ± 2.59[a]	2,300 ± 50.8[b]
	5	77.4 ± 0.20[a]	2,515 ± 25.5[a]	79.0 ± 1.83[a]	2,435 ± 6.4[a]
9.0	0	100.0 ± 1.51[c]	27 ± 0.7[i]	100.0 ± 3.18[c]	27 ± 0.7[f]
	3	79.4 ± 4.45[ab]	2,372 ± 14.0[c]	79.5 ± 0.85[a]	2,291 ± 148[b]
	5	76.2 ± 1.50[a]	2,442 ± 32.7[b]	80.2 ± 1.24[a]	2,397 ± 29.1[a]

a–i: means in the same column followed by different letters differ significantly ($p < 0.05$).

10–27 kDa range. After Flavourzyme treatment (2nd step), the protein bands at 420 kDa of EHs obtained from FoodPro-treated RSF (1st step) gradually disappeared and a novel band at 315 kDa was detected. The intensities of bands at 10–12, 15, and 27 kDa were enhanced after Flavourzyme treatment (2nd step). Therefore, the SDS-

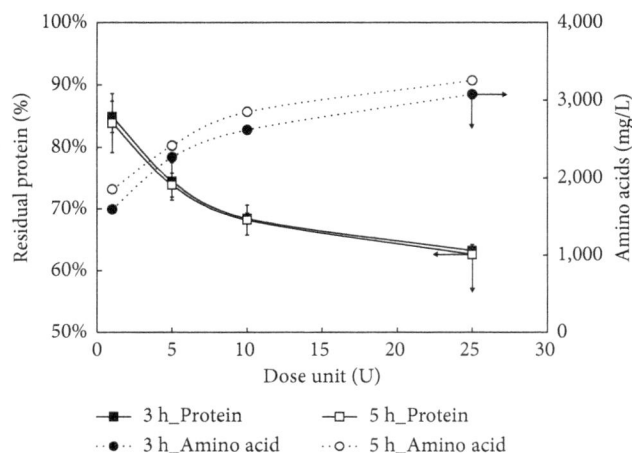

FIGURE 2: Effect of substrate (fibroin) : enzyme ratio on the proteolytic activity of FoodPro in 0.25 M citrate-phosphate buffer.

TABLE 4: Selection of the second proteolytic enzyme to enhance the efficiency of enzymatic hydrolysis of FoodPro-treated silk fibroin.

Enzyme treatment	Stage	Reaction condition			Residual protein (%)	Amino acids (mg/L)
		Time (h)	Temperature (°C)	pH		
Silk fibroin	—	0	60	—	100.0 ± 1.54^{c}	17 ± 0.7^{j}
1st step: FoodPro (FP)	1	3	60	8.0	70.6 ± 3.81^{b}	$2,040 \pm 23.7^{i}$
FP ⟶ Alcalase	2	+3	60	8.0	65.5 ± 5.67^{b}	$2,611 \pm 22.0^{d}$
FP ⟶ Alphalase	2	+3	60	7.0	65.6 ± 6.83^{b}	$2,505 \pm 28.7^{e}$
FP ⟶ Bromelain	2	+3	50	7.0	68.7 ± 9.73^{b}	$2,455 \pm 16.2^{fg}$
FP ⟶ Collupulin	2	+3	60	6.0	68.5 ± 8.32^{b}	$2,227 \pm 4.7^{h}$
2nd step FP ⟶ Flavourzyme	2	+3	50	7.0	55.1 ± 2.91^{a}	$4,519 \pm 42.1^{a}$
FP ⟶ Neutrase	2	+3	50	7.0	65.4 ± 8.68^{b}	$2,453 \pm 16.2^{fg}$
FP ⟶ Pomod 192	2	+3	50	5.0	67.1 ± 7.39^{b}	$2,689 \pm 54.7^{c}$
FP ⟶ Promod 278	2	+3	60	7.0	67.7 ± 7.61^{b}	$2,438 \pm 35.7^{g}$
FP ⟶ Promod 279	2	+3	60	5.0	65.0 ± 7.38^{b}	$2,746 \pm 18.7^{b}$
FP ⟶ Protamex	2	+3	50	7.0	66.6 ± 6.11^{b}	$2,453 \pm 25.7^{f}$

+3: additional reaction time. a–j: means in the same column followed by different letters differ significantly ($p < 0.05$).

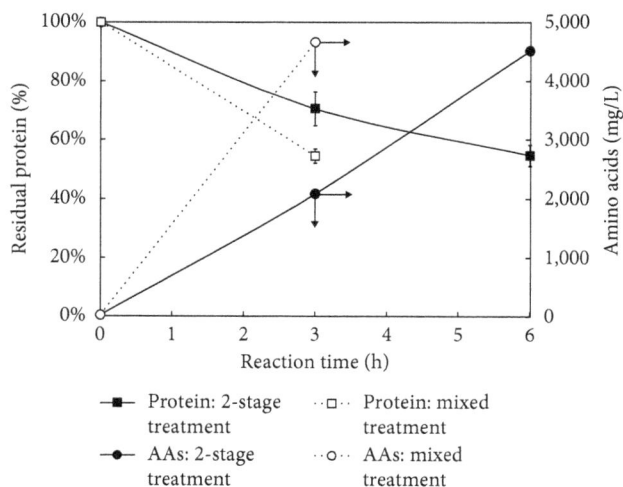

FIGURE 3: Comparison between the 2-stage and mixed treatments using FoodPro and Flavourzyme during the enzymatic hydrolysis of silk fibroin.

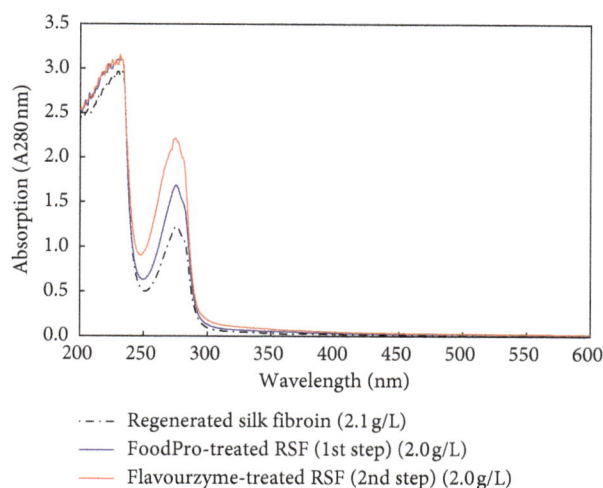

--- Regenerated silk fibroin (2.1 g/L)
— FoodPro-treated RSF (1st step) (2.0 g/L)
— Flavourzyme-treated RSF (2nd step) (2.0 g/L)

FIGURE 4: UV-visible spectra of regenerated fibroin and its enzymatic hydrolysates obtained using FoodPro in the 1st step and Flavourzyme in the 2nd step.

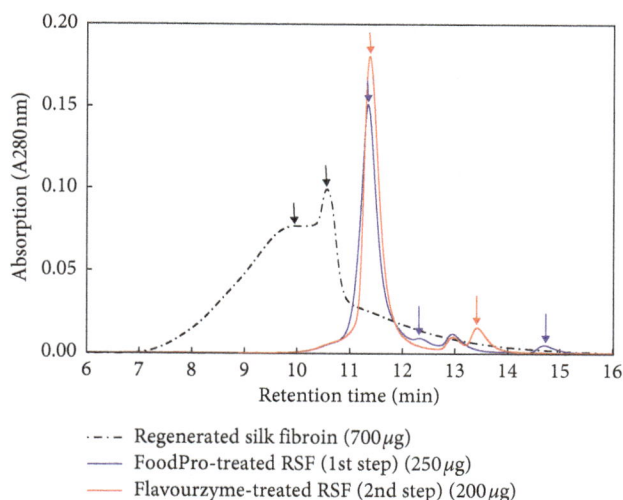

--- Regenerated silk fibroin (700 μg)
— FoodPro-treated RSF (1st step) (250 μg)
— Flavourzyme-treated RSF (2nd step) (200 μg)

FIGURE 6: Size-exclusion chromatography of the regenerated silk fibroin and its enzymatic hydrolysates obtained using FoodPro in the 1st step and Flavourzyme in the 2nd step.

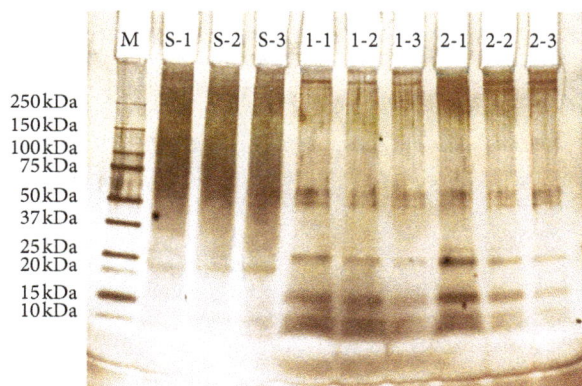

FIGURE 5: SDS-PAGE of the regenerated silk fibroin and its enzymatic hydrolysates obtained using FoodPro in the 1st step and Flavourzyme in the 2nd step. M: molecular weight marker. S-1: regenerated silk fibroin, 84 μg/well. S-2: regenerated silk fibroin, 42 μg/well. S-3: regenerated silk fibroin, 28 μg/well. 1-1: enzymatic hydrolysate obtained using FoodPro at the 1st step, 57 μg/well. 1-2: enzymatic hydrolysate obtained using FoodPro at the 1st step, 29 μg/well. 1-3: enzymatic hydrolysate obtained using FoodPro at the 1st step, 19 μg/well. 2-1: enzymatic hydrolysate obtained using Flavourzyme at the 2nd step, 45 μg/well. 2-2: enzymatic hydrolysate obtained using Flavourzyme at the 2nd step, 23 μg/well. 2-2: enzymatic hydrolysate obtained using Flavourzyme at the 2nd step, 15 μg/well.

PAGE results indicated that the aggregates and high MW proteins of RSF were dissociated and hydrolyzed in the 2-stage enzymatic treatment using FoodPro initially and then Flavourzyme.

3.7. MW Characteristics of RSF and Its EHs Using SEC. RSF samples (700 μg) and EHs (200–250 μg) were effectively separated using a 5Diol-300-II size-exclusion column with a good resolution (Figure 6). The untreated RSF exhibited

two peaks in the chromatogram. The first peak might be strongly associated to the intermolecular interaction, whereas the second peak represented the SF molecules. A previous report indicated that the elution profile of silk obtained after SEC exhibits an early-eluted peak that might represent the silk aggregates and the late-eluted peak represents the nonaggregated silk molecules [41]. This previous report suggested that Ajisawa's reagent produces more number of β-sheet structures and aggregates during dialysis owing to the high exposure of hydrophobic residues compared to the LiBr dissolution [41] and caused higher degradation of the SF heavy chains than other solvents [42]. In the SEC profile of EHs obtained from FoodPro-treated RSF, the peaks that represented untreated RSF disappeared and novel peaks were generated at low MW with enhanced intensities. In this study, the variations in absorption intensities during SEC were similar to those in UV-visible spectra analysis. These results were well matched to those of actinase-treated SF as the absorption intensity of the early peak of aggregated SF decreased with an increase in the later peak of nonaggregated SF molecules [22]. In the SEC profile of EHs, several peaks at 12.5, 13.0, and 14.8 min retention time points were observed after FoodPro treatment (1st step). After Flavourzyme treatment (2nd step), the peaks at 12.5 and 14.8 min retention time points were decreased and a novel peak at 13.5 min retention time was generated. The variations in the elution profiles of soluble SF were similar to those during the biodegradation of silk films using protease from *Streptomyces griseus* [13].

4. Conclusion

B. mori SF and its hydrolysates were investigated as functional materials in foods and alternative sources of food additives. To prepare the SF hydrolysates, enzymatic hydrolysis using commercial food-grade enzymes can replace the acid hydrolysis. The RSF solubility was relatively stable at a narrow range of pH (5.0–6.5). FoodPro and Alcalase were selected as two of the best candidate enzymes to hydrolyze SF at pH range 6.5–9.0 and temperature between 50°C and 65°C. In a single enzyme treatment using either FoodPro or Alcalase, approximately 20–25% of SF was hydrolyzed and 2 g/L AAs were released from SF. In the 2-stage enzymatic treatment using the combination of FoodPro and Flavourzyme in a sequence, approximately 45% of SF was degraded and 4.5 g/L AAs were released within 6 h of reaction time. The RSF and its EHs were characterized by performing UV-visible spectra, SDS-PAGE, and SEC analyses. In the 2-stage enzymatic treatment using FoodPro initially and then Flavourzyme, the aggregates and high MW proteins of SF were dissociated and biodegraded into the low MW proteins/peptides (10–15 kDa and 27 kDa). SF hydrolysates that are used as functional food can be produced through environment-friendly enzymatic hydrolysis using commercial food-grade proteolytic enzymes.

Conflicts of Interest

The authors declare that there are no conflicts of interest regarding the publication of this study.

Acknowledgments

This work was supported by Wonkwang University in 2017.

References

[1] K. M. Babu, *Silk: Processing, Properties and Applications*, Woodhead Publishing Ltd., Cambridge, UK, 2013.

[2] L. S. Wray, X. Hu, J. Gallego et al., "Effect of processing on silk-based biomaterials: reproducibility and biocompatibility," *Journal of Biomedical Materials Research Part B Applied Biomaterials*, vol. 99, no. 1, pp. 89–101, 2011.

[3] D. N. Rockwood, R. C. Preda, T. Yücel, X. Wang, M. L. Lovett, and D. L. Kaplan, "Materials fabrication from *Bombyx mori* silk fibroin," *Nature Protocols*, vol. 6, no. 10, pp. 1612–1631, 2011.

[4] B. Kundu, R. Rajkhowa, S. C. Kundu, and X. Wang, "Silk fibroin biomaterials for tissue regeneration," *Advanced Drug Delivery Reviews*, vol. 65, no. 4, pp. 457–470, 2013.

[5] B. Kundu, N. E. Kurland, S. Bano et al., "Silk proteins for biomedical applications: bioengineering perspectives," *Progress in Polymer Science*, vol. 39, no. 2, pp. 251–267, 2014.

[6] A. B. Li, J. A. Kluge, N. A. Guziewicz, F. G. Omenetto, and D. L. Kaplan, "Silk-based stabilization of biomacromolecules," *Journal of Controlled Release*, vol. 219, pp. 416–430, 2015.

[7] C. Vepari and D. L. Kaplan, "Silk as a biomaterial," *Progress in Polymer Science*, vol. 32, no. 8-9, pp. 991–1007, 2007.

[8] J. Melke, S. Midha, S. Ghosh, K. Ito, and S. Hofmann, "Silk fibroin as biomaterial for bone tissue engineering," *Acta Biomaterialia*, vol. 31, pp. 1–16, 2015.

[9] C. Z. Zhou, F. Confalonieri, M. Jacquet, R. Perasso, Z. G. Li, and J. Janin, "Silk fibroin: structural implications of a remarkable amino acid sequence," *Proteins: Structure, Function, and Genetics*, vol. 44, no. 2, pp. 119–122, 2001.

[10] B. Liu, Y. W. Song, L. Jin et al., "Silk structure and degradation," *Colloids and Surfaces B: Biointerfaces*, vol. 131, pp. 122–128, 2015.

[11] Y. Cao and B. Wang, "Biodegradation of silk biomaterials," *International Journal of Molecular Sciences*, vol. 10, no. 4, pp. 1514–1524, 2009.

[12] W. L. Stoppel, N. Raia, E. Kimmerling, S. Wang, C. E. Ghezzi, and D. L. Kaplan, "2.12 silk biomaterials," *Comprehensive Biomaterials II*, vol. 2, pp. 253–278, 2017.

[13] T. Arai, G. Freddi, R. Innocenti, and M. Tsukada, "Biodegradation of *Bombyx mori* silk fibroin fibers and film," *Journal of Applied Polymer Science*, vol. 91, no. 4, pp. 2383–2390, 2004.

[14] R. L. Horan, K. Antle, A. L. Collette et al., "*In vitro* degradation of silk fibroin," *Biomaterials*, vol. 26, no. 17, pp. 3385–3393, 2005.

[15] Q. Lu, B. Zhang, M. Li et al., "Degradation mechanism and control of silk fibroin," *Biomacromolecules*, vol. 12, pp. 1080–1086, 2011.

[16] C. Wongnarat and P. Srihanam, "Degradation behaviors of Thai *Bombyx mori* silk fibroins exposure to protease enzymes," *Engineering*, vol. 5, no. 1, pp. 61–66, 2013.

[17] R. You, Y. Xu, Y. Liu, X. Li, and M. Li, "Comparison of the *in vitro* and *in vivo* degradations of silk fibroin scaffolds from mulberry and nonmulberry silkworms," *Biomedical Materials*, vol. 10, no. 1, article 015003, 2015.

[18] R. You, Y. Zhang, Y. Liu, G. Liu, and M. Li, "The degradation behavior of silk fibroin derived from different ionic liquid solvents," *Natural Science*, vol. 5, no. 6, pp. 10–19, 2013.

[19] P. Srihanam and W. Simchuer, "Proteolytic degradation of silk fibroin scaffold by protease XXIII," *The Open Macromolecules Journal*, vol. 3, no. 1, pp. 1–5, 2009.

[20] Z. Luo, Q. Zhang, M. Shi, Y. Zhang, W. Tao, and M. Li, "Effect of pore size on the biodegradation rate of silk fibroin scaffolds," *Advances in Materials Science and Engineering*, vol. 2015, Article ID 315397, 7 pages, 2015.

[21] J. Brown, C. Lu, J. Coburn, and D. L. Kaplan, "Impact of silk biomaterial structure on proteolysis," *Acta Biomaterialia*, vol. 11, pp. 212–221, 2015.

[22] K. Chen, Y. Umeda, and K. Hirabayashi, "Enzymatic hydrolysis of silk fibroin," *Journal of Sericultural Science of Japan*, vol. 65, pp. 131–133, 1995.

[23] M. Sumida and V. Sutthikhum, "Fibroin and sericin-derived bioactive peptides and hydrolysates as alternative sources of food additive for promotion of human health: a review," *Research and Knowledge*, vol. 1, pp. 1–17, 2015.

[24] K. Hirao and K. Igarashi, "Teaching material study. Characteristics of silk fibroin preparations and its utilization for human food preparation," *Journal of Cookery Science of Japan*, vol. 46, pp. 54–58, 2013.

[25] K. Igarashi, K. Yoshioka, K. Mizutani, M. Miyakoshi, T. Murakami, and T. Akizawa, "Blood pressure-depressing activity of a peptide derived from silkworm fibroin in spontaneously hypertensive rats," *Bioscience, Biotechnology, and Biochemistry*, vol. 70, no. 2, pp. 517–520, 2006.

[26] K. J. Park, H. H. Jin, and C. K. Hyun, "Antigenotoxicity of peptides produced from silk fibroin," *Process Biochemistry*, vol. 38, no. 3, pp. 411–418, 2002.

[27] F. Z. Zhou, Z. Xue, and J. Wang, "Antihypertensive effects of silk fibroin hydrolysate by alcalase and purification of an ACE inhibitory dipeptide," *Journal of Agricultural and Food Chemistry*, vol. 58, no. 11, pp. 6735–6740, 2010.

[28] J. Kim, H. Hwang, J. Park, H. Y. Yun, H. Suh, and K. Lim, "Silk peptide treatment can improve the exercise performance of mice," *Journal of the International Society of Sports Nutrition*, vol. 11, no. 1, pp. 1–7, 2014.

[29] M. Watanabe, T. Kotera, A. Yura, and M. Sumida, "Enzymatic properties of fibroinase of silk gland from day one pupa of the silkworm, *Bombyx mori*," *Journal of Insect Biotechnology and Sericology*, vol. 75, pp. 39–46, 2006.

[30] M. Watanabe, K. Kamei, and M. Sumida, "Sericin digestion by fibroinase, a Cathepsin L-like cysteine proteinase, of *Bombyx mori* silk gland," *Journal of Insect Biotechnology and Sericology*, vol. 76, pp. 9–15, 2007.

[31] S. K. Kim, Y. Y. Jo, K. G. Lee et al., "Modification of the characteristics of silkworm powder by treatment with alkaline protease," *International Journal of Industrial Entomology*, vol. 31, no. 1, pp. 30–33, 2015.

[32] H. J. Chae, M. J. In, and E. Y. Kim, "Effect of solubilization conditions for molecular weight distribution of enzymatically-hydrolyzed silk peptides," *Korean Journal of Biotechnology and Bioengineering*, vol. 13, pp. 114–118, 1998.

[33] T. C. Mcilvaine, "A buffer solution for colorimetric comparison," *Journal of Biological Chemistry*, vol. 49, pp. 183–186, 1921.

[34] A. Ajisawa, "Dissolution of silk fibroin with calcium chloride-ethanol aqueous solution," *Journal of Sericultural Science of Japan*, vol. 67, pp. 91–94, 1998.

[35] O. H. Lowry, N. J. Rosebrough, A. L. Farr, and R. J. Randall, "Protein measurement with the Folin phenol reagent," *Journal of Biological Chemistry*, vol. 193, pp. 265–275, 1951.

[36] C. Cupp-Enyard, "Sigma's Non-specific protease activity assay - casein as a substrate," *Journal of Visualized Experiments*, vol. 19, p. e899, 2008.

[37] V. J. Harding and R. M. MzcLean, "The ninhydrin reaction with amines and amides," *Journal of Biological Chemistry*, vol. 25, pp. 337–350, 1916.

[38] C. Wong Po Foo, E. Bini, J. Hensman, D. P. Knight, R. V. Lewis, and D. L. Kaplan, "Role of pH and charge on silk protein assembly in insects and spiders," *Applied Physics A: Material Science and Processing*, vol. 82, no. 2, pp. 223–233, 2006.

[39] A. Matsumoto, A. Lindsay, B. Abedian, and D. L. Kaplan, "Silk fibroin solution properties related to assembly and structure," *Macromolecular Bioscience*, vol. 8, no. 11, pp. 1006–1018, 2008.

[40] M. Friedman, "Applications of the Ninhydrin reaction for analysis of amino acids, peptides, and proteins to agricultural and biomedical sciences," *Journal of Agricultural and Food Chemistry*, vol. 52, no. 3, pp. 385–406, 2004.

[41] Z. Zheng, S. Guo, Y. Liu et al., "Lithium-free processing of silk fibroin," *Journal of Biomaterials Applications*, vol. 31, no. 3, pp. 450–463, 2016.

[42] H. Yamada, H. Nakao, Y. Takasu, and K. Tsubouchi, "Preparation of undegraded native molecular fibroin solution from silkworm cocoons," *Materials Science and Engineering: C*, vol. 14, no. 1-2, pp. 41–46, 2001.

Nitrate and Nitrite Promote Formation of Tobacco-Specific Nitrosamines via Nitrogen Oxides Intermediates during Postcured Storage under Warm Temperature

Jun Wang,[1] Huijuan Yang,[1] Hongzhi Shi,[1] Jun Zhou,[2] Ruoshi Bai,[2] Mengyue Zhang,[1] and Tong Jin[1]

[1]*Henan Agricultural University, National Tobacco Cultivation & Physiology & Biochemistry Research Center, Zhengzhou 450002, China*
[2]*Beijing Cigarette Factory, Shanghai Tobacco Group Co. Ltd., Beijing 100024, China*

Correspondence should be addressed to Hongzhi Shi; shihongzhi88@163.com

Academic Editor: Yang Xu

Tobacco-specific nitrosamines (TSNAs) are carcinogenic and are present in cured tobacco leaves. This study was designed to elucidate the mechanisms of TSNAs formation under warm temperature storage conditions. Results showed that nitrogen oxides (NOx) were produced from nitrate and nitrite in a short period of time under 45°C and then reacted with alkaloids to form TSNAs. Nitrite was more effective than nitrate in promoting TSNAs formation during 45°C storage which may be due to the fact that nitrite can produce a large amount of NOx in comparison with nitrate. Presence of activated carbon effectively inhibited the TSNAs formation because of the adsorption of NOx on the activated carbon. The results indicated that TSNAs are derived from a gas/solid phase nitrosation reaction between NOx and alkaloids. Nitrate and nitrite are major contributors to the formation of TSNAs during warm temperature storage of tobacco.

1. Introduction

Tobacco-specific nitrosamines (TSNAs) are a group of important and toxic components of tobacco and tobacco smoke [1, 2]. TSNAs mainly consist of N-nitrosonornicotine (NNN), N-nitrosoanatabine (NAT), N-nitrosoanabasine (NAB), and 4-(methyl nitrosamino)-1-(3-pyridyl)-1-butanone (NNK). NNN and NNK are strong carcinogens [3, 4]. TSNAs are produced via nitrosation of tobacco alkaloids during the curing and storage of tobacco leaves (Figure 1) [5–8]. In China, both air-cured and flue-cured tobacco leaves are generally stored for approximately 18 months in warehouses before being processed for cigarette production in order to reduce the unfavorable smells [9].

Many studies reported about the factors that influence TSNAs formation during air-curing [7, 10–12]. Cui [10] found that TSNAs levels in the leaf lamina and midrib increased substantially during the fourth to seventh weeks of air-curing

stage. During this stage nitrate is reduced to nitrite via microbial activity, and the resulting nitrite, in turn, is involved in nitrosation reactions with the naturally existing alkaloids during air-curing, then leading to the formation of TSNAs (Figure 2) [11].

However, to our knowledge there are few reports available on the mechanisms of TSNAs formation during storage stage. TSNAs levels may increase several fold in comparison to the levels in freshly air-cured leaves [13, 14], but the mechanisms of TSNAs formation during storage are not clear. It has been reported that, as the storage temperature increased, TSNAs and nitrite contents increased, and the most rapid increase in TSNAs occurred during the warm temperature season [15, 16]. The interactions between temperature and abundant nitrate in cured tobacco leaf could be responsible for TSNAs formation during storage. Treatment of the tobaccos with streptomycin and rifampicin did not inhibit nitrosamine formation during storage indicating that TSNAs formation was

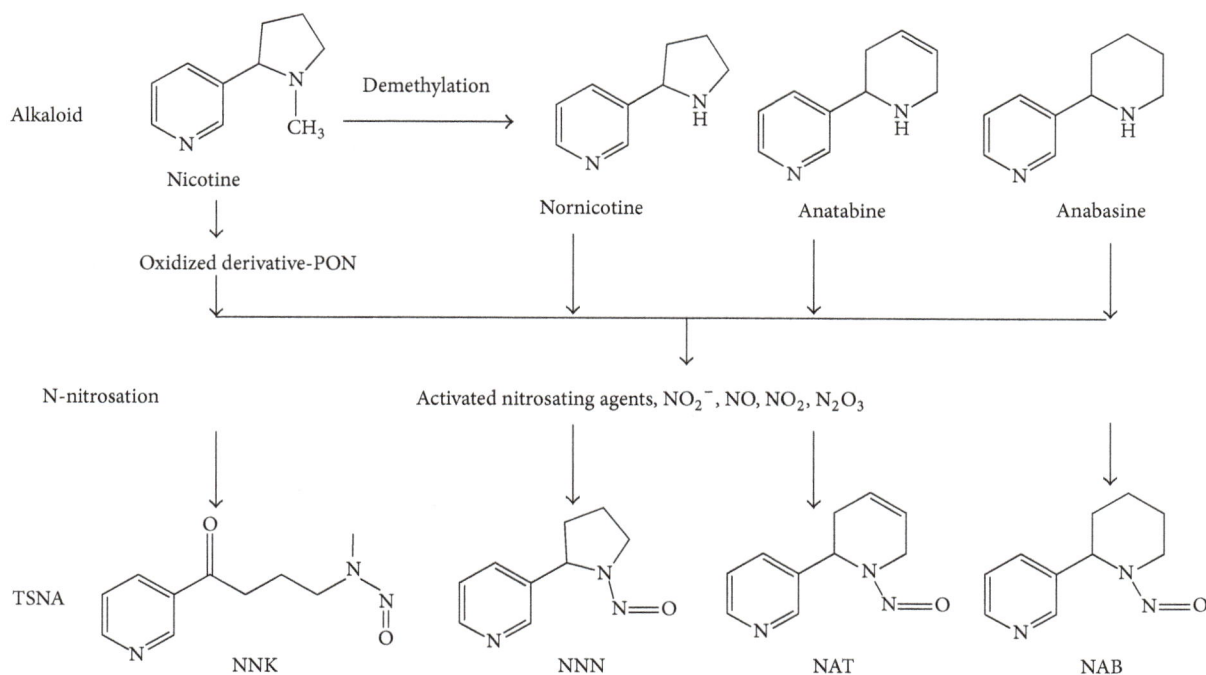

FIGURE 1: Proposed formation pathways of the major TSNAs found in cured tobacco leaves. Nitrosating agents can directly interact with nornicotine, anatabine, and anabasine to form NNN, NAT, and NAB, respectively. Nicotine is less susceptible to nitrosation; thus NNK is produced from pseudo-oxynicotine, an oxidized derivative of nicotine.

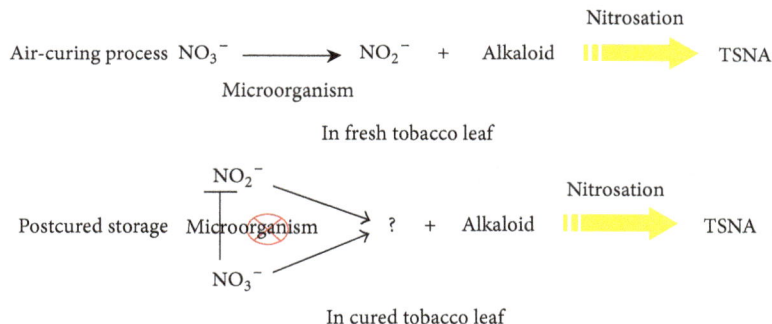

FIGURE 2: The activated nitrosating agents participate in nitrosation of alkaloid to form TSNAs during different processes of air-cured tobacco production. Nitrate (NO_3^-) is available for reduction to nitrite (NO_2^-) via microbial activity, and the nitrite, in turn, becomes involved in nitrosation reactions with the alkaloids during air-curing, leading to the formation of TSNAs. The nitrite is the most important nitrosating agent during air-curing. TSNAs formation was almost not influenced by microorganism during the storage processes.

not influenced by microorganisms during the storage process (Figure 2) [14]. How the nitrate and nitrite trigger the increase in TSNAs during warm temperature storage is still not clear (Figure 2). The objectives of this study were to elucidate the mechanisms of TSNAs formation during the storage time of cured tobacco leaf and to verify the hypothesis that nitrogen oxides produced from nitrate and nitrite in tobacco are responsible for the formation of TSNAs during storage.

2. Experiments

2.1. Plant Materials.
Tobacco samples were grown in 2015 and cured locally. Leaves from the middle stalk positions were collected. Flue-cured tobacco (*N. tabacum* cv. "Hongda") was

from Midu county, Yunnan province. Leaf samples were prepared by removing the stems, cut to strips, mixed thoroughly, sealed in plastic bags, and then stored at 4°C for the tests. The moisture content of tobacco was 12%.

2.2. Methods

2.2.1. Direct Addition of Nitrate and Nitrite to Tobacco. The treatments included four levels of $NaNO_3$ (10, 20, 30, and 40 mg/mL) and $NaNO_2$ (5, 10, 15, and 20 mg/mL), which were from 10 mL solution each, equal to 73, 146, 219, and 292 mg of NO_3^- and 33, 67, 100, and 133 mg of NO_2^-. The solutions were sprayed onto each of eight 20 g flue-cured tobacco samples. The samples were then placed to an ambient environment for air-drying. The moisture content of tobacco samples after air-drying was 13%. Then the samples were stored in an airtight vacuum desiccator (15 cm diameter) at 45°C with relative humidity of 70% for 15 d. The sample sprayed with the same volume of deionized water was used as the control. The moisture content of tobacco samples after treatment was 11.8%.

2.2.2. Separating the Tobacco and Nitrate/Nitrite Sources in a Closed System under 45°C Storage Conditions.

A vacuum desiccator with a porcelain plate was used to form a closed system in which tobacco cuts and added nitrate and nitrite could be separated during storage. Five mL of each aqueous solution of NH_4NO_3, KNO_3, $NaNO_3$, and $NaNO_2$ at 1 mol/L (equal to 310 mg of NO_3^- and 230 mg of NO_2^-) and 5 mL volume of deionized water (control) were sprayed onto the medical gauze pads (5 cm × 5 cm). After air-drying, the sprayed gauzes were placed in the bottom of the vacuum desiccator together with each 20 g sample of flue-cured tobacco. The vacuum desiccators were then tightly closed and sealed with petroleum jelly. The containers were then placed into a chamber at a temperature of 45°C and a relative humidity of 70% and stored for 15 d. The moisture content of tobacco samples after 45°C treatments was 11.1%.

2.2.3. Effect of Indirect Addition of Nitrate and Nitrite on Nitrogen Oxides Formation in a Closed System with Flue-Cured Tobacco.

Eight treatments in this experiment were divided to two groups: Group I: (1) tobacco; (2) tobacco and $NaNO_3$; (3) tobacco and $NaNO_3$ + 2.0 g activated carbon (AC); (4) tobacco and $NaNO_3$ + 10.0 g AC and Group II: (1) tobacco; (2) tobacco and $NaNO_2$; (3) tobacco and $NaNO_2$ + 2.0 g AC; (4) tobacco and $NaNO_2$ + 10.0 g AC. The amount of added $NaNO_3$ or $NaNO_2$ was 0.8 g and 0.4 g for treatments 2 to 4 (equal to 583 mg of NO_3^- and 266 mg of NO_2^-) in each group. The weight ratios of tobacco, NO_3^-, and AC were 68 : 1 : 3 and 68 : 1 : 17, respectively. And the weight ratios of tobacco, NO_2^-, and AC were 150 : 1 : 7 and 150 : 1 : 37, respectively. $NaNO_3$ and $NaNO_2$ were dissolved in 5 mL deionized water and then sprayed onto the gauze pads. After air-drying, the gauze pads were placed on the bottom of the vacuum desiccator with 40 g sample of flue-cured tobacco leaf in each desiccator. AC sample was made from wood with the particle size being 75 μm, produced by Zhengzhou Tianhe Water Purification Material Co. Ltd. in Henan province. The surface area was 901.7 m^2/g, and the total pore volume was 0.518 cm^3/g with the micropore, mesopore, and macroporous volume being 0.372 cm^3/g, 0.129 cm^3/g, and 0.016 cm^3/g, respectively. AC was activated in 100°C oven for 30 min, then was wrapped in a medical gauze, and suspended under the porcelain plate of the vacuum desiccator for the relevant treatments. All vacuum desiccators were tightly closed, sealed, and placed into a chamber at a temperature of 45°C and a relative humidity of 70% for 24 h.

The AC used in the above experiments were taken out and then were placed in the new airtight desiccator, respectively. Two grams of AC was put into another container as control. All airtight desiccators were placed into a chamber at a

temperature of 60°C and a relative humidity of 70% for 15 min and 90 min.

2.2.4. Effects of Activated Carbon (AC) on TSNAs Formation in Flue-Cured Tobacco in Response to Nitrate and Nitrite Addition.

Twelve treatments in this experiment were divided into two groups: Group I: (1) tobacco (4°C); (2) tobacco (warm temperature control 45°C); (3) tobacco and $NaNO_3$ separately; (4) tobacco and $NaNO_3$ + 1.0 g AC separately; (5) tobacco and $NaNO_3$ + 5.0 g AC separately; and (6) tobacco and $NaNO_3$ + 10.0 g AC separately and Group II: (1) tobacco (4°C); (2) tobacco (warm temperature control 45°C); (3) tobacco and $NaNO_2$ separately; (4) tobacco and $NaNO_2$ + 1.0 g AC separately; (5) tobacco and $NaNO_2$ + 5.0 g AC separately; and (6) tobacco and $NaNO_2$ + 10.0 g AC separately. The amount of $NaNO_3$ or $NaNO_2$ added to each sample of treatments 3 to 6 was 0.3 g (equal to 219 mg of NO_3^- and 200 mg of NO_2^-). The weight ratios of tobacco, NO_3^-, and AC were 91 : 1 : 4, 91 : 1 : 22, and 91 : 1 : 45, respectively. And the weight ratios of tobacco, NO_2^-, and AC were 100 : 1 : 5, 100 : 1 : 25, and 100 : 1 : 50, respectively. $NaNO_3$/$NaNO_2$ were dissolved in 5 mL deionized water and then were sprayed onto the gauze pads. After air-curing, the gauzes were placed on the bottom of the vacuum desiccators with 20 g sample of flue-cured tobacco leaf in each desiccator. All treatments were performed in the vacuum desiccator as described above. The activated AC also was used as described above. The vacuum desiccators were put into a chamber with a constant temperature of 45°C and a relative humidity of 70%, respectively. The moisture content of cured tobacco samples after treatment was 11.6%.

2.3. Chemical Analyses

2.3.1. TSNAs Measurements.

In each experiment, tobacco samples were lyophilized, ground to powder, sieved through a 0.25 mm screen, and then measured for the content of NNN, NNK, NAT, and NAB. TSNAs contents were determined at the Beijing Cigarette Factory according to the method of SPE-LC-MS/MS [12, 17, 18].

2.3.2. Nitrate, Nitrite, and Alkaloid Measurements.

NO_3-N and NO_2-N were quantified according to the method of Crutchfield and Grove [19]. The individual alkaloids were analyzed with a gas chromatograph as described by Jack and Bush [20]. Methyl tert-butyl ether was used as the extraction solvent with N-hexadecane as the internal standard.

2.3.3. Nitrogen Oxides Analysis.

The first step was diluting the standard gas with a dynamic gas calibrator (Model 146 i, Thermo Scientific, USA EPA) to give a concentration within the operational range of the instrument. The high purity nitric oxide (NO) and nitrogen dioxide (NO_2) standards in N_2 (component content: 69.8 ppm, gas sample number: L120911099, National Institute of Metrology/National Standard Material Research Center, Beijing, China) were configured into nitrogen oxides gas of a low concentration (1 ppm) by 146i calibrator. It was necessary to modify the original

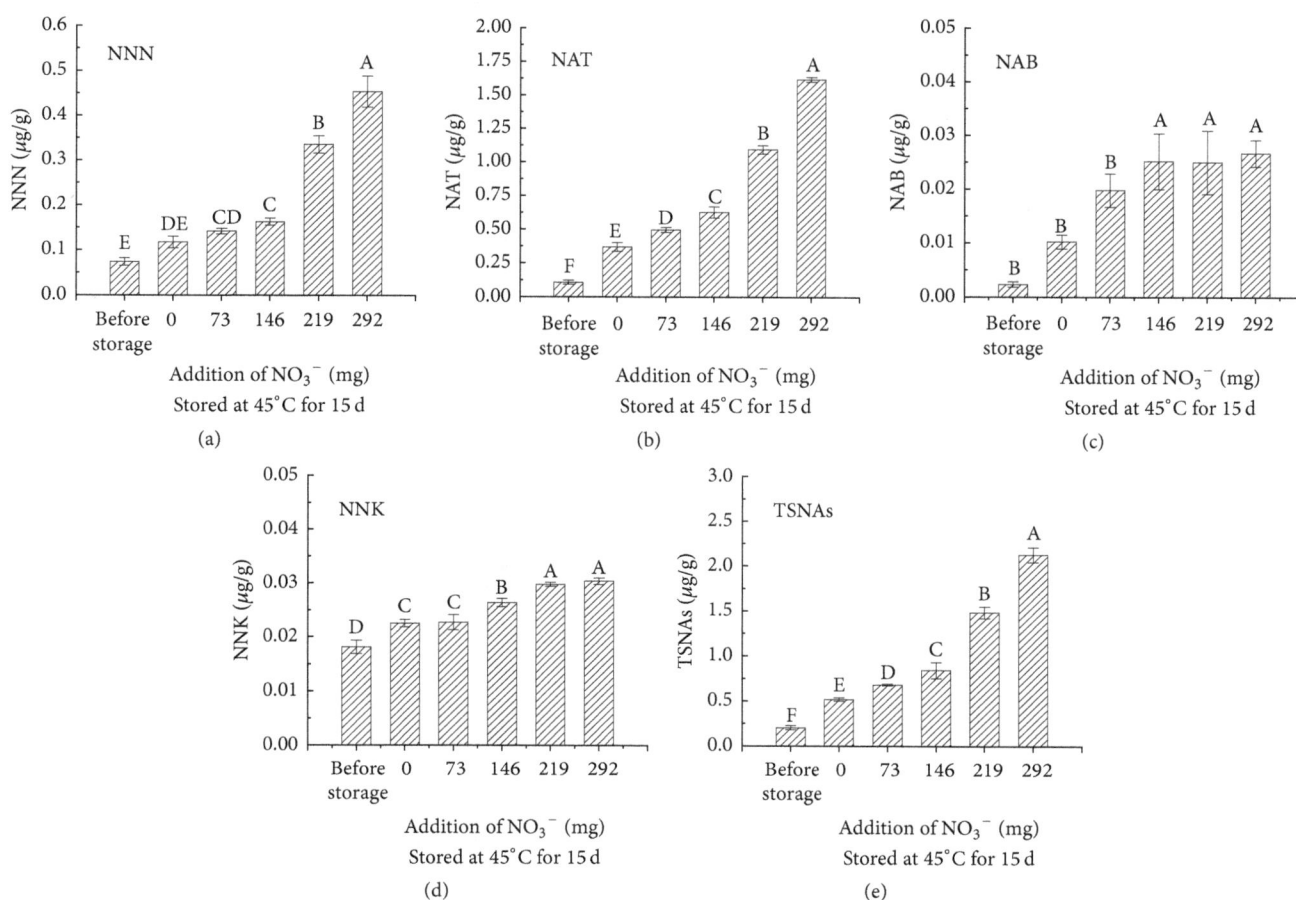

FIGURE 3: Effects of added nitrate on TSNAs formation in flue-cured tobacco stored at 45°C for 15 d. *Note.* For all experiments, each value represents the mean of three independent samples. Uppercase letters indicate significant differences between the treatments at $P < 0.05$.

procedure by configuring the standard gas again if the levels of NOx exceeded the limit of detection. The NO and NOx in the air were defaulted to the zero point by the instrument.

After experiment, the airtight vacuum desiccators were taken out and then connected with a vacuum pump (DOA-P504-BN, GAST Manufacturing, A Unit of Idex Corporation, MI, USA) to extract the gas into a gas collecting bag. After 25 s, the gas bag was pulled out and connected with the NO-NO$_2$-NO$_X$ analyzer (Model 42i, Thermo Scientific, USA EPA, Reference method RFNA-1289-074) for NOx analysis by chemiluminescence detection.

2.4. Statistical Analyses. Analysis of variance (ANOVA) and least significant difference (LSD) of TSNAs and NOx values were performed at the 0.05 level of significance. Data were statistically analyzed with SPSS 20.0. Figures were drawn with Origin 8.5. All treatments were randomly designed in triplicate.

3. Results and Discussion

3.1. Effect of Nitrate Concentration on TSNAs Formation in Flue-Cured Tobacco. The mean contents of nicotine, nornicotine, anabasine, and anatabine of tobacco used in this

experiment were 19.8, 1.2, 0.3, and 1.5 mg/g, respectively, and the NO$_3$-N and NO$_2$-N content were correspondingly 119 μg/g and 10 μg/g. With the increasing amount of NO$_3^-$ added, both individual and total TSNAs contents increased significantly (Figure 3). After the addition of 0.4 g NaNO$_3$ (equal to 292 mg of NO$_3^-$), the NAT and total TSNAs contents increased by 340% and 311%, respectively, in comparison with the control.

No significant change of NNN was observed in the tobacco samples when less than 0.2 g of NaNO$_3$ (146 mg of NO$_3^-$) was added. However, when the amount of NO$_3^-$ added increased to 219 mg (0.3 g NaNO$_3$), the NNN content was approximately double that of 0.2 g NaNO$_3$ addition. Although the NAB content was the lowest of the four individual TSNAs, a significant increase in NAB content occurred as a result of the addition of NaNO$_3$ reached to 0.2 g. The NNK content also increased significantly as nitrate addition increased.

3.2. Effect of Nitrite Concentration on TSNAs Formation in Flue-Cured Tobacco. Table 1 shows that as the concentration of added NO$_2^-$ increased, the individual and total TSNAs contents showed rapid and dramatic increases, and the relative increments were greater for the samples with greater

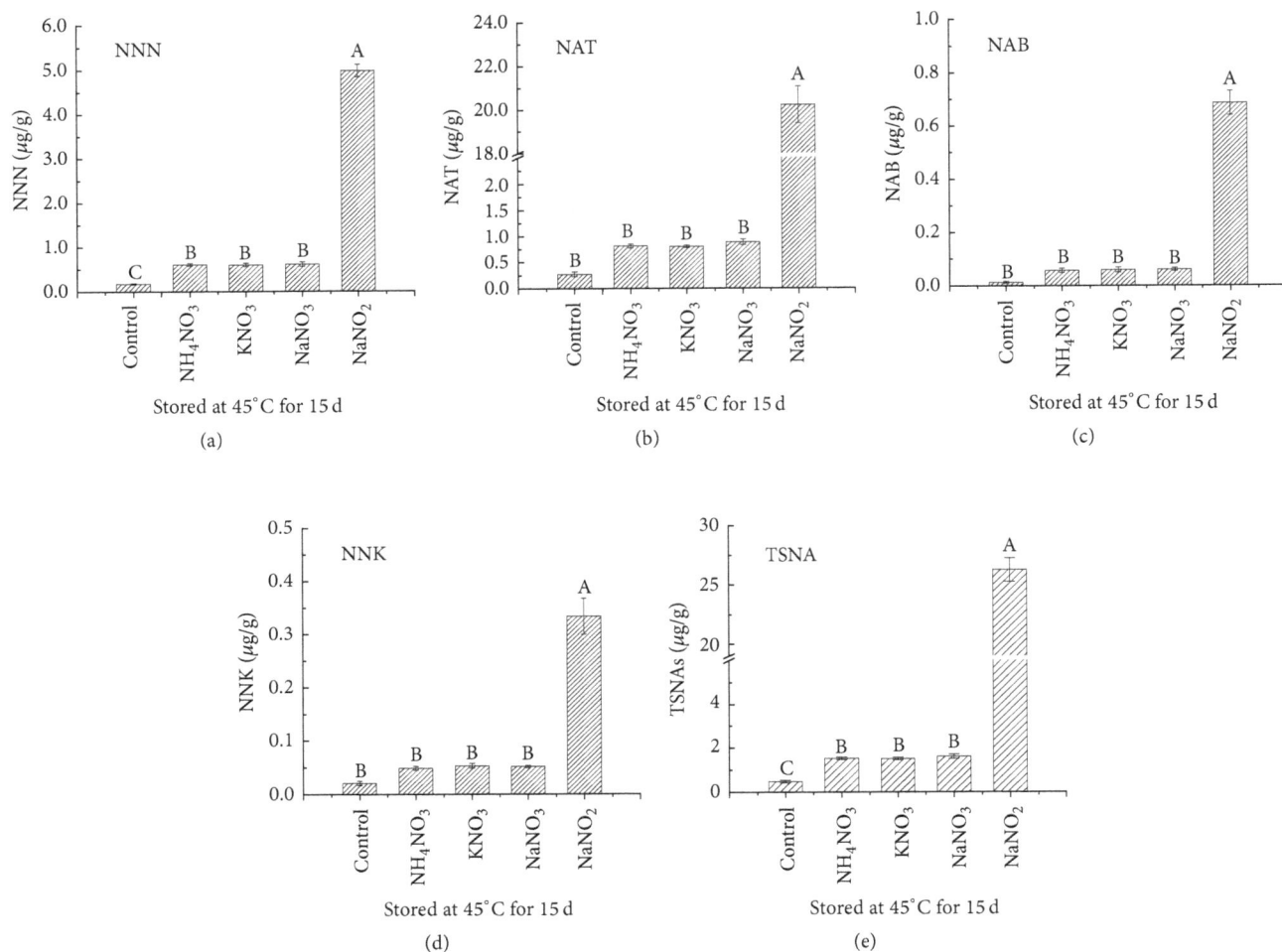

FIGURE 4: Effects of indirect addition of three nitrate compounds and sodium nitrite on TSNAs formation in flue-cured tobacco during warm temperature storage. *Note.* The 1 mol/L of nitrate, nitrite compounds (310 mg of NO_3^- and 230 mg of NO_2^-, resp.), and tobacco were added separately in the closed vessel which was then stored at 45°C for 15 d. Uppercase letters indicate significant differences between the treatments at $P < 0.05$.

amounts of added NO_2^-. When tobacco samples were treated with 133 mg of NO_2^- (0.2 g of $NaNO_2$), total TSNAs content was 117 μg/g, at a 207% increment compared with the control (0.56 μg/g). For the individual TSNA, levels of NNN, NNK, NAT, and NAB showed 257-, 38-, 203-, and 110-fold increases, respectively. The results showed that NO_2^- was more effective in promoting TSNAs formation in the cured leaf than NO_3^- during warm temperature storage.

The higher TSNAs levels in burley tobacco are partly due to the relatively higher levels of TSNAs precursors, such as alkaloids and oxide of nitrogen, that are present in the leaf tissue [21–23]. It is reported that nitrite which is derived from the bacterially mediated reduction of nitrate is considered to be the limiting factor in TSNAs formation in air-cured tobacco during air-curing [6, 24]. In this research, the addition of nitrate and nitrite in flue-cured tobacco to the levels which are equivalent to those in burley tobacco can increase the TSNAs concentration comparable

to burley tobacco especially coupled with warm temperature. Results indicated that the great amount of nitrosating species available is a major contributor to the formation of TSNAs observed in cured tobacco during storage.

3.3. Effects of Indirect Addition of Nitrate/Nitrite on TSNA Formation in Flue-Cured Tobacco during Warm Temperature Storage.
TSNAs contents increased as the storage temperature increased [15, 16], and the abundance of nitrite and nitrate could be a major contributor. To clarify how nitrate or nitrite affects the formation of TSNAs under warm temperature, indirect addition of nitrate or nitrite experiments was carried out.

After tobacco leaf treated with 1 mol/L $NaNO_2$, total TSNAs content increased almost by 54 times compared with that in the control sample (Figure 4). When tobacco leaf is placed separately with gauze pad containing 310 mg nitrate, obvious increases were observed both in individual and

TABLE 1: Effects of added nitrite on TSNAs formation in flue-cured tobacco stored at 45°C for 15 d.

Storage conditions	Addition of NO_2^- (mg)	NNN ($\mu g/g$)	NAT ($\mu g/g$)	NAB ($\mu g/g$)	NNK ($\mu g/g$)	Total TSNAs ($\mu g/g$)
Before storage	0	0.08 ± 0.01^E	0.11 ± 0.01^E	0.004 ± 0.00^E	0.03 ± 0.00^D	0.22 ± 0.00^E
	0	0.13 ± 0.01^E	0.39 ± 0.02^E	0.02 ± 0.00^E	0.03 ± 0.01^D	0.56 ± 0.03^E
	33	5.85 ± 0.28^D	18.97 ± 0.31^D	0.31 ± 0.03^D	0.09 ± 0.01^{CD}	25.22 ± 0.07^D
45°C for 15 d	67	12.85 ± 0.88^C	41.73 ± 1.95^C	0.77 ± 0.06^C	0.23 ± 0.03^C	55.57 ± 2.74^C
	100	19.22 ± 1.13^B	51.96 ± 1.54^B	1.15 ± 0.15^B	0.66 ± 0.09^B	72.99 ± 2.44^B
	133	33.66 ± 2.18^A	79.68 ± 0.96^A	2.25 ± 0.23^A	1.21 ± 0.16^A	116.80 ± 3.53^A

Note. Each value represents the mean of three independent samples, and uppercase letters indicate significant differences between the treatments at $P < 0.05$.

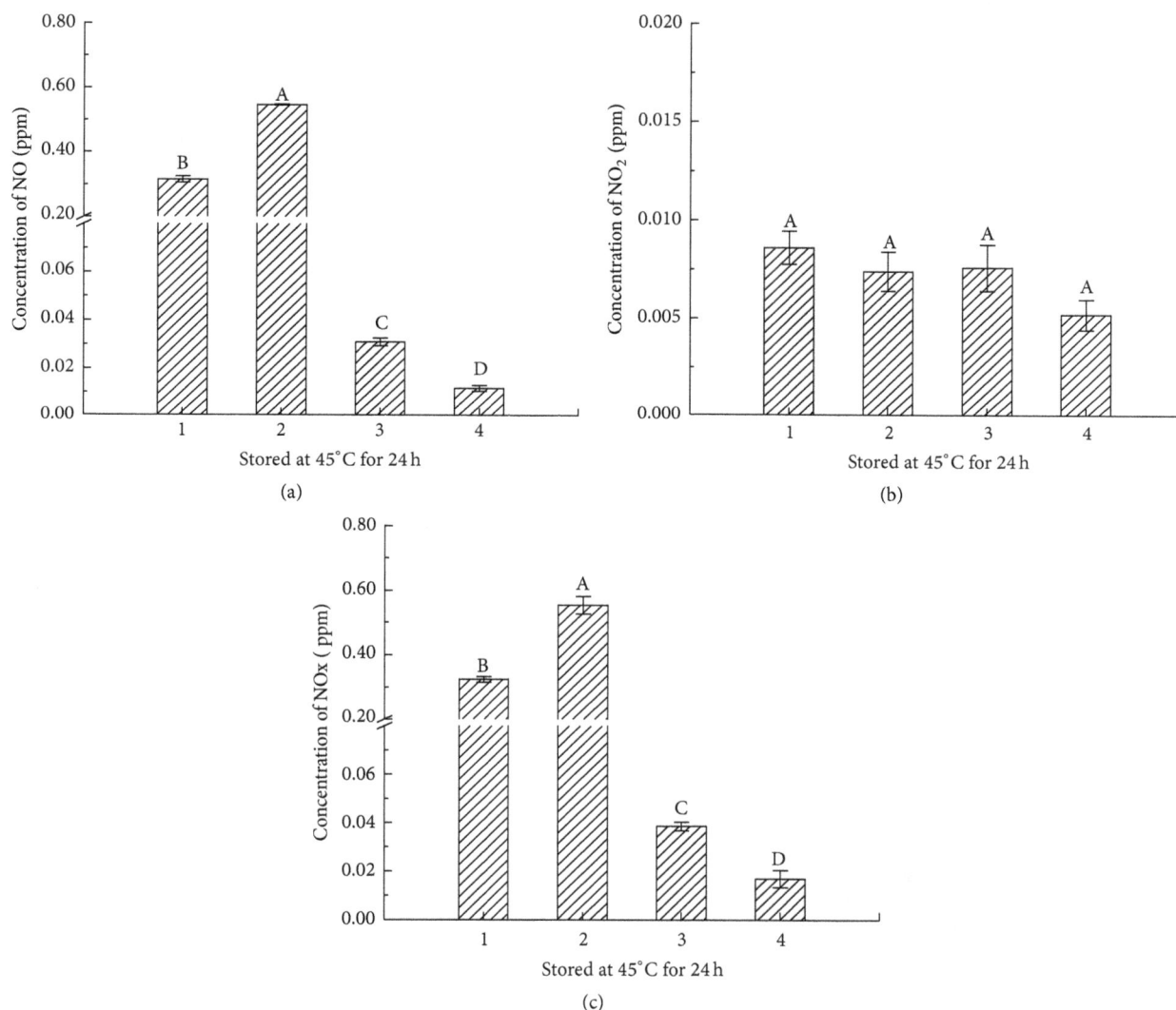

(a)

(b)

(c)

FIGURE 5: Effect of indirect addition of nitrate and activated carbon (AC) on nitrogen oxides formation in a closed system with flue-cured tobacco. *Note.* (1) tobacco, (2) tobacco + $NaNO_3$, (3) tobacco + $NaNO_3$ + AC 2.0 g, and (4) tobacco + $NaNO_3$ + AC 10.0 g, 40 g of flue-cured tobacco, and 0.8 g $NaNO_3$ (583 mg of NO_3^-). The NO_3^-, AC, and tobacco were added separately in the vacuum desiccators stored at 45°C for 24 h. Uppercase letters indicate significant differences between the treatments at $P < 0.05$.

in total TSNA content compared with the control. Data presented here suggested that even though the tobacco sample is placed separately with nitrate or nitrite, TSNA content in tobacco also increased. The formation of TSNAs in storage triggered by nitrate and nitrite is likely a gas phase reaction.

3.4. Effect of Indirect Addition of Nitrate/Nitrite on NOx Formation in a Closed System with Flue-Cured Tobacco. Flue-cured tobacco leaves could generate trace concentrations of NOx under 45°C after 24 h treatment (Figures 5 and 6). As a main component, NO accounted for more than 95% of the NOx produced from the tobacco sample. As 0.8 g of

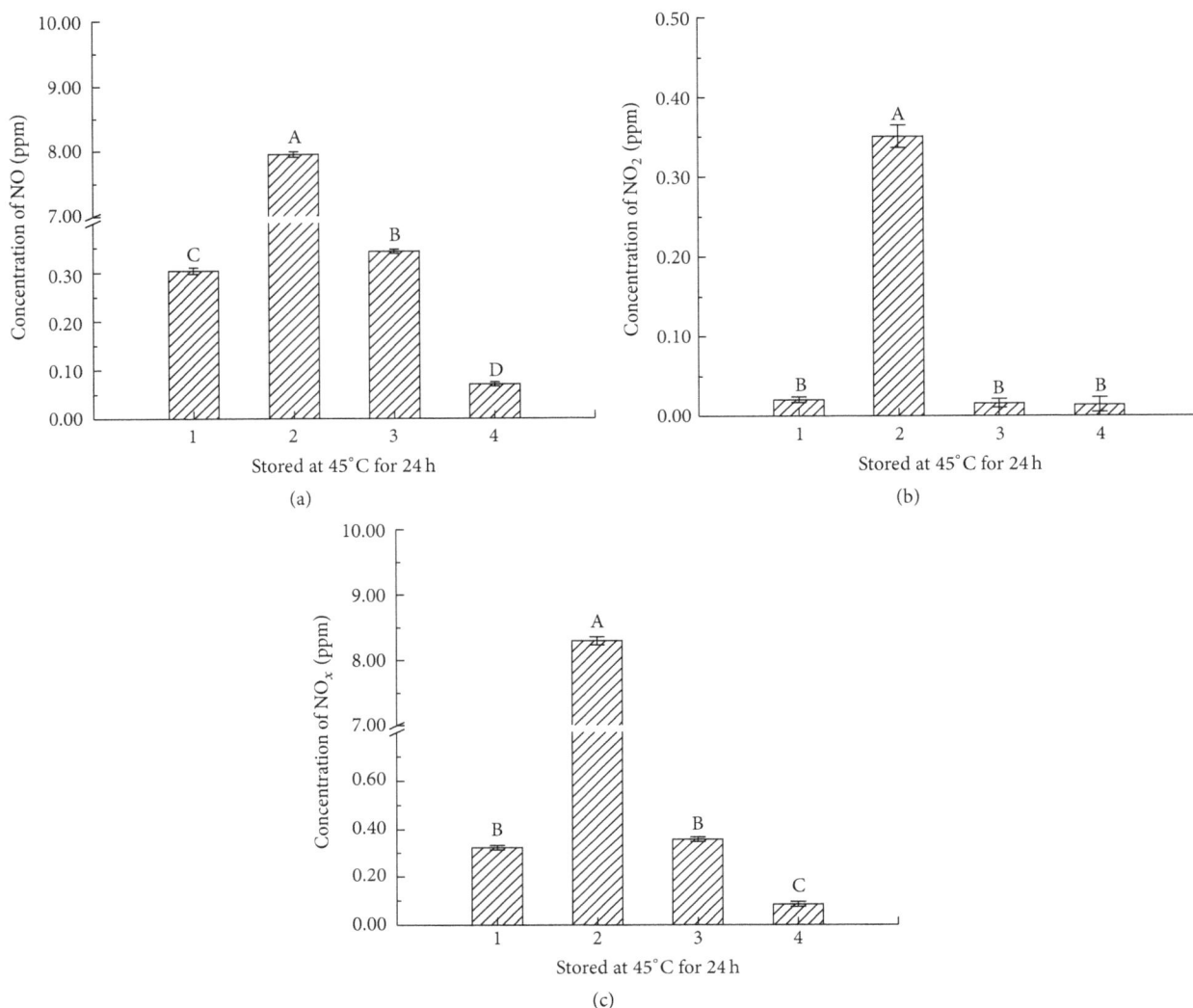

FIGURE 6: Effect of indirect addition of nitrite and activated carbon (AC) on nitrogen oxides formation in a closed system with flue-cured tobacco. *Note.* (1) tobacco, (2) tobacco + $NaNO_2$, (3) tobacco + $NaNO_2$ + AC 2.0 g, and (4) tobacco + $NaNO_2$ + AC 10.0 g, 40 g of flue-cured tobacco, and 0.4 g $NaNO_2$ (266 mg of NO_2^-). The NO_2^-, AC, and tobacco were added separately in vacuum desiccators stored at 45°C for 24 h. Uppercase letters indicate significant differences between the treatments at $P < 0.05$.

$NaNO_3$ (equal to 583 mg of nitrate) was added in the vacuum desiccator, the concentration of NO and NOx increased significantly ($P < 0.05$); NOx reached 0.6 ppm (Figure 5(c)). When 0.4 g of $NaNO_2$ (equal to 266 mg of nitrite) was added, NOx concentrations in the desiccator dramatically increased to 8.3 ppm, which was 24.6-fold greater than the control (Figure 6(c)).

It is interesting that when 2 g of AC was added to the system (the weight ratio of tobacco, NO_3^-, and AC is 68 : 1 : 3; the weight ratio of tobacco, NO_2^-, and AC is 150 : 1 : 7), a 97% decrease of NOx level was observed in comparison to the added nitrate treatment (Figure 5(c)), and almost 7.9 ppm of NOx were adsorbed by AC (Figure 6(c)). Meanwhile, similar adsorption effects were observed for NO. The concentration of NO and NOx decreased significantly ($P < 0.05$) as the addition of AC increased, indicating strong adsorption for NOx generated from nitrate and nitrite by AC.

Table 2 presents the desorption rate of NOx from the AC samples which were used for the adsorption experiments and were significantly higher than those from control group. The desorption rates of NOx and NO were obviously increased with the prolonging of treating time and the increase of NOx adsorption observed in Figures 5 and 6. Results indicated that NOx were indeed adsorbed by AC. As the weight ratio of tobacco, NO_2^-, and AC is 150 : 1 : 7, NOx decreased by 7.9 ppm (Figure 6(c)); at the same time 0.632 ppm of NOx was desorbed by the AC after treatment at 60°C for 90 min. Since the desorption rate of NOx in AC has close relationship with the temperature and time [25], further research is needed to determine the effect of desorbing temperature and time on the desorption rate of NOx adsorbed by AC.

During flue-curing process, direct-fired systems allow combustion products, specifically NOx, to mix with the air and expose the green tobacco leaves to these gases [26]. The

TABLE 2: The desorption rate of NOx at 60°C for different time.

Treatment		NO (ppm)		NOx (ppm)	
		15 min	90 min	15 min	90 min
AC (Control)		0.0094^{C}	0.0249^{C}	0.0203^{B}	0.0486^{B}
AC used in Figure 5	Treatment 3 (2 g AC)	0.0200^{B}	0.0466^{B}	0.0220^{B}	0.0505^{B}
	Treatment 4 (10 g AC)	0.0216^{B}	0.0471^{B}	0.0244^{B}	0.0511^{B}
AC used in Figure 6	Treatment 3 (2 g AC)	0.0732^{A}	0.6200^{A}	0.0864^{A}	0.6320^{A}
	Treatment 4 (10 g AC)	0.0740^{A}	0.6302^{A}	0.0887^{A}	0.6435^{A}

Note. Each value represents the mean of three independent samples, and uppercase letters indicate significant differences between the treatments at $P < 0.05$.

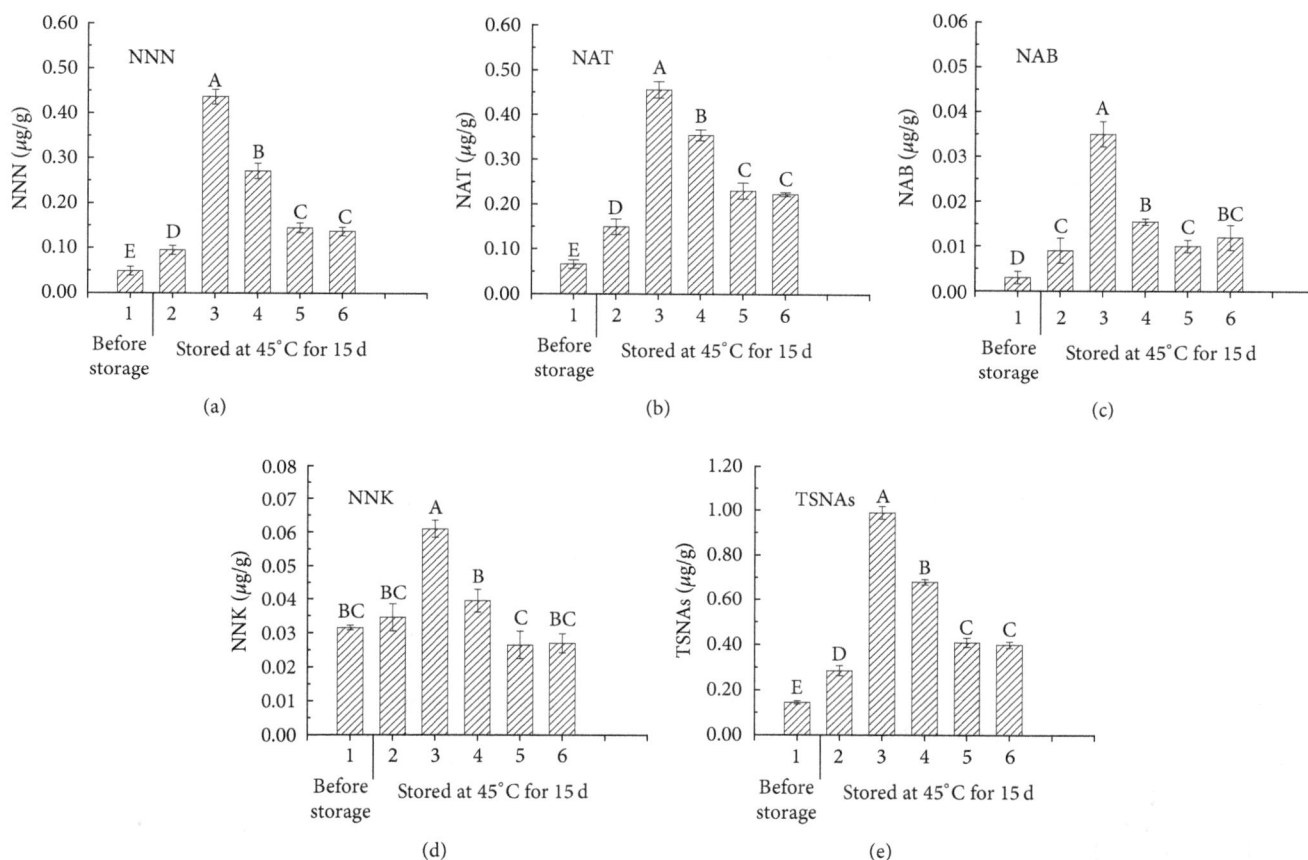

FIGURE 7: Effect of activated carbon (AC) on TSNAs formation in flue-cured tobacco in response to nitrate addition. *Note.* (1) tobacco (before storage, 4°C), (2) tobacco, (3) tobacco + $NaNO_3$, (4) tobacco + $NaNO_3$ + AC 1.0 g, (5) tobacco + $NaNO_3$ + AC 5.0 g, and (6) tobacco + $NaNO_3$ + AC 10.0 g, 20 g of flue-cured tobacco, and 0.3 g $NaNO_3$ (219 mg of NO_3^{-}). The NO_3^{-}, AC, and tobacco were added separately in the vacuum desiccators stored at 45°C for 15 d. Uppercase letters indicate significant differences between the treatments at $P < 0.05$.

previous research showed that TSNAs may be formed by nitrosation via the elevated levels of NOx in the air surrounding the leaves during the curing process [27]. However, there were few reports about the relationship between NOx and TSNAs during storage process. Data in Figure 4 suggested that TSNAs in storage is likely a gas phase reaction. This result showed that NOx can be generated from NO_2-N and NO_3-N under 45°C after 24 h. Nitrite was more effective than nitrate in the production of NOx which in turn would probably promote the formation of TSNAs.

3.5. *Effect of AC on TSNAs Formation in Flue-Cured Tobacco in Response to Nitrate and Nitrite Added.* Having a very porous structure and special surface properties, AC has been used to trap TSNAs in tobacco solution [28, 29]. Lin et al. [30] reported that AC made from coconut shells exhibited a high affinity for TSNAs. The impregnated sorbent ZnAC can remove 73% of the TSNAs in solution, offering a cost-effective candidate for industrial applications [29].

As shown in Figures 7(e) and 8(e), total TSNAs content in tobacco incubated separately with NO_2^{-} (treatment 3) at

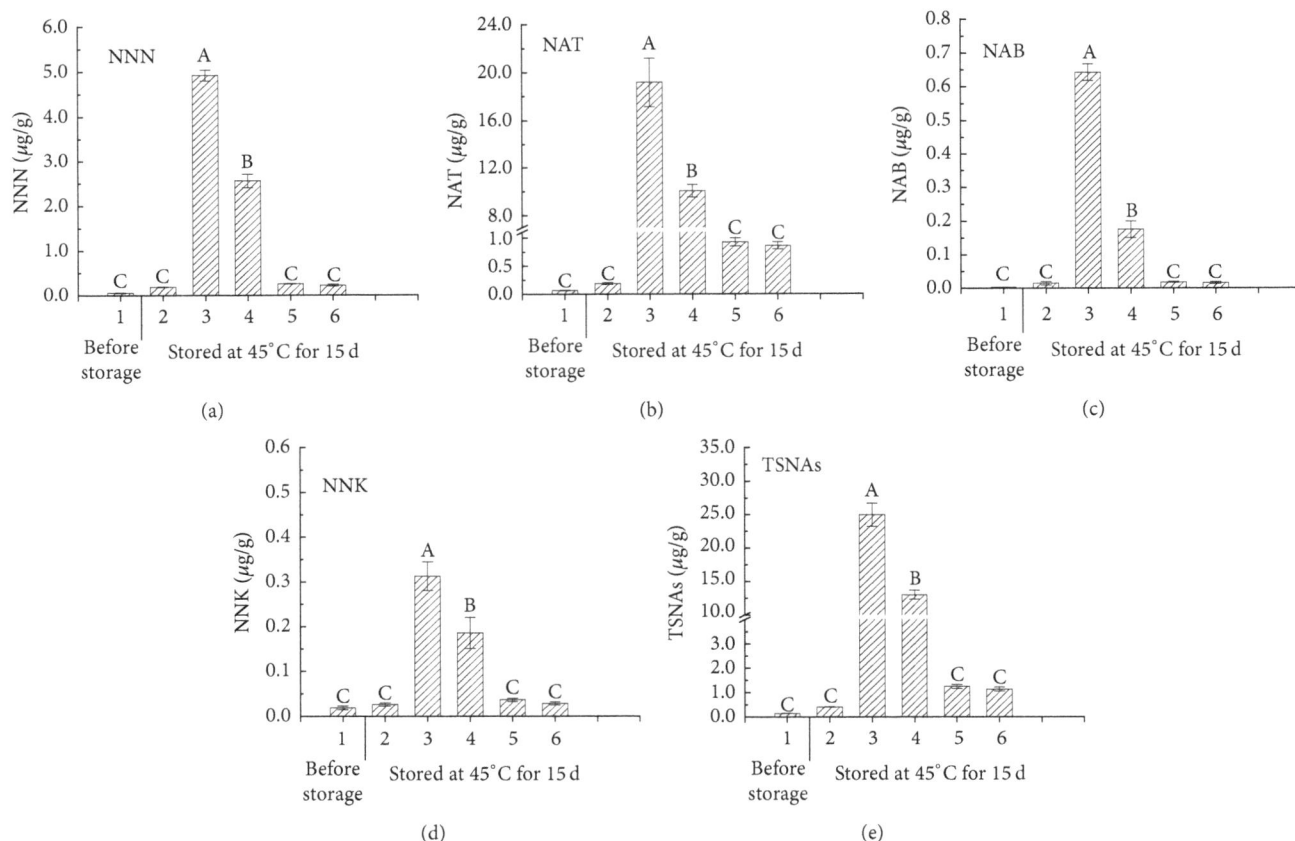

FIGURE 8: Effect of activated carbon (AC) on TSNAs formation in flue-cured tobacco in response to nitrite addition. *Note.* (1) tobacco (before storage, 4°C), (2) tobacco, (3) tobacco + NaNO$_2$, (4) tobacco + NaNO$_2$ + AC 1.0 g, (5) tobacco + NaNO$_2$ + AC 5.0 g, and (6) tobacco + NaNO$_2$ + AC 10.0 g, 20 g of flue-cured tobacco, and 0.3 g NaNO$_2$ (200 mg of NO$_2^-$). The NO$_2^-$, AC, and tobacco were added separately in the vacuum desiccators stored at 45°C for 15 d. Uppercase letters indicate significant differences between the treatments at $P < 0.05$.

45°C increased markedly to 25 μg/g, a 59-fold increase over the control (treatment 2). By treating tobacco with NO$_3^-$, however, total TSNAs increased to 1 μg/g, also significantly higher than the control. The huge difference of TSNAs content between tobacco treated with NO$_2^-$ and NO$_3^-$ could be attributed to the high concentration of NOx which were emitted from NO$_2^-$ under the same condition.

However, after adding 1 g of AC to the separating system of tobacco and NO$_2^-$, the weight ratio of tobacco, NO$_2^-$, and AC was 100 : 1 : 5, both individual and total TSNAs contents of tobacco decreased markedly, with total TSNAs content decreasing to 13 μg/g, 48% lower than in treatment 3, and NNN, NAT, NAB, and NNK decreasing by 47.9%, 47.6%, 67.3% and 41%, respectively. When the addition of AC increased to 5 g, the weight ratio of tobacco, NO$_2^-$ and AC reached to 100 : 1 : 25, total TSNAs levels further decreased to 1.2 μg/g, indicating a crucial role of AC in inhibiting TSNAs formation by adsorbing the NOx generated from NO$_2$-N and NO$_3$-N. The results also showed no significant differences in reducing TSNAs levels by increasing AC usage greater than 5 g (Figures 7 and 8). The maximum inhibitory effect of AC on TSNAs formation occurred at an AC/NO$_2^-$ ratio of 25/1. The above results indicated that AC may have the potential to be

used as adsorbent agent to reduce TSNAs formation during tobacco leaf storage. Our results strongly support the theory that TSNAs formation in storage is a gas phase reaction, and the NOx which is produced from nitrate and nitrite could intermediate the TSNAs formation during warm temperature postcured storage of tobacco. Since NOx can be produced from nitrate and nitrite under warm temperature in a short period, it can readily react with alkaloids to form TSNAs. This result also explained why the most rapid increase in TSNAs occurred during the warm temperature season [16].

Decreasing NOx level by AC adsorption significantly reduced TSNAs formation of tobacco which indicated that the removal of NOx from storage environment could be an effective way to inhibit TSNAs formation in storing tobacco leaf. Therefore, controlling the storage environment and scavenging gaseous nitrosation agents would be crucial to reduce or inhibit TSNAs formation during leaf storage.

4. Conclusion

The results proved that TSNAs are derived from a gas/solid phase nitrosation reaction between NOx and alkaloids during storage. Nitrogen oxides produced from nitrate and nitrite are responsible for the formation of TSNAs during storage under warm temperature. Presence of activated carbon in the tobacco storage containers effectively inhibited the TSNAs formation due to the adsorption of NOx on the activated carbon.

Conflicts of Interest

The authors declare that there are no conflicts of interest regarding the publication of this paper.

Authors' Contributions

Jun Wang and Huijuan Yang contributed equally to this work and should be considered co-first authors.

References

[1] J. D. Adams, S. J. Lee, N. Vinchkoski, A. Castonguay, and D. Hoffmann, "On the formation of the tobacco-specific carcinogen 4-(methylnitrosamino)-1-(3-pyridyl)-1-butanone during smoking," *Cancer Letters*, vol. 17, no. 3, pp. 339–346, 1983.

[2] D. Hoffmann, M. Dong, and S. S. Hecht, "Origin in tobacco smoke of N*ı*-nitrosonornicotine, a tobacco-specific carcinogen: brief communication," *Journal of the National Cancer Institute*, vol. 58, no. 6, pp. 1841–1844, 1977.

[3] P. C. Gupta, P. R. Murti, and R. B. Bhonsle, "Epidemiology of cancer by tobacco products and the significance of TSNA," *Critical Reviews in Toxicology*, vol. 26, no. 2, pp. 183–198, 1996.

[4] S. S. Hecht, "Biochemistry, biology, and carcinogenicity of tobacco-specific N- nitrosamines," *Chemical Research in Toxicology*, vol. 11, no. 6, pp. 559–603, 1998.

[5] H. Shi and J. Zhang, "The significant of alkaloids," in *Tobacco Alkaloids*, vol. 1st, p. 14, China Agriculture Press, Beijing, China, 2004.

[6] L. P. Bush, M. Cui, H. Shi et al., "Formation of tobacco specific nitrosamines in air-cured tobacco," *Recent Advances in Tobacco Science*, vol. 27, pp. 23–46, 2001.

[7] M. V. Djordjevic, S. L. Gay, L. P. Bush, and J. F. Chaplin, "Tobacco-specific nitrosamine accumulation and distribution in flue-cured tobacco alkaloid isolines," *Journal of Agricultural and Food Chemistry*, vol. 37, no. 3, pp. 752–758, 1989.

[8] I. Stepanov, A. Knezevich, L. Zhang, C. H. Watson, D. K. Hatsukami, and S. S. Hecht, "Carcinogenic tobacco-specific N-nitrosamines in US cigarettes: three decades of remarkable neglect by the tobacco industry," *Tobacco Control*, vol. 21, no. 1, pp. 44–48, 2011.

[9] S.-C. Shen, K.-C. Tseng, and J. S.-B. Wu, "An analysis of Maillard reaction products in ethanolic glucose-glycine solution," *Food Chemistry*, vol. 102, no. 1, pp. 281–287, 2007.

[10] M. Cui, *The source and the regulation of nitrogen oxide production for tobacco-specific nitrosamine formation during air-curing tobacco [Ph.D. thesis]*, University of Kentucky, Lexington, Ky, USA, 1998.

[11] H. R. Burton, N. K. Dye, and L. P. Bush, "Relationship between TSNA and nitrite from different air-cured tobacco varieties," *Journal of Agricultural and Food Chemistry*, vol. 42, pp. 2007–2011, 1994.

[12] X. Wei, X. Deng, D. Cai et al., "Decreased tobacco-specific nitrosamines by microbial treatment with *Bacillus amyloliquefaciens* DA9 during the air-curing process of burley tobacco," *Journal of Agricultural and Food Chemistry*, vol. 62, no. 52, pp. 12701–12706, 2014.

[13] J. L. Verrier, A. Wiernik, M. Staaf, J. l. Cadilhac, M. Onillon, and B. Vidal, "The influence of post-curing of burley tobacco and dark air-cured tobacco on TSNA and nitrite levels," in *Proceedings of the CORESTA Congress*, Shanghai, China, November 2008.

[14] R. M. Jackisch and J. H. Rovedder, "Burley tobacco post-curing management and its effect in the nitrosamine amount," in *Proceedings of the CORESTA Joint Study Group Meeting*, Krakow, Poland, October 2007, https://www.coresta.org/abstracts/burley-tobacco-post-curing-management-and-its-effect-nitro-samine-amount-1525.html.

[15] H. Saito, M. Miyazaki, and J. Miki, "Role of nitrogen oxides in tobacco-specific nitrosamine formation in burley tobacco," in *Proceedings of the 2006 CORESTA Congress*, Paris, France, October 2006, https://www.coresta.org/abstracts/role-nitrogen-oxides-tobacco-specific-nitrosamine-formation-burley-tobacco-2254.html.

[16] H. Shi, R. Wang, L. P. Bush et al., "Changes in TSNA contents during tobacco storage and the effect of temperature and nitrate level on TSNA formation," *Journal of Agricultural and Food Chemistry*, vol. 61, no. 47, pp. 11588–11594, 2013.

[17] W. Morgan, J. Reece, C. Risner et al., "A collaborative study for the determination of tobacco specific nitrosamines in tobacco," *Beiträge zur Tabakforschung International*, vol. 21, no. 3, pp. 192–203, 2014.

[18] J. Zhou, R. Bai, and Y. Zhu, "Determination of four tobacco-specific nitrosamines in mainstream cigarette smoke by gas chromatography/ion trap mass spectrometry," *Rapid Communications in Mass Spectrometry*, vol. 21, no. 24, pp. 4086–4092, 2007.

[19] J. D. Crutchfield and J. H. Grove, "A new cadmium reduction device for the microplate determination of nitrate in water, soil, plant tissue, and physiological fluids," *Journal of AOAC International*, vol. 94, no. 6, pp. 1896–1905, 2011.

[20] A. Jack and L. Bush, *The 'LC' Protocol-Appendix 3: Laboratory Procedures*, University of Kentucky, Lexington, Ky, USA, 2007, http://www.uky.edu/Ag/Tobacco/Pdf/LC-Protocol.Pdf.

[21] L. P. Bush, M. W. Cui, and H. Z. Shi, "Formation of tobacco-specific nitrosamines in air-cured tobacco," in *Proceedings of the 55th Tobacco Science Research Conference*, Greensboro, NC, USA, September 2001.

[22] H. Shi, N. E. Kalengamaliro, M. R. Krauss, W. P. Hempfling, and F. Gadani, "Stimulation of nicotine demethylation by $NaHCO_3$ treatment using greenhouse-grown burley tobacco," *Journal of Agricultural and Food Chemistry*, vol. 51, no. 26, pp. 7679–7683, 2003.

[23] H. Shi, R. Wang, L. P. Bush, H. Yang, and F. F. Fannin, "The relationships between TSNAs and their precursors in burley tobacco from different regions and varieties," *Journal of Food, Agriculture and Environment*, vol. 10, no. 3-4, pp. 1048–1052, 2012.

[24] R. S. Lewis, R. G. Parker, D. A. Danehower et al., "Impact of alleles at the *Yellow Burley* (Yb) loci and nitrogen fertilization rate on nitrogen utilization efficiency and tobacco-specific nitrosamine (TSNA) formation in air-cured tobacco," *Journal of Agricultural and Food Chemistry*, vol. 60, no. 25, pp. 6454–6461, 2012.

[25] W. J. Zhang, S. Rabiei, A. Bagreev, M. S. Zhuang, and F. Rasouli, "Study of NO adsorption on activated carbons," *Applied Catalysis B: Environmental*, vol. 83, no. 1-2, pp. 63–71, 2008.

[26] D. M. Peele, M. G. Riddick, and M. E. Edwards, "Formation of tobacco specific nitrosamines in flue-cured tobacco," *Recent Advances in Tobacco Science*, vol. 27, pp. 3–12, 2001.

[27] G. H. Ellington and M. D. Boyette, "Investigation into the correlation among nitrogen oxides and tobacco-specific nitrosamine in flue-cured tobacco," *Tobacco Science*, vol. 50, pp. 11–18, 2013.

[28] S. Tatsuoka, "Process for producing regenerated tobacco material," EP Patent:1782702 A1, 2005, http://www.freepatentsonline .com/EP1782702.html.

[29] X. D. Sun, W. G. Lin, L.-J. Wang et al., "Liquid adsorption of tobacco specific N-nitrosamines by zeolite and activated carbon," *Microporous and Mesoporous Materials*, vol. 200, pp. 260–268, 2014.

[30] W. G. Lin, B. C. Huang, B. Zhou et al., "Trapping tobacco specific N-nitrosamines in Chinese-Virginia type tobacco extracting solution by porous material," *Journal of Porous Materials*, vol. 21, no. 3, pp. 311–320, 2014.

3

Adsorptive Removal of Benzene and Toluene from Aqueous Environments by Cupric Oxide Nanoparticles: Kinetics and Isotherm Studies

Leili Mohammadi,[1] **Edris Bazrafshan,**[1] **Meissam Noroozifar,**[2] **Alireza Ansari-Moghaddam,**[1] **Farahnaz Barahuie,**[3] **and Davoud Balarak**[1]

[1]*Health Promotion Research Center, Zahedan University of Medical Sciences, Zahedan, Iran*
[2]*Analytical Research Laboratory, Department of Chemistry, University of Sistan and Baluchestan, Zahedan, Iran*
[3]*Engineering Faculty, Sistan and Baluchestan University, Zahedan, Iran*

Correspondence should be addressed to Edris Bazrafshan; ed_bazrafshan@yahoo.com

Academic Editor: Nicolas Roche

Removal of benzene and toluene, as the major pollutants of water resources, has attracted researchers' attention, given the risk they pose to human health. In the present study, the potential of copper oxide nanoparticles (CuO-NPs) in eliminating benzene and toluene from a mixed aqueous solution was evaluated. For this, we performed batch experiments to investigate the effect of solution pH (3–13), dose of CuO-NPs (0.1–0.8 g), contact time (5–120 min), and concentration of benzene and toluene (10–200 mg/l) on sorption efficiency. The maximum removal was observed at neutral pH. By using the Langmuir model, we measured the highest adsorption capacity to be 100.24 mg/g for benzene and 111.31 mg/g for toluene. Under optimal conditions, adsorption efficiency was 98.7% and 92.5% for benzene and toluene, respectively. The sorption data by CuO-NPs well fitted into the following models: Langmuir, Freundlich, Temkin, and Dubinin-Radushkevich model. The experimental information well fitted in the Freundlich for benzene and Langmuir for toluene. Based on the results, adsorption followed pseudo-second-order kinetics with acceptable coefficients. The findings introduced CuO-NPs as efficient compounds in pollutantsadsorption. In fact, they could be used to develop a simple and efficient pollutant removal method from aqueous solutions.

1. Introduction

Benzene and toluene are widely used as solvents for organic compounds, cleaning equipment, and other downstream processes in the industry. They can enter the groundwater due to leakage from storage tanks, pipes, and improper burial sites [1, 2]. US Environmental Protection Agency (EPA) has introduced these compounds as primary pollutants with adverse effects on human health. They are also classified as carcinogen with definite carcinogenic properties [3, 4].

Toluene is suspected to disrupt the central nervous system and is classified as a Group E carcinogen. Given these health effects, EPA has announced the maximum pollution level of benzene and toluene in drinking water as 5 μg/L and 2 μg/L, respectively. Also according to US Public Health Service

announcement in 1989, the amount of toluene in drinking water should not exceed 2 μg/L [5, 6]. Benzene and toluene are mostly found in effluents of chemical industries and refineries and have a high potential for pollution of surface water and groundwater. In addition, due to health effects, they are considered high-risk compounds for the environment [7]. Therefore, it is necessary to remove these compounds from water resources, particularly from surface water and groundwater.

Recently, nanoparticles (NPs) have been highly regarded by researchers for removal of pollutants. The unique properties of NPs and their high efficiency in removal of metals have made many researchers synthesize and use these substances for removal of environmental pollutants [8–10]. NPs have very good efficiency in the adsorption processes, especially adsorption of metal pollutants, due to their large effective surface and many active sites. In addition, these adsorbents have been used on a large scale to improve the capacity of adsorption of compounds with specific functional groups [11]. Nanoscale metal oxides have the potential to save water treatment costs and given their size and high absorption efficiency, they can improve the efficiency of technologies [12, 13].

Copper oxide NPs (CuO-NPs) have been used as effective adsorbents of metals including arsenic, because there is no need to correct pH or oxidize arsenic 3 to arsenic 5, and they act very well in comparison with other NPs containing anionic compounds. According to studies of Martinson and Reddy [14], CuO-NPs are known to be effective in the removal of a large number of compounds.

Benzene and toluene elimination from groundwater resources has been widely studied using various processes, mainly in biological recovery, evaporation, oxidation, and adsorption processes. In previous studies, benzene and toluene adsorption by resin, crude and refined diatomaceous earth, and organoclay have been evaluated [15].

Lu et al. found that, in the initial concentration of 60 ppm and 200 ppm of benzene and toluene, the amount of toluene adsorbed in terms of the adsorbent mass unit is greater than the amount of adsorbed benzene [16]. In addition, Aivalioti et al. [17] have shown the greater adsorption potential of diatomaceous adsorbents in the removal of benzene, compared to toluene. Daifullah and Girgis [15] have found that the adsorption of benzene in terms of the adsorbent mass unit is more than toluene.

Su et al. in a previous study reported the adsorption potential of benzene and toluene to be 212 and 225 mg/g, using sodium hypochlorite-modified nanotubes at an initial concentration of 200 ppm, contact time of 240 minutes, and adsorbent concentration of 600 mg/L. Benzene adsorption capacity was greater than toluene [18]. This finding reveals that benzene and toluene adsorption by different adsorbents is dependent on the chemical properties and porosity of the adsorbent surface. Similar results have been provided by other researchers in previous articles [15].

The aim of this research project was to conduct batch experiments to study adsorption capacity, reaction kinetics, and the effects of critical operating parameters on benzene and toluene removal by CuO-NPs. Four factors including pH, CuO-NPs dose, contact time, and initial benzene and toluene concentration were used to survey the efficiency of CuO-NPs in benzene and toluene removal from aqueous environments. Furthermore, modeling of experimental data was performed by isotherm equations.

2. Materials and Methods

2.1. Materials. Benzene, toluene, and copper nitrate $(Cu(NO_3)_2 \cdot 3H_2O)$ were supplied by Merck Co. (Germany). To prepare the stock solution, 2 mL of benzene and 2 mL of toluene (with a purity of 99.7%) were added to 1 L methanol. The homogeneous stock solution was completely sealed and stored in a volumetric flask until the daily solutions were prepared at different concentrations. A conventional method was applied to prepare CuO-NPs. First, distilled water was used to dissolve $Cu(NO_3)_2 \cdot 3H_2O$. The solution pH was set at 10 using Na_2CO_3 (1 M) through stirring. Then, the product was aged with stock liquor over 12 hours at room temperature. Filtration was performed to gather the product. Afterwards, it was rinsed with demineralized water, dried for 1 day at 60°C, and calcined at a temperature of 350°C for 4 hours [19].

2.2. Analysis. In this study, benzene and toluene concentrations were analyzed using a headspace-gas chromatograph (Agilent 7890 A, Palo Alto, CA, USA) with a flame ionization detector and a capillary column (thickness, 0.25 μm; length, 30 m; ID, 0.32 mm). For benzene and toluene measurements, the thermal program of the column was set at 40°C for 10 minutes, followed by a rise in temperature (temperature, 120°C; speed, 10°C/min) continuing for 2 minutes. The temperature of the injection site was fixed at 250°C, and 1 mL of the sample was injected into the device in the splitless mode. The gas chromatography effluent was transferred to an ionization source detector at 280°C through a 280°C transmission line. The analysis was performed in the selected ion monitoring mode. In addition, a Quanta-400F microscope, equipped with a field emission (KYKY-EM3900M, China) was used to perform scanning electron microscopy (SEM) of the prepared CuO-NPs on gold-coated samples. In order to determine the X-ray diffraction (XRD) pattern (CuK radiation; 40 kV; 30 mA), an advanced D8 diffractometer (Bruker, Germany) was used. In addition, a pH meter (WTW inoLab 7310) was employed to assess the solution pH.

2.3. Adsorption Experiments. In the present study, triplicate batch experiments were performed. Batch adsorption experiments were performed in flasks (250 mL) via magnetic stirring. We prepared 100 mL of benzene and toluene solution (initial dose, 10–200 mg/L) through dilution of the stock solution with distilled water; afterwards, it was moved to a beaker and placed on a magnetic stirrer.

We adjusted pH (range, 3–13) by using a solution containing 0.1 N of HCl or NaOH. Then, 0.1–0.8 g/L of synthesized CuO-NPs was added, and the suspension was stirred during 5–120 minutes. The samples were removed from the mixture with a micropipette after reaching the desired contact duration; then, the residual concentration of benzene and toluene was analyzed by GC method.

Adsorption kinetic experiments were performed with 250 mL beakers, containing 0.6 g of CuO-NPs and 100 mL of benzene and toluene (10, 20, 50, 100, and 200 mg/L) at pH of 7. After different intervals, we gathered and filtered the sample aliquots with syringe filters. Then, to determine the amount of the remaining contaminant, the aliquots were injected into the GC device. A similar procedure was applied for experiments on adsorption isotherms, with the exception that we used different concentrations of benzene and toluene solutions. The experiments were performed in triplicate under the same conditions to check the reproducibility of calculations; then, we measured the average values. The

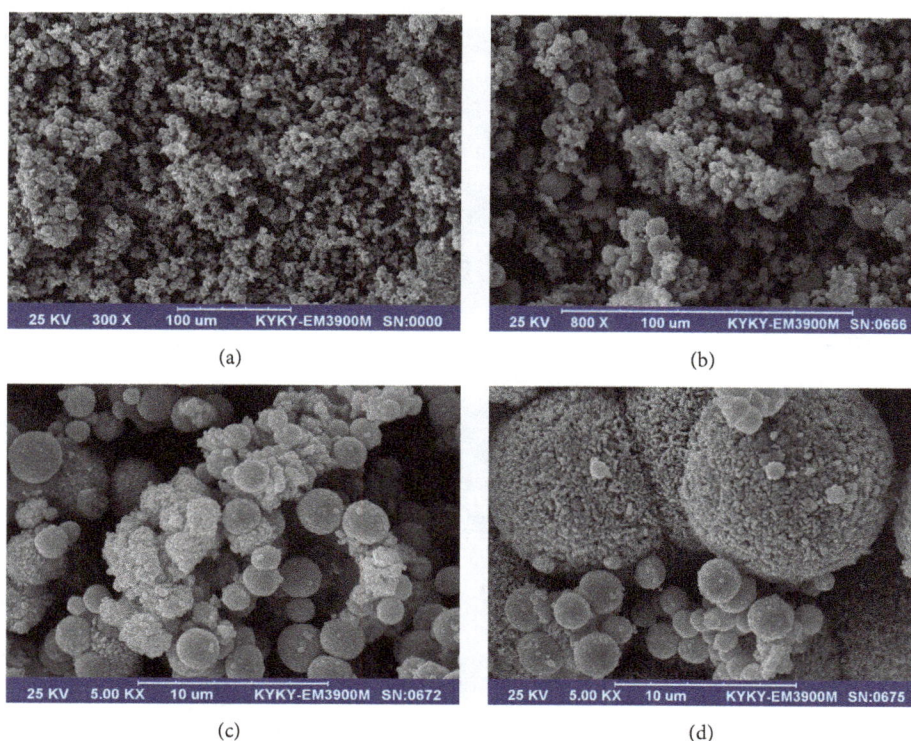

(a)

(b)

(c)

(d)

FIGURE 1: SEM images of CuO nanoparticles.

amount of the adsorbed benzene and toluene q_e (mg/g) was calculated under different conditions:

$$q_e = \frac{(C_0 - C_e) V}{M}. \tag{1}$$

In this equation, C_0 is the initial concentration, C_e denotes the liquid-phase dose of benzene and toluene at equilibrium in mg/L, V denotes the solution volume (L), and M refers to the quantity of the adsorbent (g). Moreover, to determine the benzene and toluene removal percentage, the following formula was used:

$$\text{Removal efficiency, } \% = \frac{(C_i - C_e)}{C_i} \times 100, \tag{2}$$

where C_i denotes the initial dose of the pollutant and C_e is the effluent concentration in mg/L.

We carried out the experiments in duplicate to ensure the reproducibility of the findings; then, we measured the mean of 2 calculations. The equilibrium isotherm was demonstrated by the plot of equilibrium adsorption potential versus concentration at equilibrium.

3. Results and Discussion

3.1. Adsorbent Characterization. View of adsorbent used for removing benzene and toluene in this study was analyzed by SEM and XRD device. Figure 1 shows the order specified absorbent SEM of the adsorbent. An agglomeration of nanoscale particles with spherical and flower shapes is

clearly observed. The average diameter of spherical CuO-NPs counted from the SEM images is about 148 nm. Figure 2 shows the XRD profile of the adsorbent (CuO-NPs). All 11 diffraction peaks at $2\theta = 33°C$, $36°C$, $39°C$, $49°C$, $54°C$, $59°C$, $62°C$, $66°C$, $68°C$, $73°C$, and $75°C$ were indexed to (110), (002), (111), (202), (020), (202), (113), (311), (113), (311), and (004) planes, respectively, which are related to the monoclinic crystal phase of CuO-NPs; these values are comparable with those in JCPDS file for CuO (JCPDS, card number 45-0937).

3.2. Analysis of the Effect of Solution pH. The solution pH is a critical factor in the application of nanoadsorbents as support materials in the removal of various pollutants in the adsorption process [20–25]. Under different pH environments, transference of proton might take place on the surface of metal oxides, leading to adsorption in reaction pathways [26]. To determine the optimum solution pH, it was changed from 3 to 13 with an initial benzene/toluene dosage of 50 mg/L and an adsorbent concentration (CuO-NPs) of 0.4 g per 100 mL of solution. Figure 3 shows the effect of initial solution pH on benzene/toluene adsorption. It can be seen that, by increasing pH from 3 to 7, the removal efficiency increased from about 61.76% to 67.36% for benzene and 68.34% to 73.28% for toluene and continuous increasing of pH decreased the efficiency. The greatest adsorption potential was estimated to be 84.2 and 91.6 mg/g for benzene and toluene at a pH of 7, respectively. Under these conditions, the residual concentration of benzene was 16.32 mg/L and that of toluene was 13.36 mg/L. According to these results, the optimum pH was found to be 7 for benzene/toluene

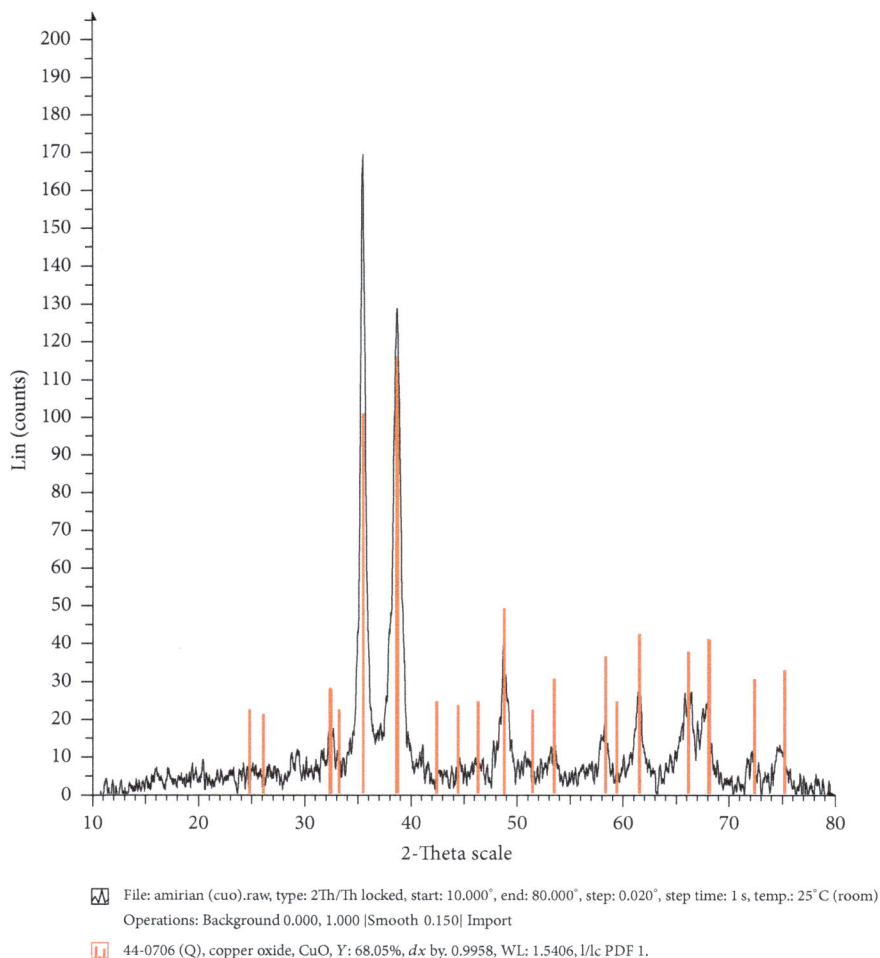

File: amirian (cuo).raw, type: 2Th/Th locked, start: 10.000°, end: 80.000°, step: 0.020°, step time: 1 s, temp.: 25°C (room)
Operations: Background 0.000, 1.000 |Smooth 0.150| Import

44-0706 (Q), copper oxide, CuO, Y: 68.05%, dx by. 0.9958, WL: 1.5406, l/lc PDF 1.

FIGURE 2: XRD patterns of CuO-NPs.

adsorption. Consequently, pH of 7 was considered optimal for the next steps.

3.3. *Effect of CuO-NP Dose on Benzene/Toluene Adsorption.* The adsorbent dose is a critical factor, as it can show the adsorbent's potential (CuO-NPs) for a specific initial concentration of benzene/toluene. Hence, to determine the impact of CuO-NPs dose on benzene/toluene adsorption, 0.1 to 0.8 g/L of the adsorbent was used in the experiments at pH of 7 and initial benzene/toluene concentration of 50 mg/L for 60 minutes. As presented in Figure 4, benzene/toluene removal efficiency is associated with CuO-NPs dosage in the solution. Based on the diagram, by increasing the concentration of the adsorbent from 0.1 to 0.6 g, the removal efficiency was enhanced from 62% to 93% for benzene and from 52% to 89% for toluene. Under these conditions, the residual concentration of benzene and toluene reached 3.29 and 5.22 mg/L, respectively.

Furthermore, as demonstrated in Figure 4, benzene/toluene removal capacity was rapidly enhanced by increasing the quantity of CuO-NPs from 0.1 to 0.6 g and from 0.6 to 0.8 g. We can explain this finding by resorting to the fact that

sorption sites continue to be unsaturated in sorption, while the available sites grow in number by raising the adsorbent dosage [21]. Furthermore, it was found that further increase in dosage beyond 8 g/L rarely affected adsorption capacity of CuO-NPs (data not showed). Consequently, 0.6 g adsorbent was considered as the optimum dose.

In addition as presented in Figure 4, the equilibrium adsorption capacity of CuO-NPs decreased from 312.4 to 77.85 mg for benzene and 263.9 to 74.63 mg for toluene. On the other hand, the adsorbed benzene/toluene per gram of adsorbent reduced fast with a rise in CuO-NPs (maximum adsorption capacity, 312.4 mg/g for benzene and 263.9 mg/g for toluene at minimum adsorbent concentration of 0.1 g/L). This observation might be associated with the fixed benzene/toluene dosage (50 mg/L), producing active sites on the surface of CuO-NPs and increasing the adsorbent dosage due to aggregation of particles [27]; similar results have been reported by other researchers [21, 22, 24, 28].

3.4. *Effects of Contact Time and Initial Benzene/Toluene Concentration.* In the adsorption process, pollutant elimination from aqueous solutions is relative to the duration of

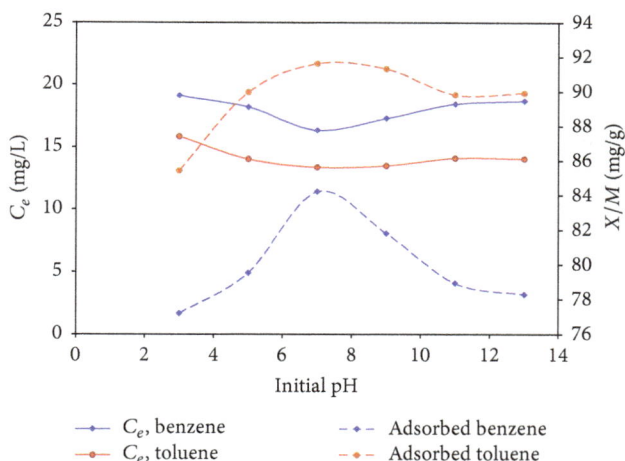

FIGURE 3: The effect of solution pH on benzene and toluene adsorption by CuO-NPs (CuO-NPs dose = 0.4 g/L, contact time = 60 min, and benzene and toluene concentration = 50 mg/L).

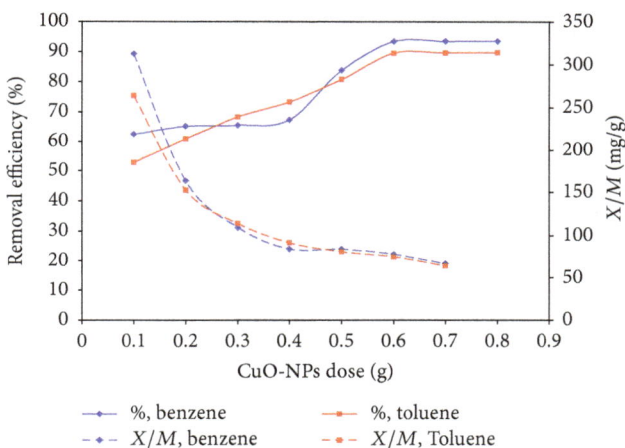

FIGURE 4: Effect of adsorbent dose on benzene and toluene adsorption by CuO-NPs (benzene/toluene concentration = 50 mg/L, initial pH = 7, and contact time = 60 min).

adsorbent-adsorbate contact [22, 29, 30]. Also, the initial concentration of solute is of remarkable importance due to its role in solid/liquid equilibrium. Hence, benzene/toluene removal by CuO-NPs was investigated as a function of contact duration. Accordingly, the effectiveness of contact duration in the adsorption of benzene/toluene on CuO-NPs was examined at a fixed pH of 7, CuO-NP dose of 0.6 g/L, and various benzene/toluene concentrations.

The analysis of the relationship between contact time and rate of adsorption in the studied pollutants by CuO-NPs at different concentrations is shown in Figures 5 and 6. As can be seen, at all concentrations tested, the highest removal efficiencies were observed in contact times of >40 minutes. Also, as presented in Figures 5 and 6, pollutant elimination improved by increasing the contact time at different initial concentrations of benzene/toluene. In fact, in the first 20 minutes, the adsorption rate was reported to be high (36–63%

for benzene and 38–59% for toluene), followed by a slower rate of adsorption and finally equilibrium within 80 minutes. Hence, due to the fact that no significant increase occurs in removal efficiency after 80 minutes, the contact time of 80 minutes was considered as the optimal adsorption time (equilibrium time). The high removal rate of benzene/toluene during the first one-third of reaction time has been attributed to the interfacial bonding of adsorbate on the empty and ready-to-adsorb binding sites of adsorbent surface. Similar findings were reported by other researchers [25, 31].

The results regarding the effect of initial benzene/toluene concentration are shown in Figures 5 and 6. As can be clearly seen, rise in the initial pollutant concentration could increase the removal efficiency and adsorption capacity. For instance, with an initial benzene concentration of 10 mg/L (contact time, 80 min), removal efficiency reached more than 87.4% with an adsorption capacity of 14.57 mg/g, whereas, at an initial benzene concentration of 200 mg/L, the corresponding values were 94.87% and 316.23 mg/g, respectively; similar trends were obtained for toluene. Consequently, the removal of benzene/toluene by CuO-NPs depends on the concentration of the pollutant. The rise in adsorption capacity (q_e) by increasing the initial benzene/toluene concentration at fixed dose of adsorbent can be described by the increased driving force, associated with the concentration gradient [32]. In fact, the initial concentration of the pollutant molecules provides an important driving force to overcome the mass transfer resistance of all molecules between the aqueous and solid phases. Bulut et al. [33] observed that the amount of dye molecules adsorbed per unit mass of biosorbent increased with an increase in initial dye concentration from 50 to 250 mg/L. It is estimated that the binding sites of sorbent stay unsaturated during the biosorption mechanism.

3.5. Adsorption Kinetics. Adsorption kinetics are of great importance in the application of adsorbents in actual situations. Several models have been employed to describe and explain the mechanisms involved in the sorption process. In the present study, to assess the adsorption kinetics of benzene and toluene on CuO-NPs, we used different models. In the pseudo-first model, it is assumed that there is a direct relation between changes in the rate of adsorption of the solved substance and time [25, 34]. The linear form of the first-order kinetics is shown in

$$\ln{(q_e - q_t)} = \ln{q_e} - k_1 t, \qquad (3)$$

where k_1 denotes the pseudo-first-order rate constant (1/minute), q_e represents the amount of adsorbed benzene and toluene at equilibrium (mg/g), q_t is the amount of adsorbed benzene and toluene at time t (mg/g), and t denotes time in minutes. Therefore, a plot of $\ln(q_e - q_t)$ versus t determines k_1 with respect to the slope and $\ln q_e$ with respect to the intercept of the plot.

The second-order kinetic model [21] is presented as follows:

$$\frac{t}{q_t} = \frac{1}{k_2 q_e^2} + \frac{t}{q_e}. \qquad (4)$$

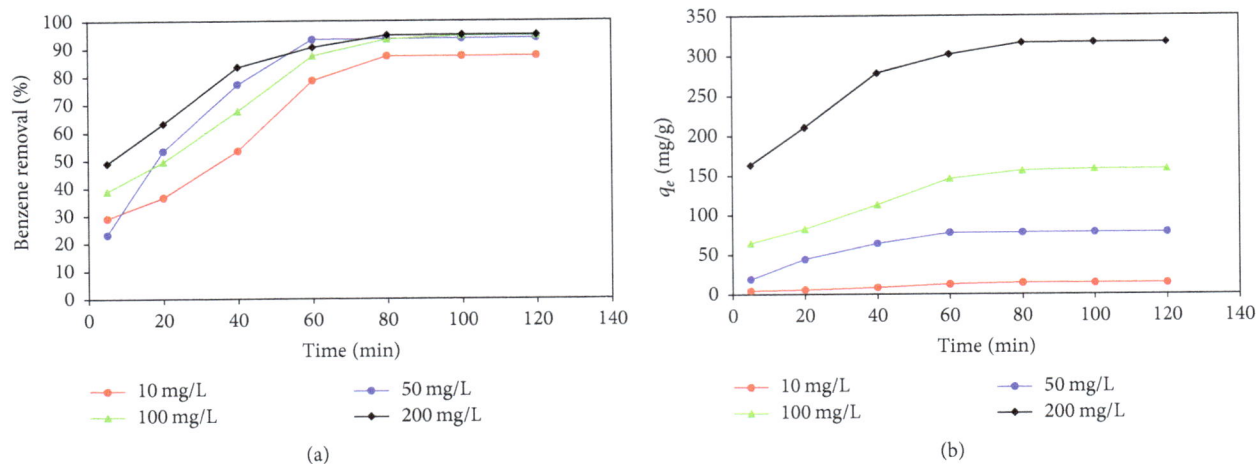

FIGURE 5: The effect of contact time on benzene adsorption by CuO-NPs ((a) percent removal efficiency, (b) adsorption capacity).

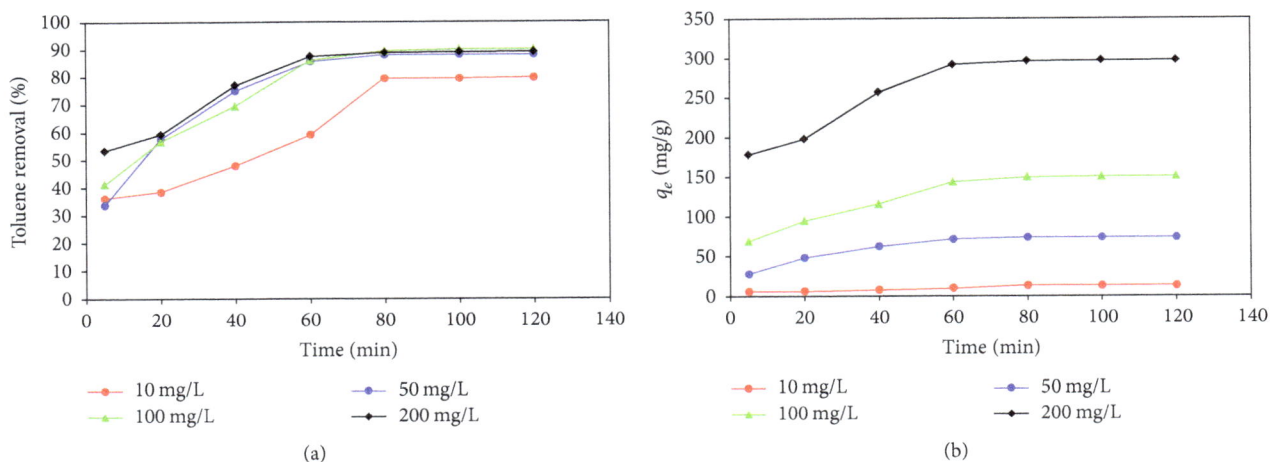

FIGURE 6: The effect of contact time on toluene adsorption by CuO-NPs ((a) percent removal efficiency, (b) adsorption capacity).

In this equation, k_2 represents the pseudo-second-order rate constant (g/mg/min), q_e denotes the amount of adsorbed benzene/toluene during equilibrium (mg/g), q_t denotes the amount of adsorbed benzene and toluene at time t (mg/g), and t refers to time in minutes. In case second-order kinetics were considered valid, a t/q_t versus t plot can indicate the linear correlation. In addition, the rate constants of k_2 and q_e can be measured with respect to the slope and intercept of the plot, respectively. This model can probably predict behaviors during the adsorption process and is in accordance with chemical sorption (stage of rate control). On the other hand, adsorption is assumed to be directed by chemical adsorption in the second-order kinetic model.

The intraparticle diffusion model is mathematically expressed as [35]

$$q_t = k_p t^{0.5} + C, \tag{5}$$

where k_p denotes the intraparticle diffusion rate constant in mg/g/min 0.5, C is the intraparticle diffusion constant in mg/g, t is time in minutes, and q_t is defined as before.

By drawing q_t in terms of $t^{0.5}$, k_p and C can be obtained. In the present research, the intraparticle diffusion model was applied to determine the adsorption mechanism. As previously described by Weber and Morris [36], it should be noted that porosity and intraparticle diffusion influence the adsorption rate in a non-flow-agitated system.

The linear plots of pseudo-first-order, pseudo-second-order, and intraparticle diffusion models for benzene and toluene are illustrated in Figures 7–9. In addition, the values of reaction constants and correlation coefficients obtained from the linear plots are presented in Table 1. It is apparent from the values of correlation coefficients that fitness of the pseudo-second-order model is better as compared to pseudo-first-order model (the linear regression correlations (R^2) from pseudo-second-order model were higher than those from pseudo-first-order model). As reported by Pillewan et al. [37], in adsorption systems following pseudo-second-order models, a common mechanism is chemical adsorption, incorporating a chemical bond between the active adsorbent sites and adsorbate valence forces.

TABLE 1: Kinetic parameters for the adsorption of benzene/toluene by CuO-NPs for different concentrations at pH = 7.

C_0, mg/L	Pseudo-first-order			Pseudo-second-order			Intraparticle diffusion		
	k_1 (1/min)	q_e (mg/g)	R^2	k_2 (g/mg/min)	q_e, mg/g	R^2	k_p (g/mg/min$^{0.5}$)	C (mg/g)	R^2
Benzene									
10	0.0235	1056	0.8629	1.5511	0.0543	0.9441	1.3165	1.4322	0.9262
50	0.0398	38.66	0.6156	0.1912	0.0108	0.9922	6.876	13.179	0.871
100	0.0465	164.35	0.9476	0.0955	0.0055	0.9823	12.151	37.5	0.9407
200	0.0515	224.08	0.9149	0.0246	0.0029	0.9971	18.694	136.61	0.9061
Toluene									
10	0.021	9.39	0.8922	1.5549	0.063	0.9342	0.9983	2.7892	0.9055
50	0.037	40.07	0.8834	0.1322	0.0123	0.9975	5.251	23.492	0.8835
100	0.0453	109.51	0.9077	0.0709	0.006	0.9926	10.097	51.504	0.9335
200	0.0499	165.67	0.9003	0.024	0.0031	0.9926	15.521	146.66	0.8969

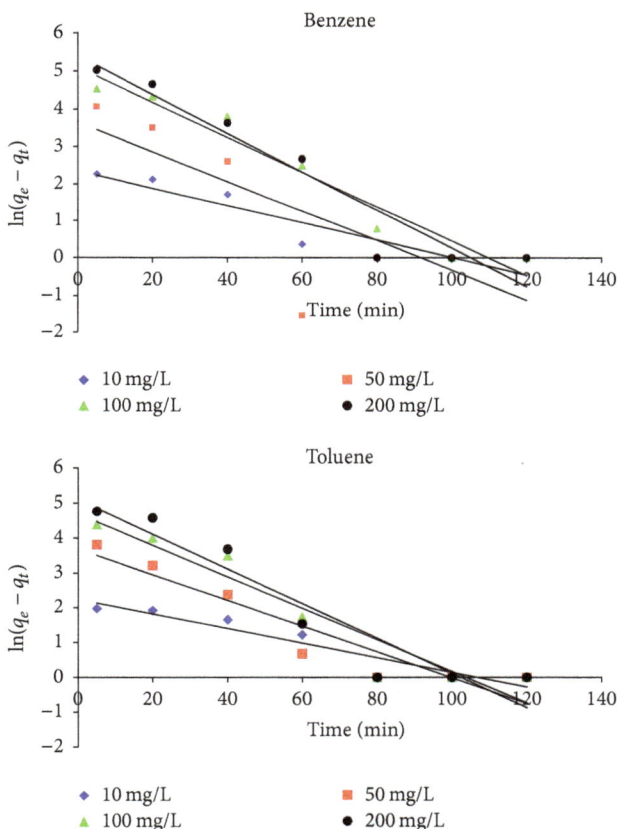

FIGURE 7: Pseudo-first-order kinetic plots for benzene/toluene adsorption on CuO-NPs at different initial concentration of benzene/toluene (adsorbent dose = 0.6 g/L, pH = 7).

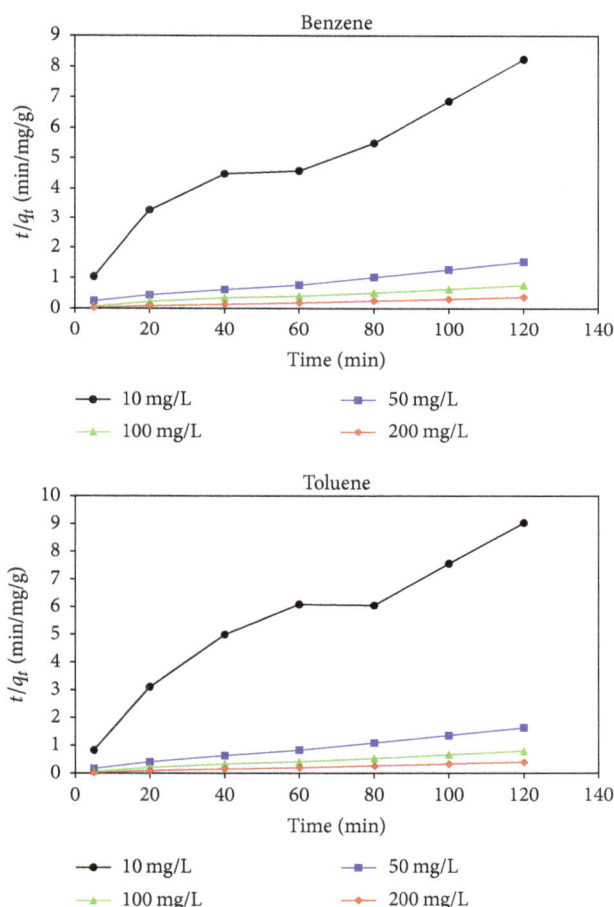

FIGURE 8: Pseudo-second-order kinetic plots for benzene and toluene adsorption on CuO-NPs at different initial concentration of benzene/toluene (adsorbent dose = 0.6 g/L, pH = 7).

Furthermore, the plot of the linear form of intraparticle diffusion model is presented in Figure 9. Also, the results of the analysis of parameters are demonstrated in Table 1. The R^2 values of the particle diffusion model are almost uniform, showing the greater contribution of adsorbate particle diffusion to constant rate analysis.

3.6. Equilibrium Adsorption Isotherm. Isotherms of adsorption are valuable curves to describe the adsorption equilibrium and provide a great deal of information to reveal the reciprocity between the adsorbent and adsorbate. On

TABLE 2: Isotherm parameters for adsorption of benzene/toluene onto CuO-NPs.

	Benzene			Toluene		
	q_m (mg/g)	k_l (L/mg)	R^2	q_m (mg/g)	k_l (L/mg)	R^2
Langmuir isotherm	100.24	0.1	0.9718	111.31	0.05	0.9826
	k_f	n	R^2	k_f	n	R^2
Freundlich isotherm	11.79	0.7	0.9836	5.88	0.76	0.9818
	k_T	B	R^2	k_T	B	R^2
Temkin isotherm	$1.79359E+31$	0.0065	0.8678	$1.70366E+52$	0.0079	0.9144
	q_m	β	R^2	q_m	β	R^2
D-R isotherm	6.08	0.0024	0.9812	5.84	0.0035	0.9782

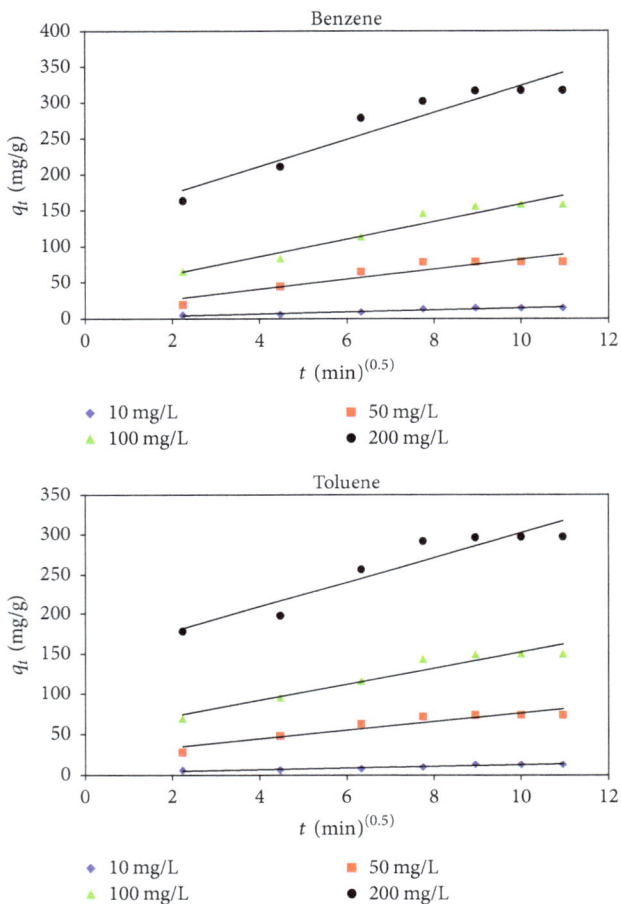

FIGURE 9: Intraparticle diffusion kinetic plots for benzene/toluene adsorption on CuO-NPs at different initial concentrations (adsorbent dose = 0.6 g/L, pH = 7).

adsorption mechanisms, surface adsorbent characteristics, adsorption affinity, and adsorption experimental data. Consequently, building an excellent correlation between equilibrium diagrams for optimization of conditions and designing adsorption systems is very important [39].

3.6.1. Langmuir Model. It is appropriate for monolayer adsorption on surfaces, consisting of a fixed quantity of identical sorption sites. On the other hand, this model includes assumptions such as monolayer adsorption, surface uniformity, and removal of the interaction of the adsorbed molecules. For monolayer adsorption, the Langmuir equation is as follows:

$$q_e = \frac{q_m k_l C_e}{1 + k_l C_e}. \quad (6)$$

In this equation, q_e denotes the quantity of adsorbed benzene/toluene per unit mass of CuO-NPs as the adsorbent (mg/g), C_e denotes the solution's concentration at equilibrium (mg/L), and q_m refers to the maximum quantity of benzene/toluene for developing a monolayer on CuO-NPs (mg/g). It is possible to reset the Langmuir model in a linear form to facilitate plotting and determine the Langmuir constants (k_l), as well as the optimal monolayer adsorption potential of the adsorbent (q_m). Also, q_m and k_l are measured based on the linear plot ($1/q_e$ versus $1/C_e$):

$$\frac{1}{q_e} = \frac{1}{q_m} + \frac{1}{q_m k_l} \frac{1}{C_e}. \quad (7)$$

3.6.2. Freundlich Model. In this model, the assumption is that adsorption takes place on heterogeneous surfaces. Based on this model, the process is defined as follows [40]:

$$q_e = k_f C_e^{1/n}. \quad (8)$$

In this equation, k_f and $1/n$ denote the characteristics of Freundlich constants, k_f represents the adsorption potential of the adsorbent, and n represents the intensity of adsorption. The equilibrium constants were examined with respect to the intercept and slope of the linear plot ($\ln q_e$ versus $\ln C_e$), respectively, according to the experimental information.

3.6.3. Temkin Model. In this model, the assumption is that reduced sorption heat is linear, and binding energy distribution is uniform (up to the highest binding energy).

the other hand, adsorption isotherms are adsorption properties and equilibrium data, which describe the reactions of pollutants with adsorbents and have a critical role in the optimization of adsorbent use [38].

As presented in Table 2, Langmuir, Freundlich, Temkin, and Dubinin-Radushkevich models were applied in order to examine the equilibrium isotherms for benzene/toluene adsorption on CuO-NPs and explain the equilibrium adsorption information. These models are applied to describe the

It considers indirect adsorbate-adsorbent interactions and suggests a linear decline in adsorption heat of all molecules inside the layer due to these interactions. Temkin model is normally as follows:

$$q_e = B \ln k_T + B \ln C_e, \qquad (9)$$

where k_T and B are measured based on a linear plot of q_e versus $\ln C_e$, k_T denotes the equilibrium binding constant (L/mg; the highest binding energy), and B is associated with the adsorption heat.

3.6.4. Dubinin-Radushkevich Model. This model is frequently applied to examine the porosity and free adsorption energy. The linear form of Dubinin-Radushkevich (D-R) model is as follows:

$$\log q_e = \ln q_m - \beta \varepsilon^2. \qquad (10)$$

In this equation, β denotes a constant associated with the mean free adsorption energy per mole of the adsorbate $(\text{mol}^2/\text{KJ}^2)$, q_m refers to the theoretical saturation potential (mg/g), and ε represents the Polanyi potential [21, 41].

The relationship between the amount of adsorbed benzene and toluene and their equilibrium concentrations was described by the four equations of adsorption isotherm, as follows: Langmuir, Freundlich, Dubinin-Radushkevich (D-R), and Temkin, and the best-fit equilibrium model was established based on the linear regression correlation coefficients, R^2. The values of R^2 from Table 2 show that the Freundlich and Langmuir models give a good fit to the sorption isotherm for benzene and toluene onto CuO-NPs, respectively.

4. Conclusion

In the present study, CuO-NPs were successfully applied for benzene/toluene removal from aqueous environments. Based on the findings, the removal percentage was dependent on the solution pH, amount of CuO-NPs as the adsorbent, initial benzene/toluene dosage, and pollutant-adsorbent contact time. The maximum removal efficiency of CuO-NPs for benzene and toluene was observed at a pH of 7. Also, based on the findings, CuO-NPs could remove 98.7–25.41% of benzene and 92.5–35.2% of toluene from solutions with an initial dosage of 10–200 mg/L and contact duration of 1 hour. By increasing CuO-NPs and duration of contact, the removal percentage of benzene and toluene improved. According to the results, the sorption data fitted into the Langmuir, Freundlich, Temkin, and Dubinin-Radushkevich models. Also, benzene well fitted into the Freundlich model, and toluene well fitted into the Langmuir model. Based on the analyses, adsorption follows pseudo-second-order kinetics.

Conflicts of Interest

The authors declare that there are no conflicts of interest regarding the publication of this manuscript.

Acknowledgments

This article is derived from the Ph.D. thesis of Ms. L. Mohammadi and all authors are grateful to the Zahedan University of Medical Sciences for the financial support of this study (Project no. 7391). Furthermore, all authors wish to thank Dr. Hossein Kamani, Dr. Ferdos Kord Mostafapour, and Ms. Sh. Sargazi for their support during the analysis of experiments.

References

[1] A. R. Bielefeldt and H. D. Stensel, "Evaluation of biodegradation kinetic testing methods and longterm variability in biokinetics for BTEX metabolism," *Water Research*, vol. 33, no. 3, pp. 733–740, 1999.

[2] L. Mohammadi, *BTEX and chlorophenols compounds removal from synthetic sollutions by heterogenous catalytic ozonation using FeO, CuO and MgO nanoparticles [Ph.D. desertaion],* Zahedan University of Medical Sciences, (in Persian), 2017.

[3] X. Fu, X. Gu, S. Lu et al., "Enhanced degradation of benzene in aqueous solution by sodium percarbonate activated with chelated-Fe(II)," *Chemical Engineering Journal*, vol. 285, pp. 180–188, 2016.

[4] N. Kang and I. Hua, "Enhanced chemical oxidation of aromatic hydrocarbons in soil systems," *Chemosphere*, vol. 61, no. 7, pp. 909–922, 2005.

[5] B. Bina, H. Pourzamani, A. Rashidi, and M. M. Amin, "Ethyl-benzene removal by carbon nanotubes from aqueous solution," *Journal of Environmental and Public Health*, vol. 2012, Article ID 817187, 8 pages, 2012.

[6] A. K. Mathur, C. B. Majumder, and S. Chatterjee, "Combined removal of BTEX in air stream by using mixture of sugar cane bagasse, compost and GAC as biofilter media," *Journal of Hazardous Materials*, vol. 148, no. 1-2, pp. 64–74, 2007.

[7] J. M. M. D. Mello, H. de Lima Brandão, A. A. U. de Souza, A. da Silva, and S. M. D. A. G. U. de Souza, "Biodegradation of BTEX compounds in a biofilm reactor-Modeling and simulation," *Journal of Petroleum Science and Engineering*, vol. 70, no. 1-2, pp. 131–139, 2010.

[8] M.-Q. Jiang, Q.-P. Wang, X.-Y. Jin, and Z.-L. Chen, "Removal of Pb(II) from aqueous solution using modified and unmodified kaolinite clay," *Journal of Hazardous Materials*, vol. 170, no. 1, pp. 332–339, 2009.

[9] A. Afkhami and R. Moosavi, "Adsorptive removal of Congo red, a carcinogenic textile dye, from aqueous solutions by maghemite nanoparticles," *Journal of Hazardous Materials*, vol. 174, no. 1–3, pp. 398–403, 2010.

[10] M. Yang, J. He, X. Hu, C. Yan, and Z. Cheng, "CuO nanostructures as quartz crystal microbalance sensing layers for detection of trace hydrogen cyanide gas," *Environmental Science and Technology*, vol. 45, no. 14, pp. 6088–6094, 2011.

[11] A. R. Türker, "New sorbents for solid-phase extraction for metal enrichment," *Clean-Soil Air Water*, vol. 35, no. 6, pp. 548–557, 2007.

[12] W.-X. Zhang, "Nanoscale iron particles for environmental remediation: an overview," *Journal of Nanoparticle Research*, vol. 5, no. 3-4, pp. 323–332, 2003.

[13] K. E. Engates and H. J. Shipley, "Adsorption of Pb, Cd, Cu, Zn, and Ni to titanium dioxide nanoparticles: Effect of particle size, solid concentration, and exhaustion," *Environmental Science and Pollution Research*, vol. 18, no. 3, pp. 386–395, 2011.

[14] C. A. Martinson and K. J. Reddy, "Adsorption of arsenic(III) and arsenic(V) by cupric oxide nanoparticles," *Journal of Colloid and Interface Science*, vol. 336, no. 2, pp. 406–411, 2009.

[15] A. A. M. Daifullah and B. S. Girgis, "Impact of surface characteristics of activated carbon on adsorption of BTEX," *Colloids and Surfaces A: Physicochemical and Engineering Aspects*, vol. 214, no. 1-3, pp. 181–193, 2003.

[16] C. Lu, F. Su, and S. Hu, "Surface modification of carbon nanotubes for enhancing BTEX adsorption from aqueous solutions," *Applied Surface Science*, vol. 254, no. 21, pp. 7035–7041, 2008.

[17] M. Aivalioti, I. Vamvasakis, and E. Gidarakos, "BTEX and MTBE adsorption onto raw and thermally modified diatomite," *Journal of Hazardous Materials*, vol. 178, no. 1-3, pp. 136–143, 2010.

[18] F. Su, C. Lu, and S. Hu, "Adsorption of benzene, toluene, ethylbenzene and p-xylene by NaOCl-oxidized carbon nanotubes," *Colloids and Surfaces A: Physicochemical and Engineering Aspects*, vol. 353, no. 1, pp. 83–91, 2010.

[19] K. Zhou, R. Wang, B. Xu, and Y. Li, "Synthesis, characterization and catalytic properties of CuO nanocrystals with various shapes," *Nanotechnology*, vol. 17, no. 15, pp. 3939–3943, 2006.

[20] E. Bazrafshan, P. Amirian, A. H. Mahvi, and A. Ansari-Moghaddam, "Application of adsorption process for phenolic compounds removal from aqueous environments: a systematic review," *Global Nest Journal*, vol. 18, no. 1, pp. 146–163, 2016.

[21] E. Bazrafshan, F. Kord Mostafapour, S. Rahdar, and A. H. Mahvi, "Equilibrium and thermodynamics studies for decolorization of Reactive Black 5 (RB5) by adsorption onto MWCNTs," *Desalination and Water Treatment*, vol. 54, no. 8, pp. 2241–2251, 2015.

[22] E. Bazrafshan, A. A. Zarei, H. Nadi, and M. A. Zazouli, "Adsorptive removal of methyl orange and reactive red 198 dyes by moringa peregrina ash," *Indian Journal of Chemical Technology*, vol. 21, no. 2, pp. 105–113, 2014.

[23] E. Bazrafshan, M. Ahmadabadi, and A. H. Mahvi, "Reactive red-120 removal by activated carbon obtained from cumin herb wastes," *Fresenius Environmental Bulletin*, vol. 22, no. 2a, pp. 584–590, 2013.

[24] E. Bazrafshan, F. K. Mostafapour, and A. H. Mahvi, "Phenol removal from aqueous solutions using pistachio-nut shell ash as a low cost adsorbent," *Fresenius Environmental Bulletin*, vol. 21, no. 10, pp. 2962–2968, 2012.

[25] F. Kord Mostafapour, E. Bazrafshan, M. Farzadkia, and S. Amini, "Arsenic removal from aqueous solutions by salvadora persica stem ash," *Journal of Chemistry*, vol. 2013, Article ID 740847, 8 pages, 2013.

[26] E. Bazrafshan, F. K. Mostafapour, A. R. Hosseini, A. Raksh Khorshid, and A. H. Mahvi, "Decolorisation of reactive red 120 dye by using single-walled carbon nanotubes in aqueous solutions," *Journal of Chemistry*, vol. 2013, Article ID 938374, 8 pages, 2013.

[27] T. Calvete, E. C. Lima, N. F. Cardoso, S. L. P. Dias, and F. A. Pavan, "Application of carbon adsorbents prepared from the Brazilian pine-fruit-shell for the removal of Procion Red MX 3B from aqueous solution—kinetic, equilibrium, and thermodynamic studies," *Chemical Engineering Journal*, vol. 155, no. 3, pp. 627–636, 2009.

[28] E. Bazrafshan, D. Balarak, A. H. Panahi, H. Kamani, and A. H. Mahvi, "Fluoride removal from aqueous solutions by cupricoxide nanoparticles," *Fluoride*, vol. 49, no. 3, pp. 233–244, 2016.

[29] E. Bazrafshan, A. A. Zarei, and F. K. Mostafapour, "Biosorption of cadmium from aqueous solutions by Trichoderma fungus: kinetic, thermodynamic, and equilibrium study," *Desalination and Water Treatment*, vol. 57, no. 31, pp. 14598–14608, 2016.

[30] P. Loganathan, S. Vigneswaran, and J. Kandasamy, "Enhanced removal of nitrate from water using surface modification of adsorbents—a review," *Journal of Environmental Management*, vol. 131, pp. 363–374, 2013.

[31] B. Saha, K. E. Taylor, J. K. Bewtra, and N. Biswas, "Laccase-catalyzed removal of diphenylamine from synthetic wastewater," *Water Environment Research*, vol. 80, no. 11, pp. 2118–2124, 2008.

[32] E. Bekhradinassab and S. Sabbaghi, "Removal of nitrate from drinking water using nano SiO2-FeOOH-Fe core-shell," *Desalination*, vol. 347, pp. 1–9, 2014.

[33] Y. Bulut, N. Gözübenli, and H. Aydin, "Equilibrium and kinetics studies for adsorption of direct blue 71 from aqueous solution by wheat shells," *Journal of Hazardous Materials*, vol. 144, no. 1-2, pp. 300–306, 2007.

[34] N. Öztürk and T. E. Bektaş, "Nitrate removal from aqueous solution by adsorption onto various materials," *Journal of Hazardous Materials*, vol. 112, no. 1-2, pp. 155–162, 2004.

[35] J. Wu and H.-Q. Yu, "Biosorption of 2,4-dichlorophenol by immobilized white-rot fungus Phanerochaete chrysosporium from aqueous solutions," *Bioresource Technology*, vol. 98, no. 2, pp. 253–259, 2007.

[36] W. J. Weber and J. C. Morris, "Advances in water pollution research: removal of biologically resistant pollutants from waste waters by adsorption," in *Proceedings of the International Conference on Water Pollution Symposium*, vol. 2, pp. 231–266, Pergamon, Oxford, UK, 1962.

[37] P. Pillewan, S. Mukherjee, T. Roychowdhury, S. Das, A. Bansiwal, and S. Rayalu, "Removal of As(III) and As(V) from water by copper oxide incorporated mesoporous alumina," *Journal of Hazardous Materials*, vol. 186, no. 1, pp. 367–375, 2011.

[38] Z. Yang, H. Yi, X. Tang et al., "Potential demonstrations of 'hot spots' presence by adsorption-desorption of toluene vapor onto granular activated carbon under microwave radiation," *Chemical Engineering Journal*, vol. 319, pp. 191–199, 2017.

[39] S.-L. Hii, S.-Y. Yong, and C.-L. Wong, "Removal of rhodamine B from aqueous solution by sorption on Turbinaria conoides (Phaeophyta)," *Journal of Applied Phycology*, vol. 21, no. 5, pp. 625–631, 2009.

[40] S. Khorramfar, N. M. Mahmoodi, M. Arami, and K. Gharanjig, "Dye removal from colored textile wastewater using tamarindus indica hull: adsorption isotherm and kinetics study," *Journal of Color Science and Technology*, vol. 3, pp. 81–88, 2009.

[41] M. M. Montazer, P. Rabbani, A. Abdolali, and A. R. Keshtkar, "Kinetics and equilibrium studies on biosorption of cadmium, lead, and nickel ions from aqueous solutions by intact and chemically modified brown algae," *Journal of Hazardous Materials*, vol. 185, no. 1, pp. 401–407, 2011.

4

Fe Isotopic Compositions of Modern Seafloor Hydrothermal Systems and Their Influence Factors

Xiaohu Li, Jianqiang Wang, and Hao Wang

Key Laboratory of Submarine Geosciences, Second Institute of Oceanography, State Oceanic Administration, Hangzhou 310012, China

Correspondence should be addressed to Xiaohu Li; lixh09@gmail.com

Academic Editor: Rafael García-Tenorio

Based on previous research on the Fe isotope compositions of various components and systems of the Earth, this study focused on the Fe isotope compositions of hydrothermal systems, including the Fe isotope variations in chalcopyrite, pyrite, and sphalerite, and their possible controlling factors. The main findings are as follows: (1) The range of Fe isotopes in hydrothermal systems at mid-ocean ridge is very large. The δ^{56}Fe values of hydrothermal fluids are characterized by significant enrichment in light Fe isotopes. (2) The δ^{56}Fe values of sulfides also exhibit lighter Fe isotope characteristics than those of hydrothermal fluids from hydrothermal vent fields at mid-ocean ridge. The vent temperature, fluid properties, and mineral deposition processes significantly affect the δ^{56}Fe values of hydrothermal sulfides. (3) Chalcopyrite is preferentially enriched in heavy Fe isotopes, whereas sphalerite and pyrite are enriched in light Fe isotopes. In addition, the δ^{56}Fe values of pyrite/marcasite display a larger range than those of chalcopyrite. This pattern is directly related to equilibrium fractionation or kinetic fractionation of Fe isotopes during the deposition of sulfides. To better understand the Fe isotope compositions of modern seafloor hydrothermal systems, the geochemical behavior and fractionation mechanisms of Fe isotopes require further in situ study.

1. Introduction

As the most abundant element on Earth, Fe is widely distributed in various components and systems of the Earth. Fe is involved in the physiochemical aspects of biological and inorganic processes in high- and low-temperature geologic environments. Fe has four stable isotopes in nature, ^{54}Fe, ^{56}Fe, ^{57}Fe, and ^{58}Fe, whose abundances are 5.84%, 91.76%, 2.12%, and 0.28%, respectively. Recently, the rapid development of a highly accurate multiple-collector inductively coupled plasma mass spectrometry (MC-ICP-MS) technique for measuring Fe isotopes has the enabled broad application of Fe isotopes in the Earth science [1, 2].

Previous research investigated the Fe isotope compositions of various components and systems of the Earth. The range of δ^{56}Fe values in meteorites is from −0.15‰ to 0.18‰, with an average δ^{56}Fe of 0.03 ± 0.06‰ (n = 71) (Figure 1). The Fe isotope composition of Earth materials is generally thought to be similar to the average of meteorites [1]. The Fe isotope composition of the mantle is controlled primarily by the Fe isotope compositions of mantle peridotites. This range

is from −1.1‰ to 0.48‰, with an average δ^{56}Fe of −0.05 ± 0.21‰ (n = 364). The Fe isotope composition of the oceanic crust is controlled by the Fe isotope compositions of basalts. This range is from −0.46‰ to 0.81‰, with an average δ^{56}Fe of 0.04 ± 0.12‰ (n = 232). The Fe isotope composition of the continental crust is controlled by the Fe isotope compositions of shale and loess/dust. The range of δ^{56}Fe values in shale is from −0.36‰ to 0.42‰, with an average δ^{56}Fe of 0.02 ± 0.17‰ (n = 24). The range of δ^{56}Fe values in loess/dust is from −0.15‰ to 0.33‰, with an average δ^{56}Fe of 0.05±0.10‰ (n = 40). Fe in marine systems is derived primarily from rivers [3, 4], atmospheric dust [5], and seafloor hydrothermal fluids [6, 7]. The removal of Fe from oceans is primarily in the form of marine sediments, carbonates, and Fe-Mn nodules and crusts [8]. The range of Fe isotope compositions of seawater is from −0.64‰ to 0.80‰, with an average δ^{56}Fe of 0.34 ± 0.25‰ (n = 497) (Figure 1). The Fe isotope output endmembers of the ocean are enriched in light Fe isotopes relative to the input endmembers. Modern seafloor hydrothermal systems are an important source of Fe in the

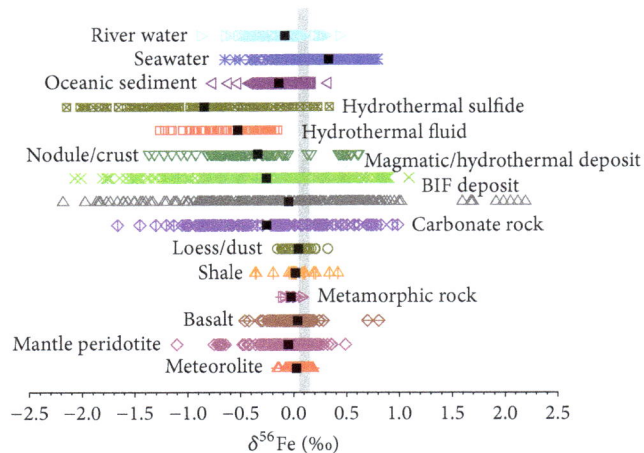

FIGURE 1: Fe isotopic composition of each repository. (The gray color column shows the Fe isotope composition of basalt with an average value of $+0.105 \pm 0.006‰$ ($n = 43$) [10]; the black rectangle shows the average value of δ^{56}Fe. All data from the literatures: river [3, 4, 11, 12]; ocean water [13–16]; oceanic sediment [6, 7, 12, 17]; hydrothermal sulfide [6, 18–20]; hydrothermal fluid [6, 7, 19–21]; Fe-Mn noddle and crust [8, 22–24]; deposit [25–31]; banded Fe formations [32–42]; carbonate rock [36, 43, 44]; loess/dust [6, 31, 45]; shale [6]; metamorphose rock [46]; basalt [32, 33, 45, 47–53]; mantle peridotite [32, 33, 45, 47–52, 54, 55]; meteorolite [24, 33, 34, 49, 50, 56–62].)

oceans. Hydrothermal fluids and ocean waters exchange a significant quantity of material and energy during seafloor hydrothermal activity. The mixing of hydrothermal fluids with seawater and their subsequent cooling by conduction significantly affect the cycling of chemical elements in the oceans [9].

Based on previous research, this study focused on the Fe isotope composition characteristics of hydrothermal systems, including composition variations of chalcopyrite, pyrite, and sphalerite, and the possible controlling factors. Therefore, the scope of this paper is mainly concerned with inorganic Fe isotope fractionation in the modern seafloor hydrothermal systems. We discuss the effects of bedrock properties, vent temperature [18], hydrothermal sulfide deposition processes [19], chemical properties of hydrothermal fluids [20], and phase separation [6] of mid-ocean ridge hydrothermal systems, and we summarize the variations in Fe isotope compositions of seafloor hydrothermal systems and the fractionation mechanisms.

2. Results and Discussion

2.1. Fe Isotopic Compositions of Modern Seafloor Hydrothermal Fluids. The range of Fe isotope compositions (δ^{56}Fe) of hydrothermal fluids along modern seafloor mid-ocean ridge is from $-1.26‰$ to $-0.14‰$ with an average value of $-0.52 \pm 0.28‰$ ($n = 69$) (Figure 1). Studies of the Fe isotope compositions of hydrothermal fluids from hydrothermal vent on the Pacific ridge [6, 19], the Atlantic ridge [6, 18, 21], and the Juan de Fuca ridge [20] revealed that the Fe isotope compositions of these vent fluids varied significantly (Figure 2). The range

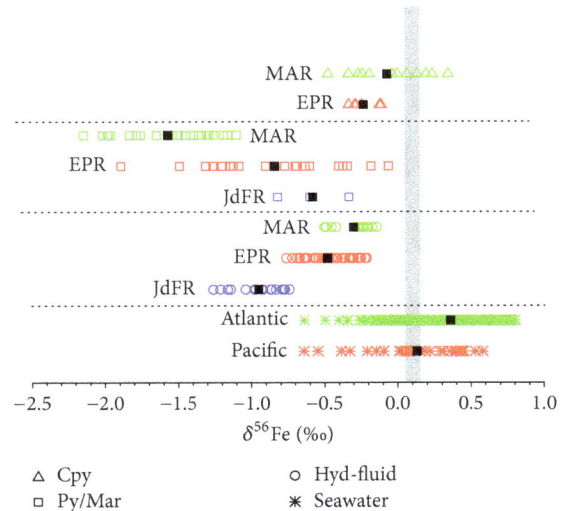

FIGURE 2: Fe isotopic composition of modern seafloor hydrothermal fluid and sulfide. (The gray color column shows the Fe isotope composition of basalt with an average value of $+0.11 \pm 0.01‰$ ($n = 43$) [10]; the black rectangle shows the average value of δ^{56}Fe. All data from the literatures is the same as Figure 1).

of Fe isotope compositions along the fast-spreading Eastern Pacific ridge is from $-0.76‰$ to $-0.25‰$, with an average δ^{56}Fe of $-0.48 \pm 0.14‰$ ($n = 29$). The range of Fe isotope compositions is from $-1.26‰$ to $-0.74‰$, with an average δ^{56}Fe of $-0.95 \pm 0.16‰$ ($n = 16$), along the moderate spreading Juan de Fuca ridge and is from $-0.50‰$ to $-0.14‰$ at the Lucky Strike vent on the slow spreading North Atlantic ridge, with an average δ^{56}Fe of $-0.30 \pm 0.11‰$ ($n = 24$). The Fe isotope composition variations among mid-ocean-ridge hydrothermal fluids may be related primarily to water-rock reactions [17], bedrock properties [18], hydrothermal sulfide deposition [19], chemical compositions of fluids [20], phase separation [6], sediments, and microbiologic effects [15, 64].

2.2. Fe Isotopic Compositions of Modern Seafloor Hydrothermal Sulfides. The ranges of Fe isotope compositions of chalcopyrite, pyrite, and sphalerite (Figure 2) are from $-0.47‰$ to $0.35‰$, $-2.14‰$ to $-0.06‰$, and $-1.08‰$ to $-0.44‰$, respectively, with average δ^{56}Fe values of $-0.12 \pm 0.24‰$ ($n = 22$), $-1.18 \pm 0.53‰$ ($n = 48$), and $-0.82 \pm 0.19‰$ ($n = 22$), respectively. The Fe isotope compositions of sulfide minerals range from heavier to lighter among chalcopyrite, pyrite/marcasite, and sphalerite, which indicates the differences in isotope fractionation mechanisms during the deposition of these sulfide minerals.

The range of Fe isotope compositions of chalcopyrite from the Atlantic ridge is from $-0.47‰$ to $0.35‰$, with an average δ^{56}Fe of $-0.07 \pm 0.27‰$ ($n = 16$) [18]. The range of Fe isotope compositions of chalcopyrite from the Eastern Pacific ridge is from $-0.33‰$ to $-0.11‰$, with an average δ^{56}Fe of $-0.23 \pm 0.08‰$ ($n = 6$) [19] (Figure 2).

The range of Fe isotope compositions of pyrite/marcasite from the Atlantic ridge is from $-2.14‰$ to $-1.10‰$, with average δ^{56}Fe of $-1.57 \pm 0.31‰$ ($n = 22$) [18]. The range of Fe isotope compositions of pyrite from the Eastern Pacific ridge is from −1.89‰ to −0.06‰, with an average δ^{56}Fe of −0.86 ± 0.43‰ (n = 25) [19]. The range of the Fe isotope compositions of the Juan de Fuca ridge is from −0.82‰ to −0.33‰, with an average δ^{56}Fe of −0.58 ± 0.20‰ (n = 3) [20].

2.3. Factors Controlling the Fe Isotopic Composition of Hydrothermal Fluids

2.3.1. Water and Rock Interaction. Large amounts of seawater permeate deeply into the oceanic crust through fracture zones at mid-ocean ridge, where the seawater is warmed by heat sources. This seawater reacts with the bedrock deep in the oceanic crust, leaches out ore-forming elements, and erupts from the seafloor, forming hydrothermal fluids. During the weathering and leaching of seafloor basalts, the leached basalts are highly depleted in Fe and have heavier Fe isotope compositions than fresh basalts [17]. Leaching experiments have been conducted on biotite granites and tholeiitic basalts at room temperature and under acidic conditions. The fractionation coefficient of the leached fluid phase relative to the original rocks is −0.5‰ to −1‰ [65]. The ranges of δ^{56}Fe values of fresh and altered basalts from the Juan de Fuca ridge are from −0.79‰ to −0.46‰ and −0.66‰ to −0.38‰, respectively [20]. These results indicate that light Fe isotopes are leached out during reactions between the seafloor hydrothermal fluids and rock.

2.3.2. Bedrock. The bedrock of mid-ocean ridge is the major source of Fe in hydrothermal systems. The bedrock of the Lucy Strike hydrothermal field at the Atlantic ridge consists of enriched mid-ocean ridge basalt (E-MORB). The basalt located at 9 to 10°N on the East Pacific Rise consists of normal mid-ocean ridge basalt (N-MORB). E-MORB exhibits a lesser degree of partial melting than does N-MORB. E-MORB has extremely strong incompatibility and is enriched in strongly incompatible elements [66]. Fe(III) is less compatible than Fe(II) during partial melting [67] and preferentially enters the melt phase. Furthermore, the Fe isotope composition of Fe(III) is heavier than that of Fe(II) [68–71]. Studies also indicated that during fractional crystallization of magma olivine is enriched in light Fe isotopes. With the continuous crystallization of olivine phenocrysts, the heavy Fe isotope enrichment in basalts increases [10]. Therefore, the Fe isotope compositions of the hydrothermal fields at the Atlantic ridge, where the bedrock is primarily E-MORB, are enriched in heavy Fe isotopes compared with those of the hydrothermal fields at approximately 9 ~ 10°N on the East Pacific Rise, where the bedrock is primarily N-MORB.

2.3.3. Biological Effect. The Juan de Fuca ridge has received hemipelagic and turbidite deposits since the late Pleistocene. Consequently, the mid-ocean ridge there is covered by sediment that is a few hundred to a few thousand meters thick [72]. Furthermore, the hydrothermal vent fluids at the Endeavour Segment have high CH_4 and NH_4^+ concentrations [73, 74], indicating significant microorganism activity

in the sediment. Deep-sea sediment is enriched in light Fe isotopes (Figure 1). Additionally, microorganisms cause preferential enrichment in light Fe isotopes [15, 64]. These hydrothermal fluids may have been affected by sediment mixing and microorganism activity during recycling and eruption, resulting in the lighter Fe isotope compositions of the hydrothermal fluids at the Juan de Fuca ridge than those of the hydrothermal fluids at the Atlantic and Pacific ridges.

It has been shown that iron stable isotopes can provide precise information on metals biogeochemical processes and help to identify and better quantify the biogeochemical Fe cycle of plant metabolism studied directly in nature [75]. Crosby et al. reported that the Fe dissimilatory reduction (DIR) promoted by *Geobacter sulfurreducens* and *Shewanella putrefaciens* strains caused Fe isotopic fractionation of approximately 2.2‰, with final Fe(II) species enriched in light isotopes [76], which demonstrated that the Fe isotope composition of Fe(II)$_{aq}$ is largely controlled by isotopic exchange with a reactive Fe(III) pool that lies in the outer layers of the ferric oxide substrate. The adsorption effect led to a particularly strong heavy iron enrichment onto cyanobacteria cells relative to Fe(II)$_{aq}$ when compared to similar experiments performed with Fe(II)$_{aq}$ [77]. The distribution of iron isotopes in sediments and sedimentary rocks is also a powerful measure of the biogeochemical cycle of Fe in the modern and ancient ocean [78].

2.3.4. Phase Separation of Hydrothermal Fluids. Experiments have demonstrated that liquid phases are strongly enriched in positive divalent ions of transition metals (such as Fe(II)) at gas-liquid equilibrium under high-temperature and high-pressure conditions [79]. Previous studies included simulated experiments on variations in the Fe isotope compositions of hydrothermal fluid systems during the phase separation period at seafloor hydrothermal vents within a temperature range of 424 to 466°C and a pressure range of 35.2 to 24.7 MPa [63]. When the solution system did not exhibit phase separation, the δ^{56}Fe value was −0.29 ± 0.03‰ (2σ). Phase separation occurred in the systems with decompression and formed gas and liquid phases. The Fe isotopes were significantly fractionated. The gas phase was enriched in heavy Fe isotopes (Figure 3(a)). During gradual decompression, the fractionation coefficient of the gas phase relative to the liquid phase was larger than zero and increased continuously until equilibrium was approached (Figure 3(b)). During the decompression, the Fe isotope composition of the liquid phase was lighter than that of the gas phase and the entire system. The Cl^- concentration of the gas phase gradually decreased (Figure 3(a)), and the chloride content decreased. The chemical properties of the entire system changed. The chemical composition difference between the gas and liquid phases affected the Fe isotope fractionation to some extent. The experimental results indicated that the fractionation coefficient of Fe isotopes between coexisting gas and liquid phases reached Δ^{56}Fe$_{(gas-liquid)}$ = +0.15 ± 0.05‰, indicating that phase separation caused the chemical composition differences between the endmembers. During the fluid phase separation period, the gas phase depleted in Cl was likely enriched in heavy Fe isotopes.

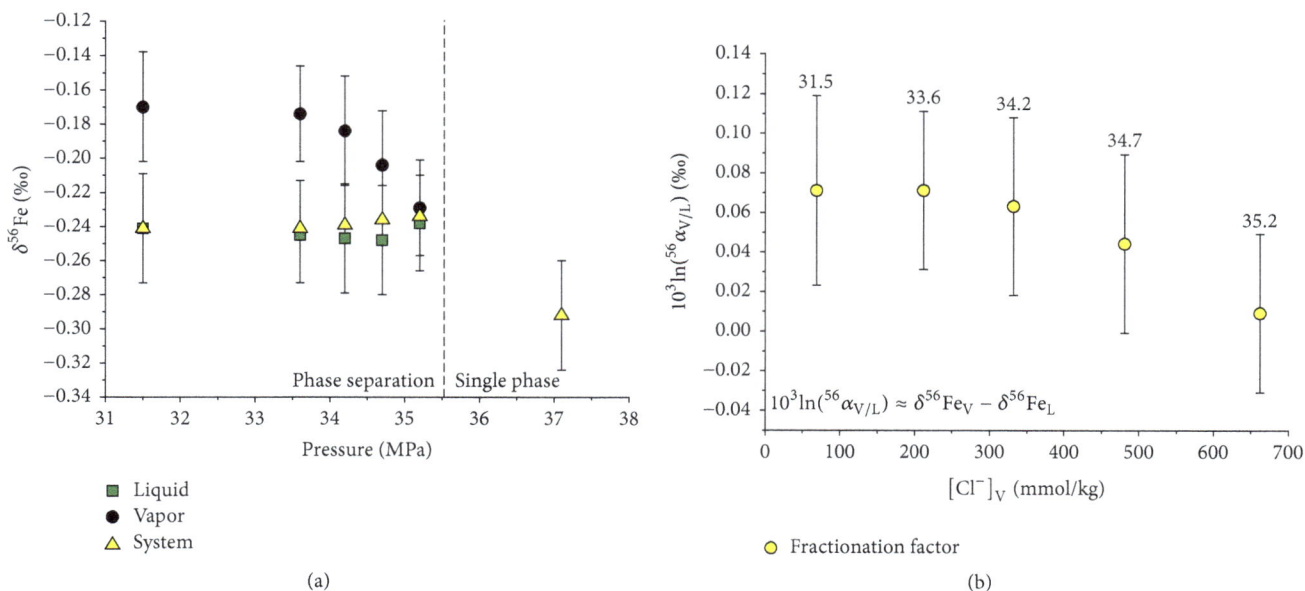

FIGURE 3: Fe isotope compositions and fractionations of the vapor, liquid, and integrated system during decompression (after [63]).

Temperature and pressure changes during phase separation under high-temperature and high-pressure conditions can cause the coordination chemistry of water-bearing metal chloride complexes to change as a function of the Cl concentration of the environment [80]. Syverson et al. proposed that $[FeCl_4]^{2-}$ was present predominantly in the liquid phase, whereas $[FeCl_2(H_2O)_2]^0$ preferentially entered the gas phase [63]. Theoretical calculations indicated that the degree of Fe isotope fractionation caused by the difference between the coordination chemistry of Fe-Cl and Fe-H_2O complexes is likely caused by an oxidization-reduction effect [81]. Water-bearing (including oxygen bonded) transition-metal complexes are enriched in heavy Fe isotopes compared with chloride transition-metal complexes [82]. In addition, previous studies indicated that because certain light Fe isotopes were lost during the collision in which the Moon was formed from the Earth, the Fe isotope compositions of the Earth and Moon are significantly heavier than those of other planets [62]. The heavier Fe isotope compositions of the Earth and the Moon could also be due to high oxygen fugacity during the magma evolution that caused the residual phase to be enriched in heavy Fe isotopes.

2.4. Factors Controlling the Fe Isotopic Compositions of Hydrothermal Sulfides

2.4.1. Vent Temperature. The hydrothermal fluid temperature of black smoker chimneys in the high-temperature hydrothermal fields at 9 to 10°N on the East Pacific Rise exceeds 300°C. The hydrothermal fluid temperature in the low-temperature hydrothermal fields is 100 to 250°C. The temperatures of the hydrothermal fluid diffusion zones are below 100°C. When the hydrothermal vent temperature is below 250°C, hydrothermal sulfides do not form chalcopyrite

[19]. The vent temperature of the Lucky Strike hydrothermal field enriched in Cu sulfides exceeds 300°C. The temperatures of vent regions rich in Fe-Zn sulfides are between 180 and 220°C [18].

The minerals formed at high temperature during hydrothermal sulfide mineralization are enriched in Se, whereas the low-temperature minerals are depleted in Se [83]. Based on the relation between the Se concentrations of hydrothermal sulfides and Fe isotope compositions (Figure 4), chalcopyrite is preferentially deposited as a high-temperature mineral and is enriched in heavy Fe isotopes, causing the hydrothermal fluids to be depleted in heavy Fe isotopes. Pyrite and sphalerite deposited later are enriched in light Fe isotopes (Figure 5(d)). The Bio9'' vent temperature shown in Figure 5(a) is as high as 383°C. Chalcopyrite is preferentially deposited as a high-temperature mineral, and its Fe and S are both heavier than those in pyrite and sphalerite, which is deposited later. In later hydrothermal fluids, heavy Fe isotopes are consumed. Therefore, the later-deposited pyrite and sphalerite are enriched in light Fe isotopes. The K-vent temperature is 203°C (Figure 5(b)). The Fe isotope compositions of the coexisting pyrite and sphalerite are essentially consistent, and both are lower than those in the hydrothermal fluid. However, pyrite is enriched in light S isotopes compared with sphalerite, possibly because sphalerite is preferentially enriched in heavy S isotopes. Studies of the Navan magmatic-hydrothermal Pb-Zn deposit in Ireland indicated a temperature of the ore-forming fluid of 150 to 170°C. During rapid sphalerite deposition, because of kinetic fractionation, the sphalerite is enriched in light $\delta^{56}Fe$ (−1.2‰) and $\delta^{66}Zn$ (−0.15‰), and the two values display a very high positive correlation [84].

The hydrothermal sulfide assemblage at the high-temperature Bio9'' vent (383°C) located at 9 to 10°N on the

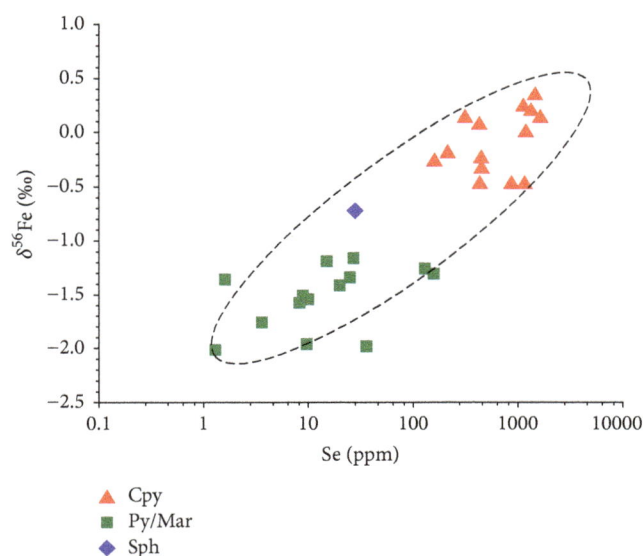

FIGURE 4: Se content versus δ^{56}Fe values of hydrothermal sulfides (after [18]).

East Pacific Rise in Figure 5 is rich in Cu, Fe, and Zn. The δ^{56}Fe range of the chalcopyrite is from $-0.29‰$ to $-0.11‰$. The range of δ^{56}Fe values of the pyrite is from $-1.26‰$ to $-1.08‰$ and δ^{56}Fe values of the sphalerite is $-0.95‰$. The hydrothermal sulfide assemblage at the low-temperature K-vent (203°C) is rich in Fe and Zn. The range of δ^{56}Fe values of the pyrite is from $-0.90‰$ to $-0.63‰$. The range of δ^{56}Fe values of the sphalerite is from $-0.54‰$ to $-0.76‰$. The inactive hydrothermal vent regions contain low-temperature diffusive fluids (~100°C), and the hydrothermal sulfide assemblage is rich in Fe and Zn. The range of δ^{56}Fe values of the pyrite is from $-1.79‰$ to $-0.18‰$. The range of δ^{56}Fe values of the sphalerite is from $-0.94‰$ to $-0.55‰$. The range of δ^{56}Fe values of the hydrothermal fluids is $-0.67‰$ to $-0.25‰$ [19]. Due to the remineralization of late-stage hydrothermal fluids, pyrite formed more strongly with hydrothermal fluids during the early-stage disequilibrium isotope fractionation exchanges of Fe and S isotopes and approaches equilibrium fractionation [18, 19].

After comparing the sulfide assemblages of high- and low-temperature hydrothermal vent regions and the corresponding sulfide δ^{56}Fe characteristics, we developed the following three conclusions: (1) Seafloor hydrothermal sulfides of Cu-Fe-Zn mineral assemblage where δ^{56}Fe$_{chalcopyrite} > \delta^{56}Fe_{pyrite}$ and δ^{56}Fe$_{pyrite} \approx \delta^{56}Fe_{sphalerite}$ indicate the mineralization of high-temperature hydrothermal vent (~ 350°C). (2) Seafloor hydrothermal sulfides exhibiting Fe-Zn mineral assemblages where δ^{56}Fe$_{pyrite} \approx \delta^{56}Fe_{sphalerite} < \delta^{56}Fe_{hydrothermal\ fluid}$ indicate the mineralization environments of low-temperature hydrothermal vent (~250°C). (3) Seafloor hydrothermal sulfides with Fe-Zn mineral assemblages where δ^{56}Fe$_{pyrite} \approx \delta^{56}Fe_{sphalerite} \approx \delta^{56}Fe_{hydrothermal\ fluid}$ in certain Fe-Zn mineral assemblages and where δ^{56}Fe$_{pyrite} < \delta^{56}$Fe$_{sphalerite}$ in others indicate the possible remineralization

of early-stage hydrothermal sulfides by low-temperature diffusive fluids in hydrothermal fields (~100°C).

2.4.2. Hydrothermal Fluid Chemistry.
Previous studies of the skarn deposit at Xinqiao, Anhui, indicated that the range of wall rock δ^{56}Fe values of the deposit is from $-0.13‰$ to $0.41‰$, with an average value of $0.07‰$, whereas the range of δ^{56}Fe values of the magnetite that formed earliest is from $-0.54‰$ to $0.20‰$ (average of $-0.11‰$). Compared with the intrusion, the magnetite was enriched in light Fe isotopes. The average δ^{56}Fe value of the pyrite that formed later is $-0.41‰$. The δ^{56}Fe characteristics of the country rock cannot explain the enrichment of Fe-bearing minerals in light Fe isotopes. The enrichment is a result of the enrichment in light Fe isotopes of the initial fluid exsolved from the magma [27]. Fe in the magma fluids and hydrothermal fluids originated from magma that existed in the form of FeCl$_2^0$ [85], and the iron chloride is enriched in light Fe isotopes, compared with magnetite and silicate melts [86]. After comparing a high-Cl vent (Pipe Organ) and a low-Cl vent (Inferno) on the Juan de Fuca ridge, we noted that if the fluid Cl concentration is low, with a higher H$_2$S/Fe value, the light Fe isotope enrichment of hydrothermal fluids during the period of sulfide deposition is greater [20]. Therefore, both the chemical property differences and changes in hydrothermal fluids can cause variations in the Fe isotope compositions of hydrothermal sulfides.

Rouxel et al. [19] proposed that the Fe isotopes of chalcopyrite display very little kinetic fractionation or are approximately at equilibrium fractionation. Studies have shown that chalcopyrite formed at high temperature in seafloor hydrothermal fluids tends to be significantly more enriched in heavy Fe isotopes than hydrothermal fluids. However, the Fe isotope composition range of chalcopyrite in the Atlantic ridge is clearly broader than that of chalcopyrite at the East Pacific Rise (Figure 2). The chalcopyrite of the East Pacific Rise is all from the Bio9″ vent. The chalcopyrite in the Atlantic ridge is from the Bairro Alto, Elisabeth, Y3, Eiffel Tower, and Marker US4 vents. The vent temperatures in the Lucky Strike hydrothermal fields are significantly different (170 to 350°C) [18]. The chemical property differences of the fluids are significantly affected by phase separation [87]. The oxidization-reduction reaction during the mixing of the vent fluids enriched in H$_2$S and seawater enriched in SO$_4^{2-}$ causes the S isotope differences of the hydrothermal fluids [87]. The ranges of δ^{34}S values of the vent fluids at the Bio9″ vent (3.1 to 3.2‰) and the Lucky Strike hydrothermal field (1.5 to 4.5‰) at the East Pacific Rise also indicate the differences in the vent fluid properties in hydrothermal fields.

2.4.3. Deposition of Hydrothermal Sulfides.
Oxidization-reduction is the primary process controlling Fe isotope fractionation. As Fe(II) is oxidized to Fe(III), Fe isotopes are significantly fractionated. The isotope composition of Fe(III) is preferentially enriched in heavy Fe isotopes [68–71]. Fe(III) rapidly forms Fe oxide and hydroxide deposits in aqueous solutions enriched in relatively heavy Fe isotopes. Under reducing conditions, Fe(II) is highly soluble. The solution

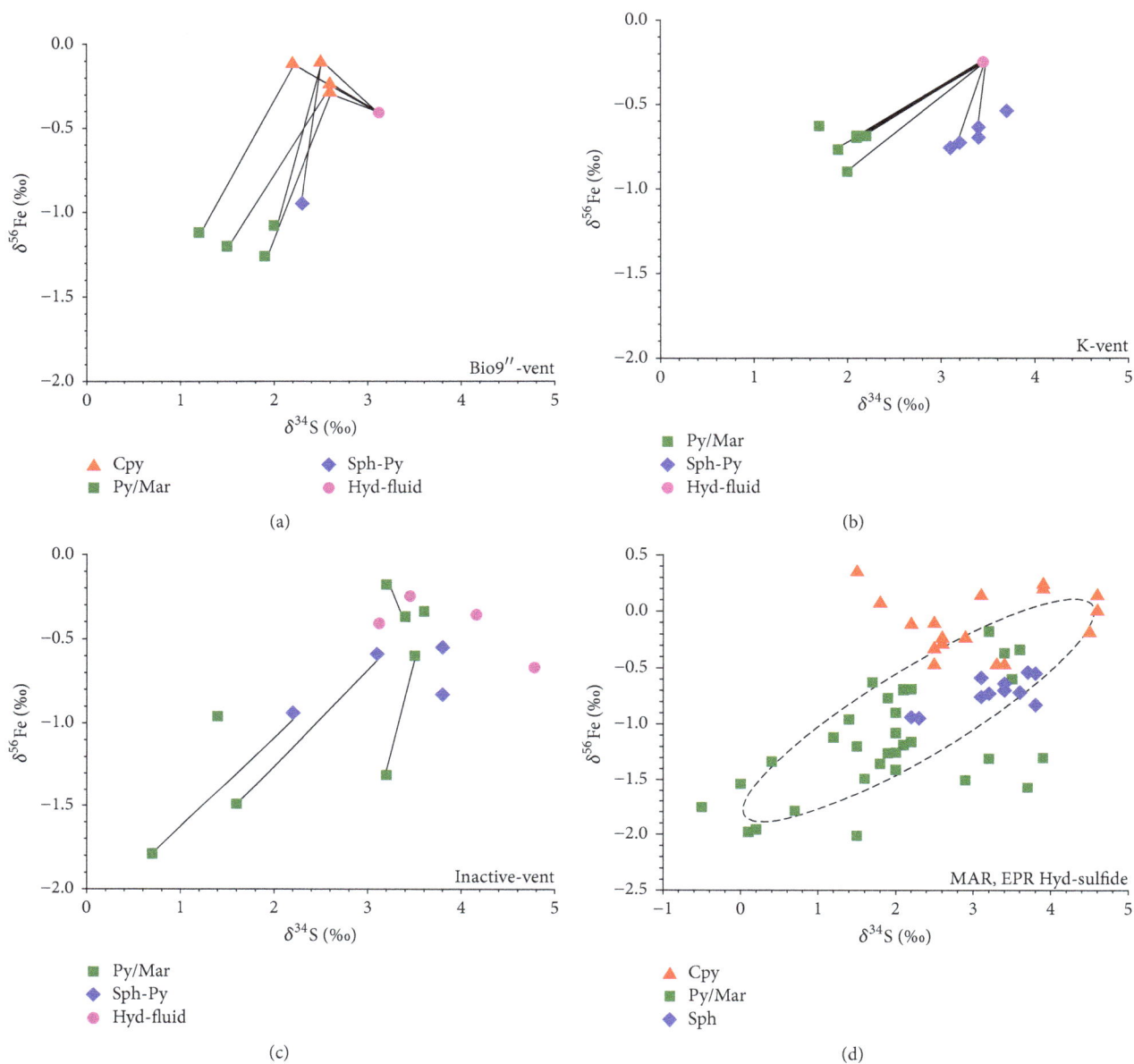

FIGURE 5: Plots of $\delta^{56}Fe$ versus $\delta^{34}S$ of hydrothermal vent fluids and sulfides (after [18, 19]).

is enriched in light Fe isotopes. Sulfide and sulfate minerals formed by these hydrothermal fluids are enriched in light Fe isotopes [88, 89].

Available data indicate that the Fe isotope composition of single magnetite in BIFs is the highest ($\delta^{56}Fe = 3.15‰$), whereas that of single siderite is the lowest ($\delta^{56}Fe = -2.05‰$) [31]. The valence state of the resulting minerals is related to oxygen fugacity and its evolution with time [44, 90], causing a significant difference in Fe isotopes in BIFs. Theoretical calculations and simulation of equilibrium fractionation during the mineral deposition of quartet pyrite, siderite, and hematite showed that Fe(III)$_{aq}$ in solution is enriched in heavier Fe isotopes, compared with Fe(II)$_{aq}$. Fe(II)$_{mineral}$ of quartet pyrite and siderite is enriched in lighter Fe

isotopes, compared with Fe(II)$_{aq}$ [91–93]. Generally, seafloor hydrothermal fluids are reducible and acidic, and Fe(III) that exists in the formation of seafloor hydrothermal sulfides rapidly reduces to Fe(II) [94].

Seafloor hydrothermal fluids form common sulfides including pyrite, chalcopyrite, and sphalerite. Pyrite formation in nature requires two processes: (1) Fe^{2+} first forms mackinawite in Fe-rich solutions (Fe$^{2+}_{aq}$ + H$_2$S$_{aq}$/HS$_{aq}^{-}$ $\rightarrow$ FeS); (2) mackinawite then dissolves and forms pyrite (FeS + S$^{2-}_{n}$/H$_2$S $\rightarrow$ FeS$_2$). Earlier research included simulated Fe isotope fractionation experiments involving mackinawite deposition at room temperature and under acidic conditions [95, 96]. At the beginning of the experiments, the $\delta^{56}Fe$ value of the mackinawite was lighter than that of the solution by approximately 0.9‰. After 7 to 38 days, the $\delta^{56}Fe_{FeS-Fe(II)}$ value was −0.9 to −0.3‰ (pH = 4) (Figure 3). The results indicated significant kinetic fractionation during deposition of the mackinawite. In a more recent study, Guilbaud et al. [97] reported that mackinawite formed under equilibrium conditions is enriched in heavier Fe isotopes compared with that formed in solution (pH = 4 or 7). The results indicated a $\delta^{56}Fe_{FeS-Fe(II)}$ value of 0.3‰ (Figure 3). Pyrite formed by mackinawite dissolution at low temperatures and under reducing conditions was significantly enriched in lighter Fe isotopes compared with mackinawite. The fractionation coefficient $\delta^{56}Fe_{FeS2-FeS}$ was from −3.0‰ to −1.7‰ (Figure 3) [97], indicating that the Fe isotopes underwent significant kinetic fractionation.

The temperature of the pyrite deposited in the late stage in the Lucky Strike hydrothermal field at the Atlantic ridge is below 200°C [18]. Kinetic fractionation under low-temperature conditions may have caused the significant enrichment of light Fe isotopes in pyrite/marcasite. The range of Fe isotope compositions of the pyrite at the East Pacific Rise is large, as well as the temperature range of this hydrothermal field (203 to 383°C). $\delta^{56}Fe_{FeS-Fe(II)}$ between pyrite/marcasite and hydrothermal fluids is −0.77 ± 0.07 to −0.58 ± 0.13‰ [19]. Certain pyrites exhibited a $\delta^{56}Fe$ value of approximately −1.5‰, indicating that the pyrites may have undergone significant kinetic fractionation during formation.

The Fe isotope compositions of certain pyrites are consistent with the $\delta^{56}Fe$ values of sphalerite (Figure 2). The average $\delta^{56}Fe$ value of sphalerite is −0.82 ± 0.19‰ (n = 22) [19]. Previous studies of the Navan Pb-Zn deposit in Ireland indicated that the range of $\delta^{56}Fe$ values of sphalerite is approximately from −2.2‰ to −0.2‰, and the Fe and Zn isotopes exhibit a very high positive correlation. The formation of iron sulfide in hydrothermal fluids is typically rapid and thus often causes kinetic fractionation, indicating that the kinetic fractionation of Fe and Zn isotopes during the formation of sphalerite causes the enrichment in light Fe and Zn isotopes [84].

Based on Figures 1 and 2, the Fe isotope compositions of certain pyrites are similar to those of chalcopyrite: they are enriched in heavy Fe isotopes. In closed hydrothermal systems, hydrothermal fluids do not mix with seawater. The hydrothermal fluids cool conductively and slowly form pyrite. The $\delta^{56}Fe$ and $\delta^{34}S$ values of sulfides and hydrothermal fluids reach equilibrium fractionation [18, 19]. Previous studies involving theoretical calculations of the fractionation coefficient between pyrite and chalcopyrite [91–93] indicated that the Fe isotope composition of pyrites coexisting with chalcopyrite under high-temperature conditions may approach or reach equilibrium fractionation. These pyrites are enriched in heavy Fe isotopes. During mineral growth, if the bond lengths of two minerals are different, then the two minerals display mass fractionation. Heavy isotopes preferentially enter the mineral with the shorter bond length and higher bond energy [98]. In theory, pyrite is enriched in heavy Fe isotopes. Therefore, the $\delta^{56}Fe$ values of coexisting chalcopyrite and pyrite are similar.

3. Conclusions

Based on our analysis of the Fe isotope compositions of various components and systems of the Earth, we note that the Fe isotope compositions differ significantly in nature. The compositions are characterized predominantly by enrichment in light Fe isotopes in hydrothermal systems. Source region differences and different geological processes are the major factors causing these Fe isotope composition variations. The $\delta^{56}Fe$ values of hydrothermal fluids are characterized by significant enrichment in light Fe isotopes. Water-rock reactions, bedrock properties, sediment and microorganism effects, and phase separation are important factors to cause variations of $\delta^{56}Fe$ values. The $\delta^{56}Fe$ values of sulfides also exhibit lighter Fe isotope characteristics relative to hydrothermal vent fluids from the mid-ocean ridge. The vent temperature, fluid properties, and mineral deposition processes significantly affect the $\delta^{56}Fe$ values of hydrothermal sulfides. Chalcopyrite is preferentially enriched in heavy Fe isotopes. Sulfides such as sphalerite and pyrite are enriched in light Fe isotopes. In addition, the $\delta^{56}Fe$ values of pyrite/marcasite display a larger variation than those of chalcopyrite. This pattern is directly related to the equilibrium fractionation or the kinetic fractionation of Fe isotopes during the deposition of sulfides. The $\delta^{56}Fe$ values of various hydrothermal sulfide mineral assemblages display large differences, indicating specific ore-forming temperature. Although researchers currently have a good understanding of the Fe isotope compositions and fractionation processes of modern seafloor hydrothermal systems, the geochemical behavior differences and the fractionation mechanisms of Fe isotopes among various minerals during the mineralization of hydrothermal sulfides from the hydrothermal fields at mid-ocean ridge require further in situ study. Furthermore, studying the coupling of iron oxidation and reduction in hydrothermal ecosystem should help to identify and better quantify the biogeochemical Fe cycle in natural environments.

Conflicts of Interest

The authors declare that they have no conflicts of interest.

Acknowledgments

This research was supported by the National Basic Research and Development Program (Grant no. 2013CB429705), the National Natural Science Foundation of China (Grant no. 41276055), the Fundamental Research Funds for Central Public Welfare Research Institutes (Grant no. JT1701), and Zhejiang Provincial Natural Science Foundation of China (Grant no. LY14D060005).

References

[1] X.-K. Zhu, Y. Wang, B. Yan et al., "Developments of non-traditional stable isotope geochemistry," *Bulletin of Mineralogy Petrology and Geochemistry*, vol. 6, pp. 651–688, 2013.

[2] N. S. Belshaw, X.-K. Zhu, Y. Guo, and R. K. O'Nions, "High precision measurement of iron isotopes by plasma source mass

spectrometry," *International Journal of Mass Spectrometry*, vol. 197, no. 1-3, pp. 191–195, 2000.

[3] J.-B. Chen, V. Busigny, J. Gaillardet, P. Louvat, and Y.-N. Wang, "Iron isotopes in the Seine River (France): Natural versus anthropogenic sources," *Geochimica et Cosmochimica Acta*, vol. 128, pp. 128–143, 2014.

[4] F. Poitrasson, L. Cruz Vieira, P. Seyler et al., "Iron isotope composition of the bulk waters and sediments from the Amazon River Basin," *Chemical Geology*, vol. 377, pp. 1–11, 2014.

[5] M. Waeles, A. R. Baker, T. Jickells, and J. Hoogewerff, "Global dust teleconnections: Aerosol iron solubility and stable isotope composition," *Environmental Chemistry*, vol. 4, no. 4, pp. 233–237, 2007.

[6] B. L. Beard, C. M. Johnson, K. L. Von Damm, and R. L. Poulson, "Iron isotope constraints on Fe cycling and mass balance in oxygenated Earth oceans," *Geology*, vol. 31, no. 7, pp. 629–632, 2003.

[7] S. Severmann, C. M. Johnson, B. L. Beard et al., "The effect of plume processes on the Fe isotope composition of hydrothermally derived Fe in the deep ocean as inferred from the Rainbow vent site, Mid-Atlantic Ridge, 36∘14′N," *Earth and Planetary Science Letters*, vol. 225, no. 1-2, pp. 63–76, 2004.

[8] S. Levasseur, M. Frank, J. R. Hein, and A. N. Halliday, "The global variation in the iron isotope composition of marine hydrogenetic ferromanganese deposits: Implications for seawater chemistry?" *Earth and Planetary Science Letters*, vol. 224, no. 1-2, pp. 91–105, 2004.

[9] J. M. Edmond, C. Measures, R. E. McDuff et al., "Ridge crest hydrothermal activity and the balances of the major and minor elements in the ocean: The Galapagos data," *Earth and Planetary Science Letters*, vol. 46, no. 1, pp. 1–18, 1979.

[10] F.-Z. Teng, N. Dauphas, S. Huang, and B. Marty, "Iron isotopic systematics of oceanic basalts," *Geochimica et Cosmochimica Acta*, vol. 107, pp. 12–26, 2013.

[11] B. A. Bergquist and E. A. Boyle, "Iron isotopes in the Amazon River system: Weathering and transport signatures," *Earth and Planetary Science Letters*, vol. 248, no. 1-2, pp. 39–53, 2006.

[12] M. S. Fantle and D. J. DePaolo, "Iron isotopic fractionation during continental weathering," *Earth and Planetary Science Letters*, vol. 228, no. 3-4, pp. 547–562, 2004.

[13] T. M. Conway and S. G. John, "Quantification of dissolved iron sources to the North Atlantic Ocean," *Nature*, vol. 511, no. 7508, pp. 212–215, 2014.

[14] T. M. Conway and S. G. John, "The cycling of iron, zinc and cadmium in the North East Pacific Ocean - Insights from stable isotopes," *Geochimica et Cosmochimica Acta*, vol. 164, pp. 262–283, 2015.

[15] A. Radic, F. Lacan, and J. W. Murray, "Iron isotopes in the seawater of the equatorial Pacific Ocean: New constraints for the oceanic iron cycle," *Earth and Planetary Science Letters*, vol. 306, no. 1-2, pp. 1–10, 2011.

[16] J. N. Fitzsimmons, G. G. Carrasco, J. Wu et al., "Partitioning of dissolved iron and iron isotopes into soluble and colloidal phases along the GA03 GEOTRACES North Atlantic Transect," *Deep-Sea Research Part II: Topical Studies in Oceanography*, vol. 116, pp. 130–151, 2015.

[17] O. Rouxel, N. Dobbek, J. Ludden, and Y. Fouquet, "Iron isotope fractionation during oceanic crust alteration," *Chemical Geology*, vol. 202, no. 1-2, pp. 155–182, 2003.

[18] O. Rouxel, Y. Fouquet, and J. N. Ludden, "Subsurface processes at the lucky strike hydrothermal field, Mid-Atlantic ridge:

Evidence from sulfur, selenium, and iron isotopes," *Geochimica et Cosmochimica Acta*, vol. 68, no. 10, pp. 2295–2311, 2004.

[19] O. Rouxel, W. C. Shanks III, W. Bach, and K. J. Edwards, "Integrated Fe- and S-isotope study of seafloor hydrothermal vents at East Pacific Rise 9-10∘N," *Chemical Geology*, vol. 252, no. 3-4, pp. 214–227, 2008.

[20] M. Sharma, M. Polizzotto, and A. D. Anbar, "Iron isotopes in hot springs along the Juan de Fuca Ridge," *Earth and Planetary Science Letters*, vol. 194, no. 1-2, pp. 39–51, 2001.

[21] S. A. Bennett, O. Rouxel, K. Schmidt, D. Garbe-Schönberg, P. J. Statham, and C. R. German, "Iron isotope fractionation in a buoyant hydrothermal plume, 5∘S Mid-Atlantic Ridge," *Geochimica et Cosmochimica Acta*, vol. 73, no. 19, pp. 5619–5634, 2009.

[22] B. L. Beard, C. M. Johnson, L. Cox, H. Sun, K. H. Nealson, and C. Aguilar, "Iron isotope biosignatures," *Science*, vol. 285, no. 5435, pp. 1889–1891, 1999.

[23] N.-C. Chu, C. M. Johnson, B. L. Beard et al., "Evidence for hydrothermal venting in Fe isotope compositions of the deep Pacific Ocean through time," *Earth and Planetary Science Letters*, vol. 245, no. 1-2, pp. 202–217, 2006.

[24] X.-K. Zhu, R. K. O'Nions, Y. Guo, and B. C. Reynolds, "Secular variation of iron isotopes in North Atlantic Deep Water," *Science*, vol. 287, no. 5460, pp. 2000–2002, 2000.

[25] S. Graham, N. Pearson, S. Jackson, W. Griffin, and S. Y. O'Reilly, "Tracing Cu and Fe from source to porphyry: In situ determination of Cu and Fe isotope ratios in sulfides from the Grasberg Cu-Au deposit," *Chemical Geology*, vol. 207, no. 3-4, pp. 147–169, 2004.

[26] G. Markl, F. von Blanckenburg, and T. Wagner, "Iron isotope fractionation during hydrothermal ore deposition and alteration," *Geochimica et Cosmochimica Acta*, vol. 70, no. 12, pp. 3011–3030, 2006.

[27] Y. Wang, X.-K. Zhu, J.-W. Mao, Z.-H. Li, and Y.-B. Cheng, "Iron isotope fractionation during skarn-type metallogeny: A case study of Xinqiao Cu-S-Fe-Au deposit in the Middle-Lower Yangtze valley," *Ore Geology Reviews*, vol. 43, no. 1, pp. 194–202, 2011.

[28] Y. Wang, X.-K. Zhu, J.-W. Mao, and Y.-B. Cheng, "Preliminary Fe isotopic study of Gushan ore magma deposit in Anhui Province," *Mineral Deposits*, vol. 33, no. 4, pp. 689–696, 2014.

[29] Y.-J. Chen, S.-G. Su, Y.-S. He et al., "Fe isotope compositions and implications on mineralization of Xishimen iron deposit in Wuan, Hebei," *Acta Petrologica Sinica*, vol. 30, no. 11, pp. 3443–3454, 2014.

[30] Y. Cheng, J. Mao, X. Zhu, and Y. Wang, "Iron isotope fractionation during supergene weathering process and its application to constrain ore genesis in Gaosong deposit, Gejiu district, SW China," *Gondwana Research*, vol. 27, no. 3, pp. 1283–1291, 2015.

[31] Y. Wang and X.-K. Zhu, "Fe isotope systematics and its implications in ore deposit geology," *Acta Petrologica Sinica*, vol. 28, no. 11, pp. 3638–3654, 2012.

[32] B. L. Beard and C. M. Johnson, "High precision iron isotope measurements of terrestrial and lunar materials," *Geochimica et Cosmochimica Acta*, vol. 63, no. 11-12, pp. 1653–1660, 1999.

[33] N. Dauphas, M. Van Zuilen, M. Wadhwa, A. M. Davis, B. Marty, and P. E. Janney, "Clues from Fe isotope variations on the origin of early Archean BIFs from Greenland," *Science*, vol. 306, no. 5704, pp. 2077–2080, 2004.

[34] N. Dauphas, N. L. Cates, S. J. Mojzsis, and V. Busigny, "Identification of chemical sedimentary protoliths using iron isotopes in

the > 3750 Ma Nuvvuagittuq supracrustal belt, Canada," *Earth and Planetary Science Letters*, vol. 254, no. 3-4, pp. 358–376, 2007.

[35] C. D. Frost, F. Blanckenburg, R. Schoenberg, B. R. Frost, and S. M. Swapp, "Preservation of Fe isotope heterogeneities during diagenesis and metamorphism of banded iron formation," *Contributions to Mineralogy and Petrology*, vol. 153, no. 2, pp. 211–235, 2007.

[36] G. Steinhoefel, I. Horn, and F. von Blanckenburg, "Micro-scale tracing of Fe and Si isotope signatures in banded iron formation using femtosecond laser ablation," *Geochimica et Cosmochimica Acta*, vol. 73, no. 18, pp. 5343–5360, 2009.

[37] G. Steinhoefel, F. von Blanckenburg, I. Horn, K. O. Konhauser, N. J. Beukes, and J. Gutzmer, "Deciphering formation processes of banded iron formations from the Transvaal and the Hamersley successions by combined Si and Fe isotope analysis using UV femtosecond laser ablation," *Geochimica et Cosmochimica Acta*, vol. 74, no. 9, pp. 2677–2696, 2010.

[38] N. Planavsky, O. Rouxel, A. Bekker, R. Shapiro, P. Fralick, and A. Knudsen, "Iron-oxidizing microbial ecosystems thrived in late Paleoproterozoic redox-stratified oceans," *Earth and Planetary Science Letters*, vol. 286, no. 1-2, pp. 230–242, 2009.

[39] N. Planavsky, O. J. Rouxel, A. Bekker, A. Hofmann, C. T. S. Little, and T. W. Lyons, "Iron isotope composition of some Archean and Proterozoic iron formations," *Geochimica et Cosmochimica Acta*, vol. 80, pp. 158–169, 2012.

[40] H. Tsikos, A. Matthews, Y. Erel, and J. M. Moore, "Iron isotopes constrain biogeochemical redox cycling of iron and manganese in a Palaeoproterozoic stratified basin," *Earth and Planetary Science Letters*, vol. 298, no. 1-2, pp. 125–134, 2010.

[41] G. P. Halverson, F. Poitrasson, P. F. Hoffman, A. Nédélec, J.-M. Montel, and J. Kirby, "Fe isotope and trace element geochemistry of the Neoproterozoic syn-glacial Rapitan iron formation," *Earth and Planetary Science Letters*, vol. 309, no. 1-2, pp. 100–112, 2011.

[42] B. Yan, X.-K. Zhu, S.-H. Tang, and M.-Y. Zhu, "Fe isotope characteristics of the Neoproterozoic BIF in Guangxi Province and its implications," *Acta Geologica Sinica*, vol. 84, no. 7, pp. 1080–1086, 2010.

[43] P. R. Craddock and N. Dauphas, "Iron and carbon isotope evidence for microbial iron respiration throughout the Archean," *Earth and Planetary Science Letters*, vol. 303, no. 1-2, pp. 121–132, 2011.

[44] A. Heimann, C. M. Johnson, B. L. Beard et al., "Fe, C, and O isotope compositions of banded iron formation carbonates demonstrate a major role for dissimilatory iron reduction in ~2.5Ga marine environments," *Earth and Planetary Science Letters*, vol. 294, no. 1-2, pp. 8–18, 2010.

[45] S. Weyer and D. A. Ionov, "Partial melting and melt percolation in the mantle: The message from Fe isotopes," *Earth and Planetary Science Letters*, vol. 259, no. 1-2, pp. 119–133, 2007.

[46] B. L. Beard and C. M. Johnson, "Inter-mineral Fe isotope variations in mantle-derived rocks and implications for the Fe geochemical cycle," *Geochimica et Cosmochimica Acta*, vol. 68, no. 22, pp. 4727–4743, 2004.

[47] B. L. Beard, C. M. Johnson, J. L. Skulan, K. H. Nealson, L. Cox, and H. Sun, "Application of Fe isotopes to tracing the geochemical and biological cycling of Fe," *Chemical Geology*, vol. 195, no. 1-4, pp. 87–117, 2003.

[48] F.-Z. Teng, N. Dauphas, and R. T. Helz, "Iron isotope fractionation during magmatic differentiation in Kilauea Iki lava lake," *Science*, vol. 320, no. 5883, pp. 1620–1622, 2008.

[49] F.-Z. Teng, N. Dauphas, R. T. Helz, S. Gao, and S. Huang, "Diffusion-driven magnesium and iron isotope fractionation in Hawaiian olivine," *Earth and Planetary Science Letters*, vol. 308, no. 3-4, pp. 317–324, 2011.

[50] S. Weyer, A. D. Anbar, G. P. Brey, C. Münker, K. Mezger, and A. B. Woodland, "Iron isotope fractionation during planetary differentiation," *Earth and Planetary Science Letters*, vol. 240, no. 2, pp. 251–264, 2005.

[51] X. Zhao, H. Zhang, X. Zhu, S. Tang, and Y. Tang, "Iron isotope variations in spinel peridotite xenoliths from North China Craton: Implications for mantle metasomatism," *Contributions to Mineralogy and Petrology*, vol. 160, no. 1, pp. 1–14, 2010.

[52] X. Zhao, H. Zhang, X. Zhu, S. Tang, and B. Yan, "Iron isotope evidence for multistage melt-peridotite interactions in the lithospheric mantle of eastern China," *Chemical Geology*, vol. 292-293, pp. 127–139, 2012.

[53] J. A. Schuessler, R. Schoenberg, and O. Sigmarsson, "Iron and lithium isotope systematics of the Hekla volcano, Iceland - Evidence for Fe isotope fractionation during magma differentiation," *Chemical Geology*, vol. 258, no. 1-2, pp. 78–91, 2009.

[54] B. L. Beard and C. M. Johnson, "Fe isotope variations in the modern and ancient earth and other planetary bodies," *Reviews in Mineralogy and Geochemistry*, vol. 55, pp. 319–357, 2004.

[55] F. Huang, Z. Zhang, C. C. Lundstrom, and X. Zhi, "Iron and magnesium isotopic compositions of peridotite xenoliths from Eastern China," *Geochimica et Cosmochimica Acta*, vol. 75, no. 12, pp. 3318–3334, 2011.

[56] X.-K. Zhu, Y. Guo, R. K. O'Nions, E. D. Young, and R. D. Ash, "Isotopic homogeneity of iron in the early solar nebula," *Nature*, vol. 412, no. 6844, pp. 311–313, 2001.

[57] E. Mullane, S. S. Russell, and M. Gounelle, "Nebular and asteroidal modification of the iron isotope composition of chondritic components," *Earth and Planetary Science Letters*, vol. 239, no. 3-4, pp. 203–218, 2005.

[58] R. Schoenberg and F. V. Blanckenburg, "Modes of planetary-scale Fe isotope fractionation," *Earth and Planetary Science Letters*, vol. 252, no. 3-4, pp. 342–359, 2006.

[59] D. C. Hezel, A. W. Needham, R. Armytage et al., "A nebula setting as the origin for bulk chondrule Fe isotope variations in CV chondrites," *Earth and Planetary Science Letters*, vol. 296, no. 3-4, pp. 423–433, 2010.

[60] F. Poitrasson and R. Freydier, "Heavy iron isotope composition of granites determined by high resolution MC-ICP-MS," *Chemical Geology*, vol. 222, no. 1-2, pp. 132–147, 2005.

[61] A. W. Needham, D. Porcelli, and S. S. Russell, "An Fe isotope study of ordinary chondrites," *Geochimica et Cosmochimica Acta*, vol. 73, no. 24, pp. 7399–7413, 2009.

[62] F. Poitrasson, A. N. Halliday, D.-C. Lee, S. Levasseur, and N. Teutsch, "Iron isotope differences between Earth, Moon, Mars and Vesta as possible records of contrasted accretion mechanisms," *Earth and Planetary Science Letters*, vol. 223, no. 3-4, pp. 253–266, 2004.

[63] D. D. Syverson, N. J. Pester, P. R. Craddock, and W. E. Seyfried, "Fe isotope fractionation during phase separation in the NaCl-H2O system: An experimental study with implications for seafloor hydrothermal vents," *Earth and Planetary Science Letters*, vol. 406, pp. 223–232, 2014.

[64] X.-K. Zhu, Y. Guo, R. J. P. Williams et al., "Mass fractionation processes of transition metal isotopes," *Earth and Planetary Science Letters*, vol. 200, no. 1-2, pp. 47–62, 2002.

[65] J. B. Chapman, D. J. Weiss, Y. Shan, and M. Lemburger, "Iron isotope fractionation during leaching of granite and basalt by hydrochloric and oxalic acids," *Geochimica et Cosmochimica Acta*, vol. 73, no. 5, pp. 1312–1324, 2009.

[66] L.-L. Zhang, *The relationship between EMORB and the expansion velocity of mid ocean ridge: implications for EMORB source and its genetic mechanism*, Graduate University of Chinese Academy of Sciences, 2012.

[67] S. Weyer, "What drives iron isotope fractionation in magma?" *Science*, vol. 320, no. 5883, pp. 1600-1601, 2008.

[68] A. D. Anbar, J. E. Roe, J. Barling, and K. H. Nealson, "Nonbiological fractionation of iron isotopes," *Science*, vol. 288, no. 5463, pp. 126–128, 2000.

[69] N. Balci, T. D. Bullen, K. Witte-Lien, W. C. Shanks, M. Motelica, and K. W. Mandernack, "Iron isotope fractionation during microbially stimulated Fe(II) oxidation and Fe(III) precipitation," *Geochimica et Cosmochimica Acta*, vol. 70, no. 3, pp. 622–639, 2006.

[70] C. M. Johnson, J. Skulan, B. L. Beard, H. J. Sun, K. Nealson, and P. S. Braterman, "Isotopic fractionation between Fe(III) and Fe(II) in aqueous solutions," *Earth and Planetary Science Letters*, vol. 195, no. 1, pp. 141–153, 2002.

[71] S. A. Welch, B. L. Beard, C. M. Johnson, and P. S. Braterman, "Kinetic and equilibrium Fe isotope fractionation between aqueous Fe(II) and Fe(III)," *Geochimica et Cosmochimica Acta*, vol. 67, no. 22, pp. 4231–4250, 2003.

[72] A. Inoue, "Two-dimensional variations of exchangeable cation composition in the terrigenous sediment, eastern flank of the Juan de Fuca Ridge," *Marine Geology*, vol. 162, no. 2-4, pp. 501–528, 2000.

[73] D. A. Glickson, D. S. Kelley, and J. R. Delaney, "Geology and hydrothermal evolution of the mothra hydrothermal field, endeavour segment, Juan de Fuca Ridge," *Geochemistry, Geophysics, Geosystems*, vol. 8, no. 6, Article ID Q06010, 2007.

[74] D. S. Kelley, J. A. Karson, D. Blackman et al., "An off-axis hydrothermal vent field near the Mid-Atlantic Ridge at 30°N," in *Nature*, vol. 412, pp. 145–149, 2001.

[75] J. Garnier, J. Garnier, C. Vieira et al., "Iron isotope fingerprints of redox and biogeochemical cycling in the soil-water-rice plant system of a paddy field," *Science of The Total Environment*, vol. 574, pp. 1622–1632, 2017.

[76] H. A. Crosby, C. M. Johnson, E. E. Roden, and B. L. Beard, "Coupled Fe(II)-Fe(III) electron and atom exchange as a mechanism for Fe isotope fractionation during dissimilatory iron oxide reduction," *Environmental Science and Technology*, vol. 39, no. 17, pp. 6698–6704, 2005.

[77] D. S. Mulholland, F. Poitrasson, L. S. Shirokova et al., "Iron isotope fractionation during Fe(II) and Fe(III) adsorption on cyanobacteria," *Chemical Geology*, vol. 400, pp. 24–33, 2015.

[78] M. Kunzmann, T. M. Gibson, G. P. Halverson et al., "Iron isotope biogeochemistry of Neoproterozoic marine shales," *Geochimica et Cosmochimica Acta*, vol. 209, pp. 85–105, 2017.

[79] G. S. Pokrovski, J. Roux, and J.-C. Harrichoury, "Fluid density control on vapor-liquid partitioning of metals in hydrothermal systems," *Geology*, vol. 33, no. 8, pp. 657–660, 2005.

[80] W. Liu, S. J. Borg, D. Testemale, B. Etschmann, J.-L. Hazemann, and J. Brugger, "Speciation and thermodynamic properties for cobalt chloride complexes in hydrothermal fluids at 35-440°C and 600bar: An in-situ XAS study," *Geochimica et Cosmochimica Acta*, vol. 75, no. 5, pp. 1227–1248, 2011.

[81] P. S. Hill and E. A. Schauble, "Modeling the effects of bond environment on equilibrium iron isotope fractionation in ferric aquo-chloro complexes," *Geochimica et Cosmochimica Acta*, vol. 72, no. 8, pp. 1939–1958, 2008.

[82] J. R. Black, A. Kavner, and E. A. Schauble, "Calculation of equilibrium stable isotope partition function ratios for aqueous zinc complexes and metallic zinc," *Geochimica et Cosmochimica Acta*, vol. 75, no. 3, pp. 769–783, 2011.

[83] G. Auclair and Y. Fouquet, "Distribution of selenium in high-temperature hydrothermal sulfide deposits at 13 degree north, east pacific rise," *The Canadian Mineralogist*, vol. 25, no. 4, pp. 577–587, 1987.

[84] D. Gagnevin, A. J. Boyce, C. D. Barrie, J. F. Menuge, and R. J. Blakeman, "Zn, Fe and S isotope fractionation in a large hydrothermal system," *Geochimica et Cosmochimica Acta*, vol. 88, pp. 183–198, 2012.

[85] I. M. Chou and H. P. Eugster, "Solubility of magnetite in supercritical chloride solutions," *American Journal of Science*, vol. 277, no. 10, pp. 1296–1314, 1977.

[86] A. Heimann, B. L. Beard, and C. M. Johnson, "The role of volatile exsolution and sub-solidus fluid/rock interactions in producing high 56Fe/54Fe ratios in siliceous igneous rocks," *Geochimica et Cosmochimica Acta*, vol. 72, no. 17, pp. 4379–4396, 2008.

[87] J. L. Charlou, J. P. Donval, E. Douville et al., "Compared geochemical signatures and the evolution of Menez Gwen (35°50N) and Lucky Strike (37°17N) hydrothermal fluids, south of the Azores Triple Junction on the Mid-Atlantic Ridge," *Chemical Geology*, vol. 171, no. 1-2, pp. 49–75, 2000.

[88] C. M. Johnson, B. L. Beard, and E. E. Roden, "The iron isotope fingerprints of redox and biogeochemical cycling in modern and ancient earth," *Annual Review of Earth and Planetary Sciences*, vol. 36, pp. 457–493, 2008.

[89] K. G. Taylor and K. O. Konhauser, "In earth surface systems: A major player in chemical and biological processes," *Elements*, vol. 7, no. 2, pp. 83–88, 2011.

[90] W. Li, A. D. Czaja, M. J. Van Kranendonk, B. L. Beard, E. E. Roden, and C. M. Johnson, "An anoxic, Fe(II)-rich, U-poor ocean 3.46 billion years ago," *Geochimica et Cosmochimica Acta*, vol. 120, pp. 65–79, 2013.

[91] V. B. Polyakov and S. D. Mineev, "The use of Mossbauer spectroscopy in stable isotope geochemistry," *Geochimica et Cosmochimica Acta*, vol. 64, no. 5, pp. 849–865, 2000.

[92] V. B. Polyakov, R. N. Clayton, J. Horita, and S. D. Mineev, "Equilibrium iron isotope fractionation factors of minerals: Reevaluation from the data of nuclear inelastic resonant X-ray scattering and Mössbauer spectroscopy," *Geochimica et Cosmochimica Acta*, vol. 71, no. 15, pp. 3833–3846, 2007.

[93] V. B. Polyakov and D. M. Soultanov, "New data on equilibrium iron isotope fractionation among sulfides: Constraints on mechanisms of sulfide formation in hydrothermal and igneous systems," *Geochimica et Cosmochimica Acta*, vol. 75, no. 7, pp. 1957–1974, 2011.

[94] D. Rickard and J. W. Morse, "Acid volatile sulfide (AVS)," *Marine Chemistry*, vol. 97, no. 3-4, pp. 141–197, 2005.

[95] I. B. Butler, C. Archer, D. Vance, A. Oldroyd, and D. Rickard, "Fe isotope fractionation on FeS formation in ambient aqueous solution," *Earth and Planetary Science Letters*, vol. 236, no. 1-2, pp. 430–442, 2005.

[96] R. Guilbaud, I. B. Butler, R. M. Ellam, and D. Rickard, "Fe isotope exchange between Fe(II)aq and nanoparticulate

mackinawite (FeSm) during nanoparticle growth," *Earth and Planetary Science Letters*, vol. 300, no. 1-2, pp. 174–183, 2010.

[97] R. Guilbaud, I. B. Butler, R. M. Ellam, D. Rickard, and A. Oldroyd, "Experimental determination of the equilibrium Fe isotope fractionation between Feaq2+ and FeSm (mackinawite) at 25 and 2∘C," *Geochimica et Cosmochimica Acta*, vol. 75, no. 10, pp. 2721–2734, 2011.

[98] J. R. O'Neil, "Theoretical and experimental aspects of isotopic fractionation," *Reviews in Mineralogy*, vol. 16, pp. 1–40, 1986.

5

Electrocoagulation-Adsorption to Remove Anionic and Cationic Dyes from Aqueous Solution by PV-Energy

J. Castañeda-Díaz, T. Pavón-Silva, E. Gutiérrez-Segura, and A. Colín-Cruz

Facultad de Química, Universidad Autónoma del Estado de México, Paseo Colón y Tollocan s/n, 50000 Toluca, MEX, Mexico

Correspondence should be addressed to T. Pavón-Silva; th.pavon@gmail.com

Academic Editor: Davide Vione

The cationic dye malachite green (MG) and the anionic dye Remazol yellow (RY) were removed from aqueous solutions using electrocoagulation-adsorption processes. Batch and continuous electrocoagulation procedures were performed and compared. Carbonaceous materials obtained from industrial sewage sludge and commercial activated carbons were used to adsorb dyes from aqueous solutions in column systems with a 96–98% removal efficiency. The continuous electrocoagulation-adsorption system was more efficient for removing dyes than electrocoagulation alone. The thermodynamic parameters suggested the feasibility of the process and indicated that the adsorption was spontaneous and endothermic (ΔS = 0.037 and −0.009 for MG and RY, resp.). The ΔG value further indicated that the adsorption process was spontaneous (−6.31 and −10.48; T = 303 K). The kinetic electrocoagulation results and fixed-bed adsorption results were adequately described using a first-order model and a Bohart-Adams model, respectively. The adsorption capacities of the batch and column studies differed for each dye, and both adsorbent materials showed a high affinity for the cationic dye. Thus, the results presented in this work indicate that a continuous electrocoagulation-adsorption system can effectively remove this type of pollutant from water. The morphology and elements present in the sludge and adsorbents before and after dye adsorption were characterized using SEM-EDS and FT-IR.

1. Introduction

The contamination of surface water and groundwater with dyes is a serious environmental problem and a threat to human and aquatic life. Several studies have reported that more than 100,000 dyes are commercially available with an estimated annual production of over 7×10^5–1×10^6 tons [1]. Industries that produce textiles, cosmetics, paper, leather, light-harvesting arrays, agricultural products, photoelectrochemical cells, pharmaceuticals, and food processing also produce large volumes of wastewater that are polluted with a high concentration of dyes and other components [2].

Malachite green (MG) is a cationic dye that is mainly used to dye textiles and paper; MG is also used as a fungicide and antiparasitic agent in fishkeeping [3] and causes injuries to humans and animals after inhalation and ingestion. Its entry into the environment causes reduced human fertility and generates carcinogenic, mutagenic, and respiratory

hazards [4, 5]. Remazol yellow (RY) dye, or reactive yellow 105, is an anionic dye used for textile dyeing due to its ease of use, colour stability, and resistance to washing [6]. The manufacture of azo dyes can impact the environment because of the presence of toxic amines in the effluent [7]. Moreover, dyes must be eliminated from wastewater before discharging it into water bodies. Due to the complex aromatic structure and synthetic origins of dyes, they are stable to heat, oxidizing agents, photodegradation, and biodegradation [8]. Several conventional methods can be applied to remove dyes from wastewater, including biological, physical, and chemical methods [9]. For example, coagulation is effective for sulphur and dispersive dyes. Acid, direct, vat, and reactive dyes coagulate but do not settle, while cationic dyes do not coagulate. Furthermore, chemical coagulation causes additional pollution (due to the undesired reactions in treated water) and produces large amounts of sludge [10, 11]. Due to the large variability in the composition of textile wastewater, most

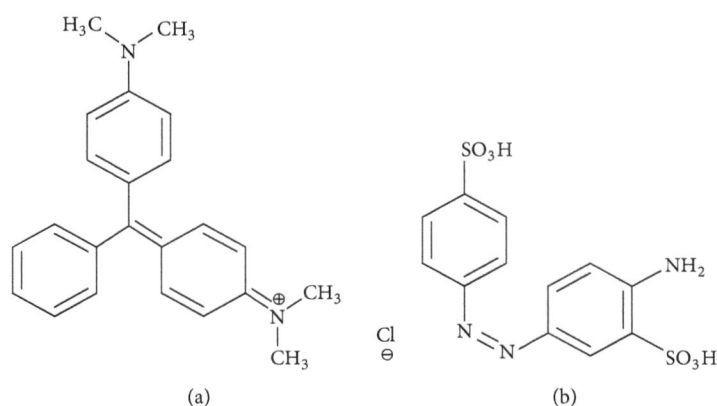

FIGURE 1: Chemical structure of (a) malachite green and (b) Remazol yellow dyestuff.

conventional methods are expensive and are now becoming inadequate. Thus, there is an urgent need to develop more efficient and cost-effective techniques for the treatment of wastewater [12]. A combination of various techniques for the treatment of effluents can lead to higher removal efficiencies than a single treatment method.

Electrocoagulation technology is a simple, reliable, cost-effective, and promising technique for treating various wastewaters without additional chemicals and thus reducing the volume of produced sludge; this technology requires only a low-intensity electrical current and can therefore be operated using green processes, such as solar cells, windmills, and fuel cells [2]. The coagulating agent is generated in situ during the electrooxidation of a sacrificial anode; the hydroxide formed on the mineral surfaces in situ is 100 times higher than on preprecipitated hydroxides when metal hydroxides are used as the coagulant. The flocs formed during electrocoagulation are relatively large and contain less bound water; therefore, they can easily be removed by filtration. The use of electrocoagulation with effluents containing low levels of dissolved solids is limited by the required minimum solution conductivity. During the removal of organic compounds from effluent-containing chlorides, toxic chlorinated organic compounds can form, and the sacrificial anodes must be replaced periodically; this method has high electricity costs and sludge production [10, 13]. Therefore, electrocoagulation in combination with other treatment methods is a safe and effective method for dye removal.

Adsorption is a widely used, effective technology for dye removal treatment with applicability in wastewater treatment. Adsorption can also remove soluble and insoluble organic pollutants; the removal capacity of this method can reach 99.9%. Adsorption is the accumulation of a substance at a surface or interface. In water treatment, the process occurs at an interface between a solid adsorbent and contaminated water. The adsorbed pollutant is called the adsorbate, and the adsorbing phase is called the adsorbent [14]. The most efficient methods are technologies based on the adsorption of activated carbon from water, which appears to be the best method for removing dyes. However, this process is expensive and difficult to regenerate after use [15]. Alternatively, sewage

sludge-derived carbonaceous materials that act as sorbents have the advantage of providing economic value to waste as a material for producing adsorbents for the removal of pollutant materials. In addition, this application could solve the problem of sewage sludge pollution. Therefore, the feasibility of achieving cationic and anionic dye removal from aqueous solutions using a combined electrocoagulation and adsorption system composed of commercial activated carbon and sewage sludge-derived carbonaceous material as adsorbents was examined. The effects of the operating conditions on the performance of the electrocoagulation process in a batch reactor were evaluated; the combined process using electrocoagulation followed by adsorption in a sequenced continuous step was evaluated and compared. To the best of our knowledge, no such studies have been carried out previously. Finally, the Brunauer–Emmett–Teller (BET) surface area of the adsorbent and the sludge and dye-loaded adsorbents were characterized using FT-IR and SEM.

2. Materials and Methods

2.1. Materials. The carbonaceous material was obtained from industrial sewage sludge. The sludge feedstock and pyrolysis procedure have been described elsewhere [16]. Sludge pyrolysis was performed at 500°C for 60 min with a nitrogen flow rate of 350 mL min^{-1}. The carbonaceous material was treated with a 10% hydrochloric acid solution at 20°C for 24 h. Afterwards, it was washed five times with distilled water and dried at 70°C for 2 h. Commercial granular activated carbon (CAC) from Clarimex®, Mexico, was used for comparison. The materials were milled, sieved, and washed with deionized water to eliminate the fine particles. The grains with diameters between 0.42 mm and 0.84 mm were selected for both adsorbents.

2.2. Chemicals. MG chloride, a cationic dye (Figure 1(a)) (molecular formula: $C_{23}H_{25}N_2Cl$, molecular weight: 365 g mol^{-1}, and maximum light absorption: $\lambda_{max} = 617$ nm), is a commercial salt (Hycel, Mexico) and was used without further purification. RY is an anionic dye that is soluble in water and has a molar mass of 606 g mol^{-1} and maximum light absorption at $\lambda_{\max}$ = 269 nm. RY is a commercial salt (Dystar, Mexico), and its chemical structure is shown in Figure 1(b). MG and RY were selected and studied to compare the effectiveness of electrocoagulation-adsorption according to the type of dyestuff present. MG and YR dye solutions were used for preparing synthetic dye wastewater and were diluted to the desired concentration of 100 mg L^{-1}.

2.3. Electrochemical Cell.

2.3. Electrochemical Cell. Experiments were performed in batch and continuous modes with a 5.5-L rectangular reactor made of Perspex glass; the unit consisted of an electrochemical reactor and sludge separator. The volumes of the reactor and sludge separator were 2.5 and 3 L, respectively. The two compartments were connected by a triangular groove, and the wastewater passed from the first compartment into the second by overflow. The solution was continuously circulated in the system using iron plates with a peristaltic pump (Masterflex L-S®) at a flowrate of 65 mL min^{-1}. Sodium chloride was added to the batch runs to achieve a solution conductivity of 69–78 mS cm^{-1}.

Four plates ($100 \times 70 \times 1$ mm^3) were used as electrodes with 70-cm^2 area. The gaps between the anodes and cathodes were maintained at 1 cm. Each anode was placed between two cathodes to improve the current distribution. Magnetic stirrers were used to maintain a constant composition of feed wastewater. A solar cell supplied the system with 1-2 A of current intensity (35.7 A m^{-2}–71.4 A m^{-2}). Samples (10 mL) were periodically taken from the reactor, and the dye concentrations were measured. This procedure was repeated twice. The dye concentration of the samples was calculated using a calibration curve that was prepared previously. The effects of the operating parameters were studied, and the optimal conditions of electrocoagulation for decolourization were determined.

The electrodes were treated with HCl (1 M) for cleaning prior to use to avoid passivation. After each run, the corroded parts of the anodes and electroreduced substances on the cathodes were removed with a revolving metal brush. The surfaces of the electrodes were replenished with sandpaper prior to each new experiment.

2.4. Fixed-Bed Experiments.

2.4. Fixed-Bed Experiments. The adsorption process was conducted using 17 cm bed heights in glass columns with a 2.2 cm internal diameter. The weights of the adsorbents were 36.5 and 33.9 g of carbonaceous material. The fixed-bed volumes for both adsorbents were 64.62 cm^3. The used carbonaceous material was a residual sludge from a wastewater treatment plant, and we sought to determine and compare the applicability of an absorbent after a pyrolysis treatment and that of commercial activated carbon. Glass wool was placed in the bottom of the column to support the adsorbent. A volumetric flow rate of 40 mL min^{-1}, corresponding to a hydraulic charge of 152 m^3 m^{-2} day^{-1}, was calculated according to a previous report [17]. Liquid samples were withdrawn at different time points.

2.5. Analytical Procedure.

2.5. Analytical Procedure. The dye concentration was measured using a UV/vis spectrophotometer (Perkin Elmer

model LAMBDA 125) at a wavelength corresponding to the maximum absorbance of each dye. The samples were filtered with Whatman paper number 1 every 5 min for analysis. The colour removal efficiency (R, %) after each treatment was calculated using the following formula:

$$R\,(\%) = \frac{(C_0 - C_e)}{C_0}100. \tag{1}$$

where C_0 and C_e are the concentrations of dye before and after electrocoagulation, respectively. The faradic yield of the metal dissolution (Φ) was estimated according to Faraday's law:

$$\Phi = \frac{MIt}{nF}. \tag{2}$$

where Φ is the amount of iron dissolution (g), M is the molecular weight of the iron (55.85 g mol^{-1}), n is the number of electron moles (n = 2), and F is the Faraday constant (F = 96,487 C mol^{-1}). The conductivity and pH were measured using a multiparameter instrument (Conductronic pH-15), and a voltmeter (Fluke 179) was used to measure the voltage during the electrocoagulation process.

The rate of dye removal during electrocoagulation can be represented as follows:

$$\frac{dC}{dt} = -kC^m. \tag{3}$$

where C represents the dye concentration, m represents the order of the reaction, k represents the reaction rate coefficient, and t represents the time. For a first-order reaction, the above equation becomes the following:

$$\ln\frac{[C]}{C_0} = -kt. \tag{4}$$

where C_0 is the initial dye concentration [18].

The rate of adsorption and the adsorbed amount of cationic and anionic dye on the materials were calculated using a breakthrough curve obtained by plotting C_e/C_0 (effluent concentration/influent concentration) versus time. The experimental results were adjusted to the Bohart-Adams model using Origin 8.0 software. The expression used is as follows [11]:

$$\frac{C_e}{C_0} = \exp\left(k_{AB}C_0t - k_{AB}N_0\frac{Z}{F}\right). \tag{5}$$

where k_{AB} is the kinetic constant (Lmg^{-1} min^{-1}), F is the linear flow rate (cm min^{-1}), Z is the bed depth of the column (cm), and N_0 is the adsorption capacity (mg L^{-1}). The experimental data were fitted to this model; the R^2 parameter indicates the correlation between the experimental points and predicted values [19].

The kinetics of the electrocoagulation removal process were coupled with the fixed-bed adsorption to determine an effective percentage removal.

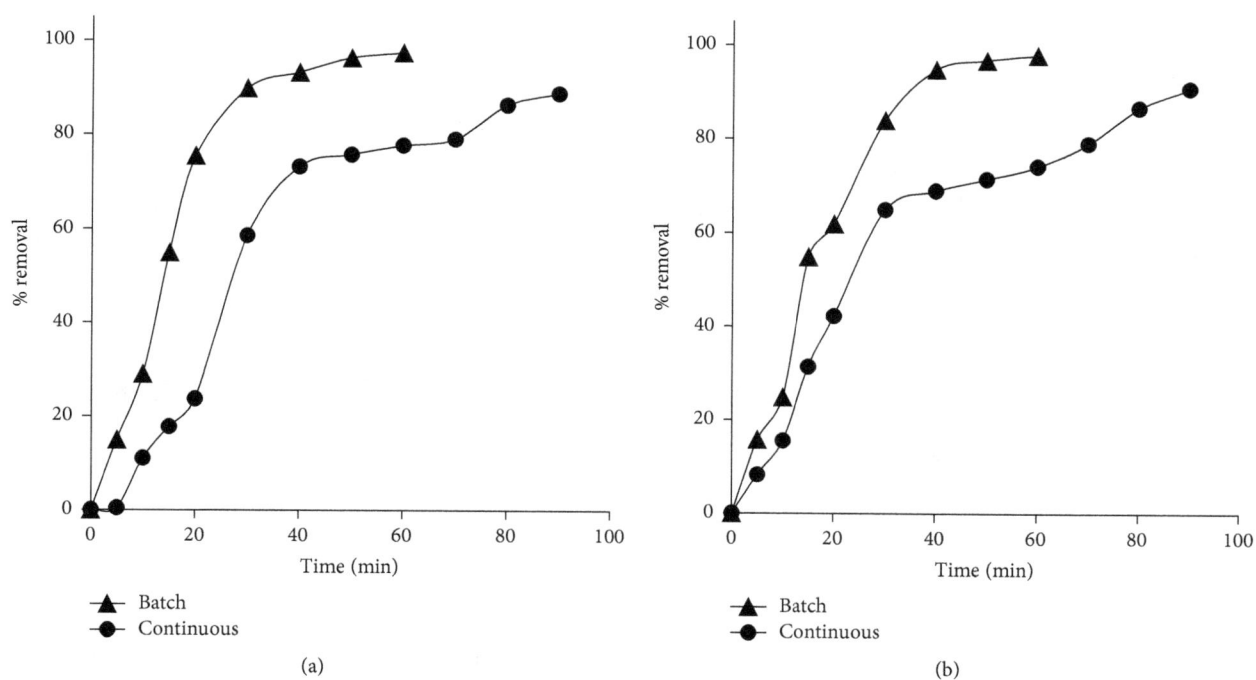

(a)

(b)

FIGURE 2: Comparative performance of the batch and the continuous-mode systems in treating (a) malachite green and (b) Remazol yellow with 100 mg/L and current input of 35.7 A/m^2.

2.6. Characterizations. For the FT-IR studies, the sludge generated during electrocoagulation before and after dye adsorption was analysed using an FT-IR spectrometer (Nicolet Avatar 360). The samples were dispersed in spectroscopic grade KBr to record the spectra (sample: KBr = 1 : 100); the IR spectra were recorded over the range of 4000–400 cm^{-1}.

For the SEM observations, the commercial activated carbon and carbonaceous samples, before and after dye adsorption and sewage sludge electrocoagulation, were mounted directly on the holders and observed at 10 and 20 kV using a JEOL 8810 LV microscope. The microanalysis was performed using an EDS system (Oxford).

To determine the point of zero charge, the carbonaceous material (10 mg) was placed in opaque vials with a 0.01 M NaCl solution that was previously adjusted to a specific pH value between 1 and 12 (1 unit intervals) using 0.1 M HCl or NaOH solutions. After 24 h of contact, the samples were centrifuged and decanted, and the pH was determined in the liquid phase (HI 2550 pH/ORP meter and EC/TDS/NaCl meter HANNA analyser).

Specific surface areas were determined using the N$_2$ BET nitrogen adsorption method with a Micrometrics® Gemini 2360 surface area analyser. The dry and degassed samples were analysed using a multipoint N$_2$ adsorption-desorption method at room temperature.

2.7. Adsorption Batch Experiments. One-hundred milligrams of carbonaceous material and 10 mL of aliquots of different concentrations of RY or MG (at concentrations of 20, 40, 60, 80, 100, and 150 mg L^{-1}) were studied over the course of 5 h in a water bath (303, 313, and 323 K) using a previously

established method [16]. Later, the samples were decanted to separate the phases. All adsorption experiments were performed in duplicate. In addition, dye solutions with varying pH values (4, 6, 8, and 10) were tested. The pH of the solution was adjusted using 1 M HCl and 1 M NaOH. Each mixture was vacuum filtered, and the dye concentrations were determined in the liquid phase. The amount of dye on the corresponding material q_e (mg/g) was calculated using a mass balance relationship: $q_e = (C_0 - C_e)V/W$, where C_0 (mg L^{-1}) and C_e (mg L^{-1}) are the initial and equilibrium liquid-phase concentrations of the dye, respectively; V (L) is the volume of the solution; and W (g) is the weight of the corresponding material.

3. Results and Discussion

3.1. Batch and Continuous Modes of Operation. The batch mode was found to significantly enhance the efficiency of the electrochemical treatment process in terms of the dye removal, as demonstrated in Figure 2. The maximum dye removal for both dyes using the batch-mode and continuous-mode electrocoagulation was achieved at 60 min and 90 min, respectively.

Notably, a continuous system operates under steady-state conditions with a fixed pollutant concentration and effluent flow rate. In contrast, the dynamic nature of a batch reactor enables studying a range of operating conditions and is more suited to research studies. Continuous systems are better suited for industrial processes with large effluent volumes, whereas batch reactors are suited for laboratory- and pilot plant-scale applications. The continuous mode of operation

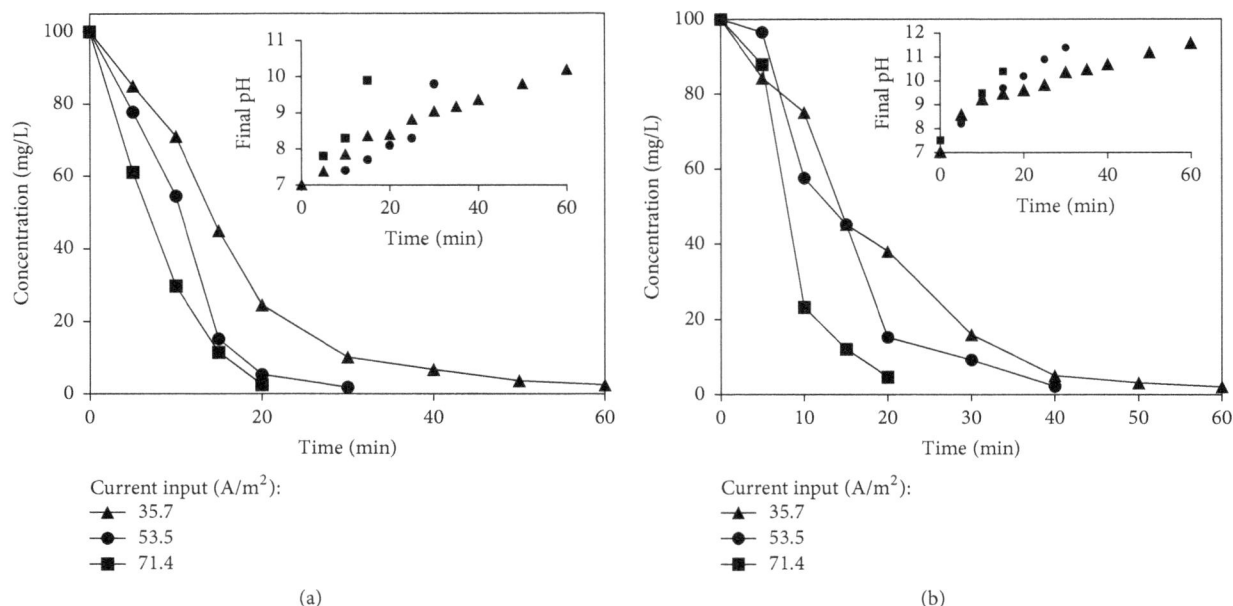

FIGURE 3: Effect of current density and final pH on the (a) Remazol yellow removal and (b) malachite green with 100 mg/L. ▲: 35.7 A/m^2, ●: 53.5 A/m^2, and ■: 71.4 A/m^2.

is preferred due to its better control compared to the batch mode of operation [10]. We studied the removal of reactive blue 21 dye by electrocoagulation using iron electrodes in batch and continuous modes. The results showed that the colour and chemical oxygen demand removal in the batch reactor were higher than those in the continuous-flow electrocoagulator because the distribution of electroactive species in the batch reactor was more effective [20]. Moreover, the electrocoagulation process was conducted in both continuous and batch modes and was determined to be an efficient and viable process for achieving a high degree of total petroleum hydrocarbon removal from petroleum-contaminated groundwater [21]. However, on a large scale in the field, the treatment process must be run in continuous mode [10].

3.2. Operation of the Continuous-Mode Electrocoagulation Process.
To establish a favourable current density value for the continuous mode, electrocoagulation processes were performed. Three values of the current density, 35.7, 53.5, and 71.4 A m^{-2}, at a fixed inlet concentration of $C_0 = 100$ mg L^{-1} with pH values between 5 and 6 were studied.

According to [22], the pollutant removal efficiency decreases with increasing initial dye concentration at a constant current density due to the formation of flocs formed from metal hydroxides that are unable to settle. Therefore, in our case, having a variable current density with a continuous system, it was decided to use an initial concentration of 100 mg L^{-1} by predicting a residue less than 20 mg L^{-1} for the coupled sorption process. Residual concentrations of 10 mg L^{-1} have been reported for sorption processes [4, 23, 24].

The initial pH values of these two dyes were nearly neutral. Therefore, chemical modification was not necessary since previous studies [25] have reported that pH values between 6 and 7 are suitable and it was compatible with Fe electrodes. Importantly, since the tests were continuous, a significant pH increase was not expected, regardless of whether the water was introduced from a surface water table or was reused; hence, the pH was expected to remain neutral, and we did not expect to apply acid or base to the treated water [26].

The data for a 60 min operation are presented in Figure 3.

As shown in Figure 3, the concentration of MG and RY decreased significantly with the increased current density. For both dyes, the time required to reach the plateau region also decreased from 60 min to less than 20 min when the current density was increased from 35.7 to 71.4 A m^{-2}. Consequently, the number of metallic hydroxides produced increased with electrocoagulation at 71.4 A m^{-2}. Therefore, the optimum current density was chosen as 35.7 A m^{-2}. This current density was used for further experiments with the continuous-mode system. The current density determines the coagulant dosage within an electrocoagulation process, adjusts the bubble production, and, hence, affects the growth of flocs in both batch and continuous reactors [9, 27].

According to Faraday's law (see (1)), as the current density increases, the ion concentrations produced by the electrodes also increase, thereby increasing the flow production and improving the dye removal efficiency [26].

The amount of the electrode that dissolved was theoretically calculated as 1.04, 1.56, and 2.08 g, following Faraday's law, corresponding to current densities of 35.7, 53.5, and 71.4 A m^{-2}, respectively. Similar effects of the current density on the electrocoagulation process have also been reported for the treatment of other organic and inorganic contaminants, such as textile wastewater [9] and heavy metals from waste fountain solutions (copper, zinc, and nickel) [28].

Furthermore, solar cells were used to supply power as a sustainable and low-cost option for treatment. Photovoltaic modules have a long life and low maintenance cost and can produce high direct currents when exposed to sunlight [29].

The electrolytic reactions during electrocoagulation with Fe electrodes include metal dissolution (anode) and water reduction (cathode). Two mechanisms for the production of metal hydroxides have been proposed [22, 30]. During the process studied herein, the pH increased from 5.3 to 11 (Figure 3) due to the production of hydroxides. The OH^- production from the reaction increased the pH during electrolysis. Consequently, insoluble $Fe(OH)_2$ precipitates were formed at pH > 5.5 and remained in equilibrium with Fe^{2+} up to a pH of 9.5 or with monomeric species such as $Fe(OH)^+$, $Fe(OH)_2$, and $Fe(OH)_3$ at higher pH values. Once the insoluble flocs of $Fe(OH)_3$ were produced, they could remove the dissolved dyes by surface complexation or electrostatic attraction [31]. The relative stability of the pH after this process was likely due to the formation of insoluble $Fe(OH)_3$ flocs and other metal hydroxides. The influent pH is an important parameter that influences the performance of the electrocoagulation process. When the pH of the dye solutions was between 5.5 and 8.5, maximum removal was achieved. As the initial pH value was approximately 6.5 for both dye solutions, additional chemicals were not required to change the initial pH values.

The formed $Fe(OH)_{n(s)}$ can remove dye molecules by surface complexation or electrostatic attraction. During surface complexation, it was assumed that the pollutant could act as a ligand to bind hydrous iron moieties using precipitation and adsorption mechanisms [22, 30].

Additionally, the removal mechanism of the azo dye may have actually complexed with the iron hydroxide-forming ionic bonds as follows [18]:

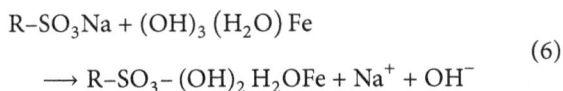

$$R–SO_3Na + (OH)_3(H_2O)Fe$$

$$\longrightarrow R–SO_3–(OH)_2 H_2OFe + Na^+ + OH^- \tag{6}$$

Figure 4 presents plots of the first-order kinetic model fitted to the experimental data. Figure 4 reveals a reasonably good fit for the first-order kinetic model for both dyes ($r^2 > 0.93$). Based on the linear regression equation fitted to the experimental data, the first-order removal rate constants for MG and RY were 0.0035 and 0.0034 min^{-1}, respectively.

Similar results were found during the removal of phosphate from landscape water using an electrocoagulation process with aluminium electrodes powered directly by photovoltaic solar modules; the first model was adequately applied to the experimental data [29]. Based on these results, we confirmed that electrocoagulation has the ability to simultaneously remove cationic and anionic dyes from aqueous solutions.

3.3. Batch Adsorption Experiments.
Adsorption equilibrium experiments were performed to describe the dye adsorption

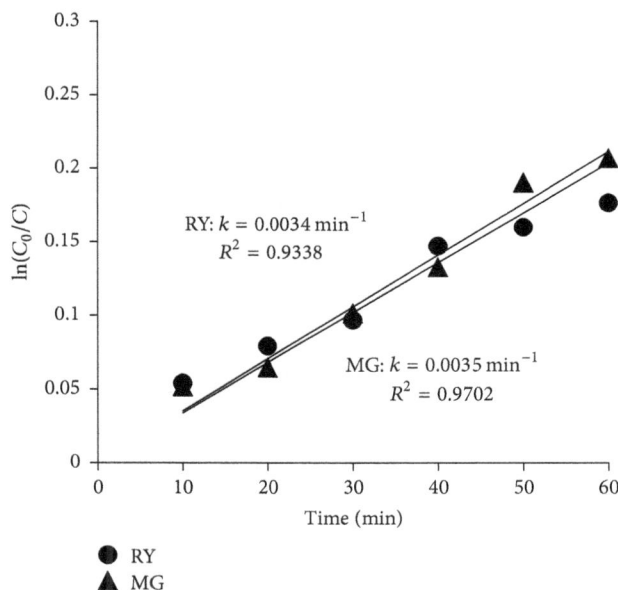

FIGURE 4: Plot of the first-order equation for (electrocoagulation) removal kinetics of malachite green (MG) and Remazol yellow (RY) in aqueous solution.

behaviour of the carbonaceous material. Two models, the Langmuir and Freundlich isotherms, were employed to analyse MG and RY adsorption data, and the respective constants of each model were estimated using nonlinear regression analysis [32].

Langmuir:

$$q_e = \frac{q_o b C_e}{1 + b C_e}. \tag{7}$$

Freundlich:

$$q_e = K_F C_e^{1/n}, \tag{8}$$

where C_e is the dye concentration at equilibrium (mg L^{-1}), q_e is the adsorption capacity at equilibrium (mg g^{-1}), q_{omax} is the maximum adsorption capacity (mg g^{-1}), b is the adsorption energy constant (L mg^{-1}), K_F is the Freundlich adsorbent capacity, and $1/n$ is the heterogeneity factor. Table 1 shows the parameters obtained after applying these models to the experimental data, and Figure 5 shows the experimental data and Langmuir model. The correlation coefficient for the Langmuir isotherm was greater than that for the Freundlich isotherm for RY adsorption, while the opposite behaviour was observed for MG adsorption. The Langmuir model assumes monolayer adsorption onto a surface containing a finite number of adsorption sites via a uniform adsorption mechanism without transmigration of the adsorbate along the surface plane. The Freundlich equation is based on adsorption onto a heterogeneous surface [33].

In the MG adsorption case, the experimental data were adjusted to both models with similar correlation values ($R^2 =$

TABLE 1: Adsorption isotherm parameters of Remazol yellow and malachite green by carbonaceous material.

| Dye | Adsorption isotherms | | | | | |
| | Langmuir | | | Freundlich | | |
	q_{max} (mg/g)	b	r^2	K_F	$1/n$	r^2
Remazol yellow	11.69	0.015	0.9926	0.5593	0.55	0.9832
Malachite green	39.76	0.086	0.9706	6.16	0.45	0.9745

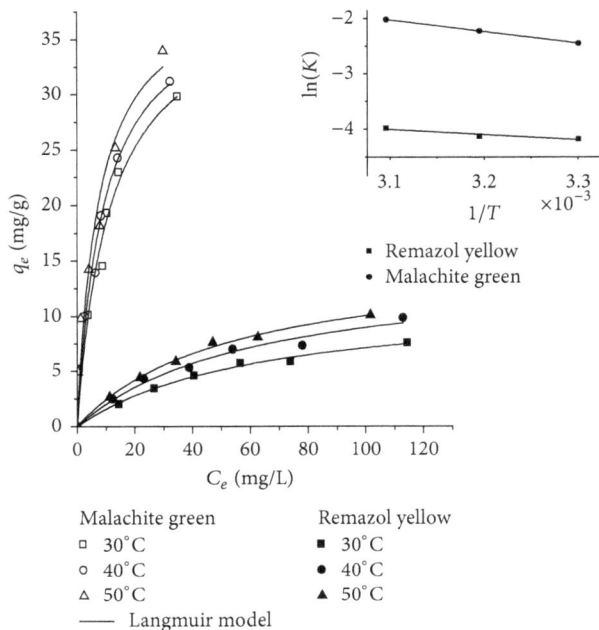

FIGURE 5: Langmuir model and Van't Hoff plot for the adsorption of the Remazol yellow and malachite green onto carbonaceous material.

TABLE 2: Thermodynamic parameters for the adsorption of Remazol yellow and malachite green by carbonaceous material.

| Dye | ΔH (kJ·mol^{-1}) | ΔS (kJ·mol^{-1}) | ΔG (kJ·mol^{-1}) | | |
			303 K	313 K	323 K
Remazol yellow	7.76	−0.009	10.48	10.57	10.66
Malachite green	17.53	0.037	6.31	5.94	5.57

of dye to interact with the adsorbent surface in solution [24]; the dyes could be positively or negatively charged to favour the selective adsorption using basic and acid materials, respectively.

To estimate the effect of the temperature on the dye adsorption onto the carbonaceous material, three basic thermodynamic parameters, the enthalpy (ΔH), entropy (ΔS), and Gibbs free energy (ΔG), were calculated using a relationship between the equilibrium constant (K_L) and temperature (T), which is expressed using the Van't Hoff equation as follows [41, 42]:

$$\ln K_L = \frac{\Delta S}{R} - \left(\frac{\Delta H}{R}\right)\left(\frac{1}{T}\right)$$
$$\Delta G = \Delta H - T\Delta S,$$
(9)

where K_L, R, and T are the Langmuir equilibrium constant, the universal gas constant (8.314 J·mol^{-1} K^{-1}), and the absolute temperature in K, respectively.

The values of the thermodynamic parameters, enthalpy (ΔH), Gibbs energy change (ΔG), and entropy change (ΔS), of the adsorption process for RY and MG are shown in Figure 5 and Table 2. The positive value for the Gibbs energy change (ΔG) shows that the adsorption process was not spontaneous [42]. The positive value of ΔH indicates the endothermic nature of the adsorption process. The positive values of ΔS for MG indicate that the randomness at the solid-liquid interface during the adsorption process increased and reflect the affinity of the adsorbent material for MG. In contrast, for RY, the negative standard entropy change (ΔS) suggests that the randomness at the solid/solution interface decreased during adsorption. Similar behaviour was observed during the adsorption of three Bezathren dyes using sodic bentonite [43].

3.4. Fixed-Bed Adsorption Experiments. Figure 6 shows the fit of the experimental results to the Bohart-Adams model, and Table 3 summarizes the parameters calculated for the adsorption of MG and RY by the commercial activated carbon and carbonaceous material.

0.97). The Langmuir model is based on the assumption of a homogeneous adsorbing surface and the independence of the adsorption sites. However, carbonaceous material is a heterogeneous product of sewage sludge carbonization, and the chemistry composition of the material used herein was heterogeneous based on the EDS analyses.

The maximum adsorption capacities of RY (q_e = 11.69 mg g^{-1}) and MG (q_e = 39.76 mg g^{-1}) on the carbonaceous material were determined. A similar result was found for the adsorption of RY onto the carbonaceous material prepared from sewage sludge (12.72 mg g^{-1}) [16], while the adsorption of MG onto the carbonaceous material was superior to that on the commercial activated carbon in the batch system [40].

The pH of a solution is an important parameter during a sorption process. The initial pH was inversely related to the adsorption capacity at equilibrium for both dyes. When the initial pH of the dye solution increased from 4 to 12 for RY, the adsorption capacity decreased; with an increase in the pH, the adsorption capacity of MG increased (figure not shown). The pH values substantially affected the availability

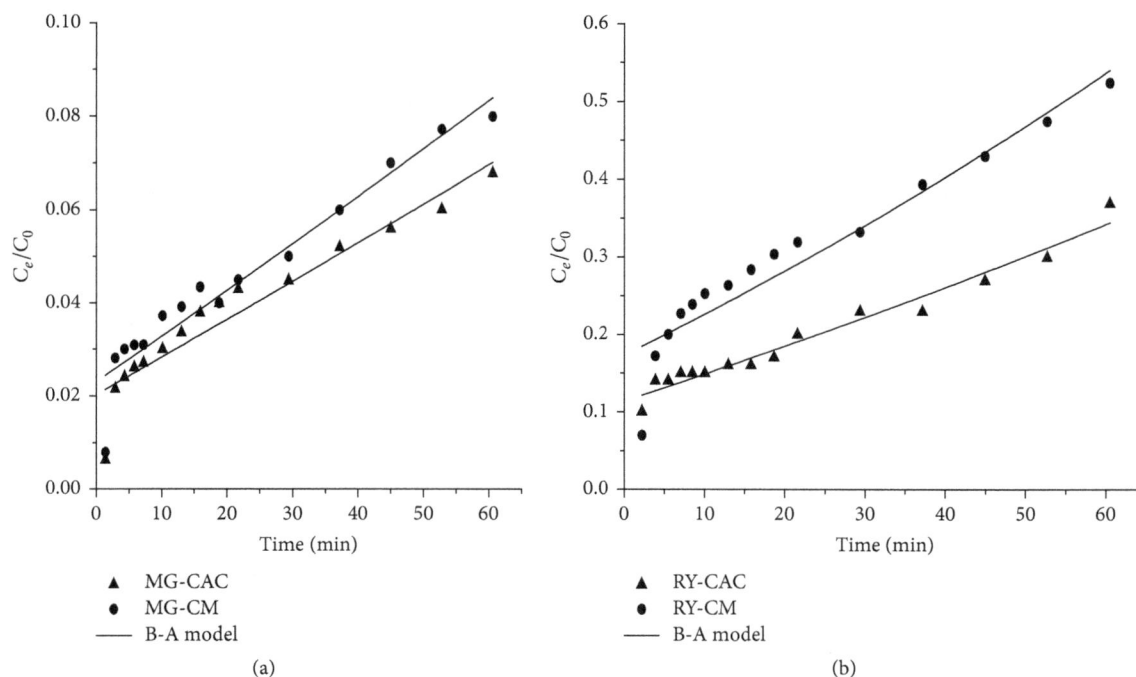

FIGURE 6: Comparison of fitted curves and experimental data using the Bohart-Adams model for the adsorption of (a) malachite green and (b) Remazol yellow using CAC and CM (h = 17 cm).

The removal of RY using carbonaceous material from pyrolyzed sewage sludge treated with 10% HCl was evaluated using adsorption equilibria. The results indicated that the main mechanism involved in the adsorption of the dye was chemisorption on a heterogeneous material [16]. However, MG removal has not been reported for batch adsorption.

For both materials, the results revealed that the maximum adsorption capacities were higher for the cationic dye (MG) than for the anionic dye (RY), even though the initial concentration was higher for the first dye. The treatment of the carbonaceous material with 10% HCL likely played a major role in the removal of the cationic dye. For the MG adsorption, the carbonaceous material breakthrough curve showed similar behaviour to that obtained for the CAC breakthrough curve (Figure 6(a)).

The adsorption rate was faster for RY than for MG. As seen in Table 3, the R^2 values were higher than 0.90 for both dyes and adsorbents. Thus, this model showed good agreement between the predicted and experimental data up to C_e/C_0 = 0.5 for all experiments. This trend can be explained by the prevalence of external mass transfer during the initial adsorption in the column [44].

A comparison of the two adsorbents (Table 3) indicated that the values of the maximum adsorption capacities for the CAC system were higher than those for the carbonaceous material system.

Usually, the breakthrough point is defined as the time when the effluent concentration (C_e) reaches a percentage of the influent concentration (C_0) that is considered unacceptable, for example, 10% (C_e/C_0 = 0.1). For the anionic dye, the breakthrough point was less than 5 min, while for the cationic

TABLE 3: Bohart-Adams–fitting parameters for the adsorption of malachite green and Remazol yellow using commercial activated carbon and carbonaceous material.

	Malachite green		Remazol yellow	
	CAC	CM	CAC	CM
C_0 (mg L^{-1})	10.34	10.34	6.10	6.10
k_{BA} (10^{-5}) (Lmg^{-1} min^{-1})	7.77	9.45	56.47	84.93
N_o (mg L^{-1})	7796.32	6396.67	971.27	602.42
R^2	0.9057	0.9279	0.9630	0.9019

dye it was more than 60 min. These results indicate that both of the adsorbent materials exhibited a high affinity for the MG dye.

Figure 7 compares the removal efficiencies of MG and RY by electrocoagulation as a single treatment method and electrocoagulation-adsorption using the CAC and carbonaceous material; this figure shows the effectiveness of the coupled treatment during dye removal.

The removal increased significantly, and the increase in the removal demonstrated the feasibility and synergic effect of the coupled process. The dye removal efficiencies ranged from 96% to 99%; the high dye removal efficiency was due to the combined electrocoagulation and adsorption on the CAC and its high surface area. Furthermore, both adsorbents showed a greater affinity for the cationic dye (MG) than for the anionic dye (RY). These results describe the electrocoagulation-adsorption performance in a continuous

TABLE 4: Comparison of dye removal performances of various treatment technologies.

Treatment method	Dye	Material	Initial concentration	Removal efficiency/q_e	Reference
Biological	MG	*T. versicolor* laccase-basic exchange resin D201	50 mg L^{-1}	35–40%	[34]
Catalysis	MG	ZnS, ZnS:Fe	15 mg L^{-1}	98.3/99.0%	[5]
	RY	ClO$_2$ catalytic oxidation	200 mg L^{-1}	94.03%	[35]
Electrolysis	MG	BDD electrodes	20 mg L^{-1}	91%	[36]
Adsorption	MG	Microorganisms based compost	50–1000 mg L^{-1}	96.8%/136.6 mg g^{-1}	[37]
	MG	Shell-treated *Zea mays*	10–200 Mg L^{-1}	NA/81.5 mg g^{-1}	[38]
	RY	Fe–Ni nanoscale oxides	100 mg L^{-1}	83%/157 mg g^{-1}	[39]
	RY	Fe–Cu nanoscale oxides	100 mg L^{-1}	70%/117.6 mg g^{-1}	[39]
	RY	Montmorillonite KSF	100 mg L^{-1}	NA/8.62 mg g^{-1}	[6]
	RY	Carbonaceous material-1% HCl	100 mg L^{-1}	NA/12.72 mg g^{-1}	[16]
	RY	Apple pulp-TiO$_2$	10 mg L^{-1}	86.97%/NA	[23]
Electrocoagulation-adsorption	MG	Fe electrodes/commercial activated carbon	100 mg L^{-1}	99.29%/7796.32 mg L^{-1}	This work
	MG	Fe electrodes/carbonaceous material	100 mg L^{-1}	99.17%/6396.67 mg L^{-1}	This work
	RY	Fe electrodes/commercial activated carbon	100 mg L^{-1}	99.77%/971.27 mg L^{-1}	This work
	RY	Fe electrodes/carbonaceous material	100 mg L^{-1}	96.79%/602.42 mg L^{-1}	This work

NA: not available.

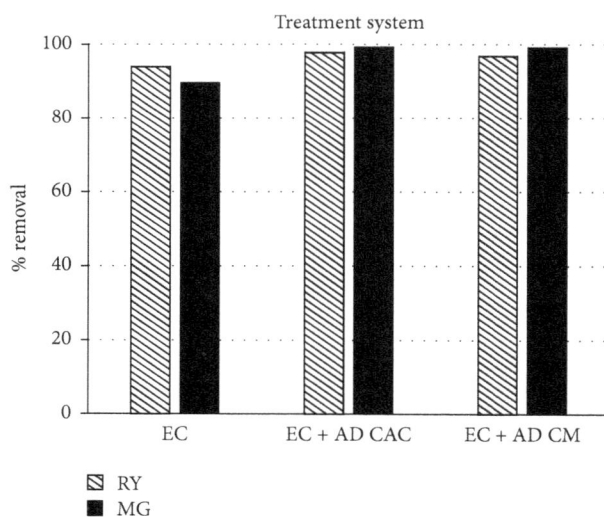

FIGURE 7: Malachite green and Remazol yellow removal efficiencies with electrocoagulation and electrocoagulation-adsorption CAC and electrocoagulation-adsorption CM.

system to explore the possibility of using the system for treating real textile effluents.

Previous studies have compared electrocoagulation-adsorption to other conventional and advanced methods (Table 4). Abundant information is available regarding MG and RY adsorption but not MG and RY electrocoagulation.

It is difficult to compare the results of this work with those from the literature because the dye removal efficiency was determined under different experimental conditions and was dependent on the type of treatment, initial dye concentration, electrode type, and chemical composition of the adsorbent. Notably, in Table 4, the electrocoagulation-adsorption resulted in a higher removal efficiency compared to the other treatment methods. The carbonaceous material used in this study shows relatively good adsorption capacities compared with CAC.

3.5. *Characterization.* The SEM image of the solid sludge precipitates (the product of RY electrocoagulation) is presented in Figure 8(a). This SEM image shows continuous flake-shaped aggregates with diameters of 10–100 μm. The elemental composition of the MG and RY sludge precipitate samples (Table 5) exhibited carbon (C), oxygen (O), and iron (Fe). The chemical composition of the sewage shows that the quantity of iron notably increased (>40.51%) when the samples were treated with electrocoagulation. The presence of silicon (Si) was attributed to the glass slit used for the EDS analysis of the sample.

Figure 8(b) shows the morphology of the activated carbon; a porous structure can be observed, and Figure 8(c) shows the morphology of the carbonaceous material, which is similar to the morphology reported elsewhere [45].

The chemical compositions of the CAC and carbonaceous material before and after MG and RY adsorption are presented in Table 3. The amount of carbon (C) in the adsorbent

TABLE 5: Elemental analysis of the dye sludge produced in the electrocoagulation process and commercial activated carbon and carbonaceous material (CM), before and after malachite green and Remazol yellow adsorption by EDS.

Element	Weight percent SEWAGE-MG	Weight percent SEWAGE-YR	Weight percent CAC	Weight percent CAC-MG	Weight percent CAC-YR	Weight percent CM	Weight percent CM-MG	Weight percent CM-YR
C	15.37 ± 5.44	19.40 ± 8.12	82.2 ± 1.3	76.66 ± 1.32	83.94 ± 1.24	43.83 ± 7.14	44.34 ± 4.22	47.44 ± 8.10
O	36.53 ± 1.93	36.13 ± 7.86	12.3 ± 1.1	20.41 ± 0.35	15.07 ± 1.48	21.31 ± 1.45	29.01 ± 3.81	18.94 ± 1.70
Na	—	—	0.6 ± 0.1	—	—	1.56 ± 0.89	1.02 ± 0.33	1.03 ± 0.21
Mg	—	—	0.2 ± 0.0	—	—	0.96 ± 0.22	1.5 ± 0.27	0.86 ± 0.33
Al	—	—	—	—	—	3.26 ± 0.06	1.00	3.23 ± 0.47
Si	2.68 ± 0.11	2.91 ± 0.96	0.3 ± 0.1	0.98 ± 0.23	—	11.98 ± 0.27	5.17 ± 0.17	14.56 ± 2.66
P	—	—	4.0 ± 0.3	1.95 ± 0.58	0.99 ± 0.26	1.80 ± 0.42	0.85 ± 0.12	1.24 ± 0.32
Ca	—	—	—	—	—	6.48 ± 3.91	10.31 ± 1.29	4.82 ± 1.69
S	—	0.22	—	—	—	1.61 ± 1.10	1.15 ± 0.31	1.08 ± 0.05
Cl	—	0.78 ± 0.46	—	—	—	1.65 ± 1.60	0.55 ± 0.35	0.22 ± 0.10
K	—	—	—	—	—	1.00 ± 0.09	0.95 ± 0.55	1.12 ± 0.34
Fe	45.43 ± 5.64	40.55 ± 13.02	0.4 ± 0.1	—	—	2.02 ± 0.74	2.10 ± 1.48	2.46 ± 1.32
Ti	—	—	—	—	—	0.55 ± 0.10	0.68 ± 0.26	0.76 ± 0.44
Cr	—	—	—	—	—	0.67 ± 0.18	0.62 ± 0.20	0.76 ± 0.23
Zn	—	—	—	—	—	1.13 ± 0.31	0.95 ± 0.25	1.42 ± 0.85

FIGURE 8: SEM image of the (a) sludge produced in the Remazol yellow-EC process and materials: (b) CAC and (c) CM before adsorption.

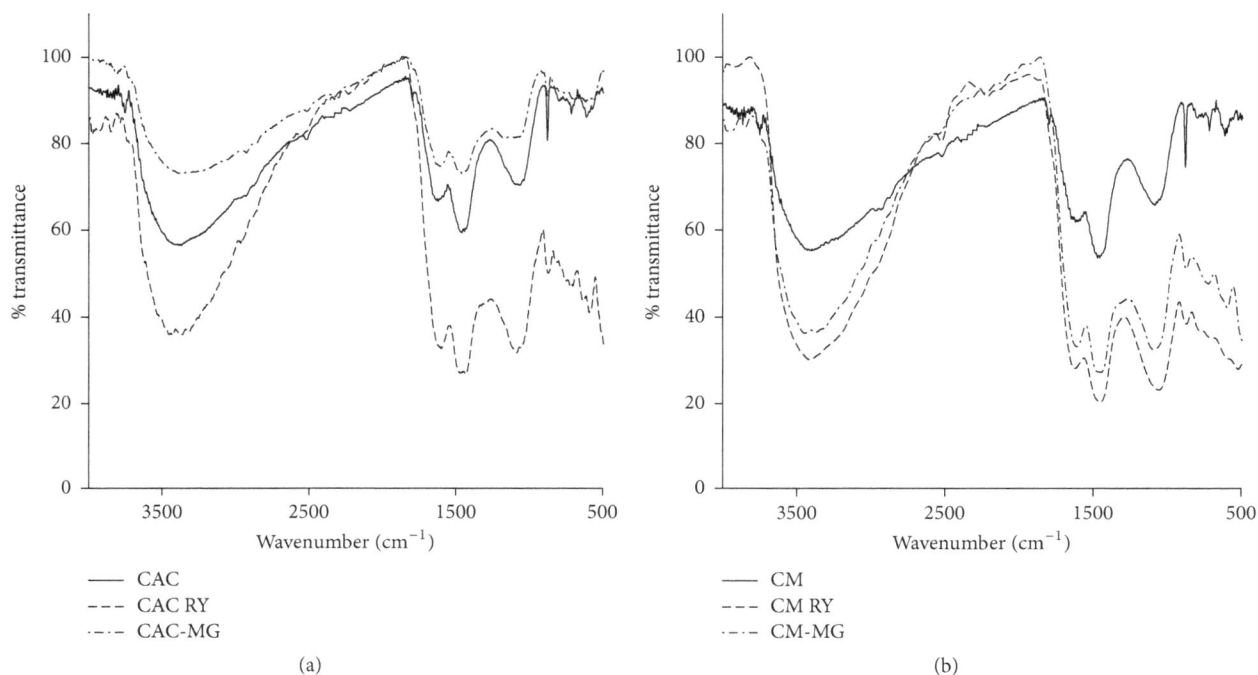

(a)

(b)

FIGURE 9: FT-IR characterization of (a) CAC and (b) CM before and after malachite green and Remazol yellow adsorption.

increased when the samples were in contact with the dye solution, which indicated that dye adsorption occurred on the materials.

The point of zero charge of the carbonaceous material was 7.1, and the FT-IR spectra of the MG and RY sludge (figure not shown) presented a broad and intense band at $3317 \, \text{cm}^{-1}$ that was attributed to the stretching vibrations of –OH. The variable band at $1630 \, \text{cm}^{-1}$ was attributed to the stretching vibration of the C=C and C=N aromatic groups. The bands at $1369 \, \text{cm}^{-1}$ indicated the presence of CH_2, and the bands in the 1600 to $1000 \, \text{cm}^{-1}$ region corresponded to the C–C and C–N groups [46]. Similar results have been reported elsewhere [39].

The FT-IR spectra of the carbonaceous material and CAC before and after dye adsorption at $4000–400 \, \text{cm}^{-1}$ are presented in Figure 9.

Bands appearing at 3367, 2908, 1620, 1458, 1057, and $876 \, \text{cm}^{-1}$ in Figure 9(a) were assigned to OH stretching, C–H stretching in the alkane, COO^- anion stretching, OH bending, C–O stretching in the ester or ether, and N–H deformation in the amines, respectively. Of these groups, the carboxylic acid and hydroxyl groups played a major role in the dye removal. Figure 9(b) shows a large number of bands, indicating the complex nature of the adsorbent carbonaceous material. The bands appearing at 3402, 2974, 1597, 1458, 1080, and $876 \, \text{cm}^{-1}$ were attributed to the presence of OH groups, C–H stretching in the alkanes, C=O stretching in the carboxylate groups, and C–O–H stretching. The bands below $1200 \, \text{cm}^{-1}$ were considered to be the fingerprint region. The MG- and RY-loaded adsorbents showed either a shift or reduction in most adsorption bands, suggesting that the functional groups played an important role in adsorption.

The zero charge point of a carbonaceous material is determined when the pH_{eq} is equal or quite similar to $pH_{initial}$. The zero charge point obtained in this study (7.0) is of the same order of magnitude as the value found by Márquez-Mendoza et al. [47]; thus, the zero charge point appears to be dependent on the origin of the material.

The BET specific surface areas of the adsorbents were 1471.22 and 100.79 $m^2 g^{-1}$, and the total pore volumes were 0.7278 and 0.1983 $cm^3 g^{-1}$ for the CAC and carbonaceous material, respectively. The BET specific surface area may have been responsible for the adsorption behaviour of the materials.

4. Conclusions

During electrocoagulation, the removal efficiency achieved in the batch system was higher than that achieved in the continuous electrochemical process. The continuous electrocoagulation-adsorption processes were more efficient for dye removal than electrocoagulation alone. The cationic and anionic dye removal efficiencies achieved during adsorption with the CAC and carbonaceous material were very similar at 96–98%.

The thermodynamic parameters ΔG, ΔH, and ΔS indicated that the adsorption process was spontaneous and endothermic. The SEM-EDS and FT-IR analyses of the sludge produced during electrocoagulation and of the adsorbent after adsorption revealed that carbon, oxygen, and the elemental components of the dye were the main byproducts formed after electrocoagulation and adsorption. The chemical composition of the sewage indicated that the quantity of iron notably increased when the samples were treated with electrocoagulation. Thus, the results presented in this work indicate that a continuous electrocoagulation-adsorption process can effectively remove cationic and anionic dyes from water.

Conflicts of Interest

The authors declare that they have no conflicts of interest.

Acknowledgments

The authors acknowledge financial support from the CONA-CYT scholarship (Grant no. 306269) for Josué Alonso Castañeda Díaz and Universidad Autónoma del Estado de México, Project 3688/2014/CIB.

References

[1] A. Blanco-Flores, A. Colín-Cruz, E. Gutiérrez-Segura et al., "Efficient removal of crystal violet dye from aqueous solutions by vitreous tuff mineral," *Environmental Technology*, vol. 35, no. 12, pp. 1508–1519, 2014.

[2] E. Brillas and C. A. Martínez-Huitle, "Decontamination of wastewaters containing synthetic organic dyes by electrochemical methods. An updated review," *Applied Catalysis B: Environmental*, vol. 166-167, pp. 603–643, 2015.

[3] D. Podstawczyk, A. Witek-Krowiak, K. Chojnacka, and Z. Sadowski, "Biosorption of malachite green by eggshells: mechanism identification and process optimization," *Bioresource Technology*, vol. 160, pp. 161–165, 2014.

[4] M. Ghaedi, E. Shojaeipour, A. M. Ghaedi, and R. Sahraei, "Isotherm and kinetics study of malachite green adsorption onto copper nanowires loaded on activated carbon: artificial neural network modeling and genetic algorithm optimization," *Spectrochimica Acta Part A: Molecular and Biomolecular Spectroscopy*, vol. 142, pp. 135–149, 2015.

[5] H. R. Rajabi, O. Khani, M. Shamsipur, and V. Vatanpour, "High-performance pure and Fe^{3+}-ion doped ZnS quantum dots as green nanophotocatalysts for the removal of malachite green under UV-light irradiation," *Journal of Hazardous Materials*, vol. 250-251, pp. 370–378, 2013.

[6] A. P. DíazGómez-Treviño, V. Martínez-Miranda, and M. Solache-Ríos, "Removal of remazol yellow from aqueous solutions by unmodified and stabilized iron modified clay," *Applied Clay Science*, vol. 80-81, pp. 219–225, 2013.

[7] M. T. Yagub, T. K. Sen, S. Afroze, and H. M. Ang, "Dye and its removal from aqueous solution by adsorption: a review," *Advances in Colloid and Interface Science*, vol. 209, pp. 172–184, 2014.

[8] V. K. Gupta, R. Kumar, A. Nayak, T. A. Saleh, and M. A. Barakat, "Adsorptive removal of dyes from aqueous solution onto carbon nanotubes: a review," *Advances in Colloid and Interface Science*, vol. 193-194, pp. 24–34, 2013.

[9] B. Merzouk, B. Gourich, A. Sekki, K. Madani, Ch. Vial, and M. Barkaoui, "Studies on the decolorization of textile dye wastewater by continuous electrocoagulation process," *Chemical Engineering Journal*, vol. 149, no. 1–3, pp. 207–214, 2009.

[10] V. Khandegar and A. K. Saroha, "Electrocoagulation for the treatment of textile industry effluent—a review," *Journal of Environmental Management*, vol. 128, pp. 949–963, 2013.

[11] E. Gutiérrez-Segura, M. Solache-Ríos, A. Colín-Cruz, and C. Fall, "Comparison of cadmium adsorption by inorganic adsorbents in column systems," *Water, Air, and Soil Pollution*, vol. 225, no. 6, pp. 1943–1956, 2014.

[12] E. Yuksel, E. Gurbulak, and M. Eyvaz, "Decolorization of a reactive dye solution and treatment of a textile wastewater by electrocoagulation and chemical coagulation: Techno-economic comparison," *Environmental Progress and Sustainable Energy*, vol. 31, no. 4, pp. 524–535, 2012.

[13] V. K. Gupta and Suhas, "Application of low-cost adsorbents for dye removal—a review," *Journal of Environmental Management*, vol. 90, no. 8, pp. 2313–2342, 2009.

[14] I. Ali, M. Asim, and T. A. Khan, "Low cost adsorbents for the removal of organic pollutants from wastewater," *Journal of Environmental Management*, vol. 113, pp. 170–183, 2012.

[15] N. Abidi, E. Errais, J. Duplay et al., "Treatment of dye-containing effluent by natural clay," *Journal of Cleaner Production*, vol. 86, pp. 432–440, 2015.

[16] J. Torres-Pérez, M. Solache-Ríos, and A. Colín-Cruz, "Sorption and desorption of dye remazol yellow onto a Mexican surfactant-modified clinoptilolite-rich tuff and a carbonaceous material from pyrolysis of sewage sludge," *Water, Air, and Soil Pollution*, vol. 187, no. 1–4, pp. 303–313, 2008.

[17] G. Tchobanoglous, F. L. Burton, and H. D. Stensel, *Wasteweater Engineering*, McGraw Hill Inc, New York, NY, USA, 2003.

[18] I. A. Şengil and M. Özacar, "The decolorization of C.I. Reactive black 5 in aqueous solution by electrocoagulation using sacrificial iron electrodes," *Journal of Hazardous Materials*, vol. 161, no. 2-3, pp. 1369–1376, 2009.

[19] S. Sánchez-Rodríguez, J. Trujillo-Reyes, E. Gutiérrez-Segura, M. Solache-Ríos, and A. Colín-Cruz, "Removal of indigo carmine by a Ni nanoscale oxides/Schoenoplectus acutus composite in batch and fixed bed column systems," *Separation Science and Technology*, vol. 50, no. 11, pp. 1602–1610, 2015.

[20] N. Ardhan, T. Ruttithiwapanich, W. Songkasiri, and C. Phalakornkule, "Comparison of performance of continuous-flow and batch electrocoagulators: A case study for eliminating reactive blue 21 using iron electrodes," *Separation and Purification Technology*, vol. 146, pp. 75–84, 2015.

[21] G. Moussavi, R. Khosravi, and M. Farzadkia, "Removal of petroleum hydrocarbons from contaminated groundwater using an electrocoagulation process: batch and continuous experiments," *Desalination*, vol. 278, no. 1–3, pp. 288–294, 2011.

[22] N. Daneshvar, A. Oladegaragoze, and N. Djafarzadeh, "Decolorization of basic dye solutions by electrocoagulation: an investigation of the effect of operational parameters," *Journal of Hazardous Materials*, vol. 129, no. 1–3, pp. 116–122, 2006.

[23] N. Ozbay and A. S. Yargic, "Factorial experimental design for Remazol Yellow dye sorption using apple pulp/apple pulp carbon-titanium dioxide co-sorbent," *Journal of Cleaner Production*, vol. 100, pp. 333–343, 2015.

[24] K. Ahmed, F. Rehman, C. T. G. V. M. T. Pires, A. Rahim, A. L. Santos, and C. Airoldi, "Aluminum doped mesoporous silica SBA-15 for the removal of remazol yellow dye from water," *Microporous and Mesoporous Materials*, vol. 236, pp. 167–175, 2016.

[25] C. Barrera-Diaz, *Aplicaciones electroquímicas al tratamiento de aguas residuales*, Editorial Reverte, 2016.

[26] N. Daneshvar, A. R. Khataee, A. R. Amani Ghadim, and M. H. Rasoulifard, "Decolorization of C.I. Acid Yellow 23 solution by electrocoagulation process: Investigation of operational parameters and evaluation of specific electrical energy consumption (SEEC)," *Journal of Hazardous Materials*, vol. 148, no. 3, pp. 566–572, 2007.

[27] S. Zodi, B. Merzouk, O. Potier, F. Lapicque, and J.-P. Leclerc, "Direct red 81 dye removal by a continuous flow electrocoagulation/flotation reactor," *Separation and Purification Technology*, vol. 108, pp. 215–222, 2013.

[28] M. Prica, S. Adamovic, B. Dalmacija et al., "The electrocoagulation/flotation study: the removal of heavy metals from the waste fountain solution," *Process Safety and Environmental Protection*, vol. 94, pp. 262–273, 2015.

[29] S. Zhang, J. Zhang, W. Wang, F. Li, and X. Cheng, "Removal of phosphate from landscape water using an electrocoagulation process powered directly by photovoltaic solar modules," *Solar Energy Materials and Solar Cells*, vol. 117, pp. 73–80, 2013.

[30] U. T. Un and E. Aytac, "Electrocoagulation in a packed bed reactor-complete treatment of color and cod from real textile wastewater," *Journal of Environmental Management*, vol. 123, pp. 113–119, 2013.

[31] C. A. Martínez-Huitle and E. Brillas, "Decontamination of wastewaters containing synthetic organic dyes by electrochemical methods: a general review," *Applied Catalysis B: Environmental*, vol. 87, no. 3-4, pp. 105–145, 2009.

[32] E. J. Lara-Vásquez, M. Solache-Ríos, and E. Gutiérrez-Segura, "Malachite green dye behaviors in the presence of biosorbents from maize (Zea mays L.), their Fe–Cu nanoparticles composites and Fe–Cu nanoparticles," *Journal of Environmental Chemical Engineering*, vol. 4, no. 2, pp. 1594–1603, 2016.

[33] Ch. Djilani, R. Zaghdoudi, F. Djazi et al., "Adsorption of dyes on activated carbon prepared from apricot stones and commercial activated carbon," *Journal of the Taiwan Institute of Chemical Engineers*, vol. 53, pp. 112–121, 2015.

[34] X. Zhang, S. Zhang, B. Pan, M. Hua, and X. Zhao, "Simple fabrication of polymer-based Trametes versicolor laccase for decolorization of malachite green," *Bioresource Technology*, vol. 115, pp. 16–20, 2012.

[35] X. Bi, P. Wang, C. Jiao, and H. Cao, "Degradation of remazol golden yellow dye wastewater in microwave enhanced ClO_2 catalytic oxidation process," *Journal of Hazardous Materials*, vol. 168, no. 2-3, pp. 895–900, 2009.

[36] F. Guenfoud, M. Mokhtari, and H. Akrout, "Electrochemical degradation of malachite green with BDD electrodes: Effect of electrochemical parameters," *Diamond and Related Materials*, vol. 46, pp. 8–14, 2014.

[37] T. B. Pushpa, J. Vijayaraghavan, S. J. Sardhar-Basha, V. Sekaran, K. Vijayaraghavan, and J. Jegan, "Investigation on removal of malachite green using EM based compost as adsorbent," *Ecotoxicology and Environmental Safety*, vol. 118, pp. 177–182, 2015.

[38] A. A. Jalil, S. Triwahyono, M. R. Yaakob et al., "Utilization of bivalve shell-treated Zea mays L. (maize) husk leaf as a low-cost biosorbent for enhanced adsorption of malachite green," *Bioresource Technology*, vol. 120, pp. 218–224, 2012.

[39] J. Trujillo-Reyes, V. Sánchez-Mendieta, M. José Solache-Ros, and A. Colín-Cruz, "Removal of remazol yellow from aqueous solution using Fe–Cu and Fe–Ni nanoscale oxides and their carbonaceous composites," *Environmental Technology*, vol. 33, no. 5, pp. 545–554, 2012.

[40] R. Gopinathan, A. Bhowal, and C. Garlapati, "Thermodynamic study of some basic dyes adsorption from aqueous solutions on activated carbon and new correlations," *The Journal of Chemical Thermodynamics*, vol. 107, pp. 182–188, 2017.

[41] N. Atar, A. Olgun, and S. Wang, "Adsorption of cadmium (II) and zinc (II) on boron enrichment process waste in aqueous solutions: Batch and fixed-bed system studies," *Chemical Engineering Journal*, vol. 192, pp. 1–7, 2012.

[42] T. Chen, B. Li, L. Fang, D.-S. Chen, W.-B. Xu, and C.-H. Xiong, "Response surface methodology for optimizing adsorption performance of gel-type weak acid resin for Eu(III)," *Transactions of Nonferrous Metals Society of China*, vol. 25, no. 12, pp. 4207–4215, 2015.

[43] I. Belbachir and B. Makhoukhi, "Adsorption of Bezathren dyes onto sodic bentonite from aqueous solutions," *Journal of the Taiwan Institute of Chemical Engineers*, vol. 75, pp. 105–111, 2017.

[44] M. A. Shavandi, Z. Haddadian, M. H. S. Ismail, and N. Abdullah, "Continuous metal and residual oil removal from palm oil mill effluent using natural zeolite-packed column," *Journal of the Taiwan Institute of Chemical Engineers*, vol. 43, no. 6, pp. 934–941, 2012.

[45] E. Gutiérrez-Segura, A. Colín-Cruz, C. Fall, M. Solache-Ríos, and P. Balderas-Hernández, "Comparison of Cd–Pb adsorption on commercial activated carbon and carbonaceous material from pyrolysed sewage sludge in column system," *Environmental Technology*, vol. 30, no. 5, pp. 455–461, 2009.

[46] J. B. Lambert, H. F. Shurvell, D. A. Lightner, and R. G. Cooks, *Organic Structural Spectroscopy*, Prentice Hall Inc, Upper Saddle River, NJ, USA, 1998.

[47] S. S. Márquez-Mendoza, M. Jiménez-Reyes, M. Solache-Ríos, and E. Gutiérrez-Segura, "Fluoride removal from aqueous solutions by a carbonaceous material from pyrolysis of sewage," *Water, Air, and Soil Pollution*, vol. 223, no. 5, pp. 1959–1971, 2012.

6

Dissolution and Solubility Product of Cd-Fluorapatite [Cd$_5$(PO$_4$)$_3$F] at pH of 2–9 and 25–45°C

Ju Lin,[1] Zongqiang Zhu ⓘ,[2] Yinian Zhu ⓘ,[1] Huili Liu ⓘ,[1] Lihao Zhang ⓘ,[2] and Zhangnan Jiang[1]

[1]*College of Environmental Science and Engineering, Guilin University of Technology, Guilin 541004, China*
[2]*Guangxi Key Laboratory of Environmental Pollution Control Theory and Technology, Guilin University of Technology, Guilin 541004, China*

Correspondence should be addressed to Yinian Zhu; zhuyinian@glut.edu.cn

Academic Editor: Henryk Kozlowski

Dissolution of the synthetic cadmium fluorapatite [Cd$_5$(PO$_4$)$_3$F] at 25°C, 35°C, and 45°C was experimentally examined in HNO$_3$ solution, pure water, and NaOH solution. The characterization results confirmed that the cadmium fluorapatite nanorods used in the experiments showed no obvious variation after dissolution. During the dissolution of Cd$_5$(PO$_4$)$_3$F in HNO$_3$ solution (pH = 2) at 25°C, the fluoride, phosphate, and cadmium ions were rapidly released from solid to solution, and their aqueous concentrations had reached the highest values after dissolution for <1 h, 1440 h, and 2880 h, respectively. After that, the total dissolution rates declined slowly though the solution Cd/P molar ratios increased incessantly from 1.55~1.67 to 3.18~3.22. The solubility product for Cd$_5$(PO$_4$)$_3$F (K_{sp}) was determined to be 10$^{-60.03}$ (10$^{-59.74}$~10$^{60.46}$) at 25°C, 10$^{-60.38}$ (10$^{-60.32}$~10$^{60.48}$) at 35°C, and 10$^{-60.45}$ (10$^{-60.33}$~10$^{60.63}$) at 45°C. Based on the log K_{sp} values obtained at an initial pH of 2 and 25°C, the Gibbs free energy of formation for Cd$_5$(PO$_4$)$_3$F (ΔG_f^0) was calculated to be −4065.76 kJ/mol (−4064.11~4068.23 kJ/mol). The thermodynamic parameters for the dissolution process were computed to be 342515.78 J/K·mol, −85088.80 J/mol, −1434.91 J/K·mol, and 2339.50 J/K·mol for ΔG^0, ΔH^0, ΔS^0, and ΔC_p^0, correspondingly.

1. Introduction

Apatite [Ca$_5$(PO$_4$)$_3$(F,OH)] forms a large family of minerals due to many isomorphous substitutions, which play a very important role in numerous industrial, medical, and environmental processes [1–3]. Apatite minerals are the raw materials to produce phosphatic fertilizers and usually contain many harmful minor elements. Among them, cadmium is one of the most toxic heavy elements in natural environment and can cause animal osteoporosis and osteomalacia due to its possible concentration in mammal's hard tissues through food chains [4–6]. Cd^{2+} (0.095 nm radius) and F$^-$ have the probability of substitution for Ca^{2+} (0.100 nm radius) and OH$^-$ in the vertebral animals' bones and teeth that are principally composed of calcium hydroxyapatite. Additionally, the ion exchangeability of apatite offers a possibility for the remediation of heavy metal-contaminated soils and the removal of hazardous heavy metals from industrial wastewaters. Apatite minerals have been applied to stabilize many heavy metals including cadmium [7–10].

The substitution of Cd^{2+} for Ca^{2+} in the apatite results in the formation of its isomorph, cadmium fluorapatite (Cd-FAP) [11]. Therefore, it is essential to understand its fundamental physicochemical properties, predominantly the solubility of cadmium apatite and its dissolution mechanism. Although many experimental works on the dissolution mechanism and kinetics of apatites in aqueous solution had already been executed [7, 12–16], much of them only concentrated on calcium hydroxyapatite and fluorapatite. Unfortunately, the thermodynamic data and dissolution kinetics for cadmium fluorapatite in aqueous solution under different conditions are now deficient, even though its dissolution and following release of cadmium, phosphate, and fluoride ions into water play an important role in cycling of these components.

No researches have been done on the dissolution and solubility of cadmium fluorapatite [Cd-FAP, $Cd_5(PO_4)_3F$], for which no thermodynamic data could be obtained in literatures to evaluate the environmental risk of Cd in relation with fluorapatite. In the present work, cadmium fluorapatite was synthesized by the precipitation method and characterized with different instruments. The dissolution mechanism of the synthetic cadmium fluorapatite was examined at different initial pHs and temperatures. As comparison, a similar test was also performed using the pure synthetic calcium fluorapatite (Ca-FAP). Furthermore, the aqueous concentrations of cadmium, calcium, phosphate, and fluoride ions from the experiment were used to estimate the solubility products and Gibbs free energies of formation.

2. Experimental Methods

2.1. Solid Preparation and Characterization

2.1.1. Solid Preparation. The pure cadmium fluorapatite [Cd-FAP, $Cd_5(PO_4)_3F$] was synthesized in the similar method as that used in our previous research [11]. The synthetic detail for the Cd-FAP preparation was dependent on the following precipitation: $5Cd^{2+} + 3PO_4^{3-} + F^- = Cd_5(PO_4)_3F$. In an 1 L polypropylene bottle, cadmium nitrate [$Cd(NO_3)_2 \cdot 4H_2O$] was firstly dissolved in ultrapure water to prepare 0.5 L of 0.2 mol/L Cd^{2+} solution. 0.1 L of 0.2 mol/L NaF solution and 0.3 L of 0.2 mol/L $(NH_4)_2HPO_4$ solution were then rapidly mixed with the cadmium solution in the bottle to form white suspension. The pH of the resulting mixed solution with a Cd : P : F molar ratio of 5 : 3 : 1 was adjusted to 9.00 by using NH_4OH solution, which was then stirred for 10 min at room temperature and aged at 100°C for 48 h. Finally, the white solid of cadmium fluorapatite obtained was carefully washed using ultrapure water and dried at 70°C for 48 h.

2.1.2. Characterization. A quantity of 0.01 g of the synthetic cadmium fluorapatite (Cd-FAP) was dissolved in 0.025 L of 1 M HNO_3 solution and diluted to 0.1 L in a volumetric flask with ultrapure water. The total cadmium, phosphor, and fluor contents were analyzed by a PerkinElmer inductively coupled plasma optical emission spectrometer (ICP-OES, Optima 7000DV) or a Dionex ion chromatography system (IC, ICS-2100). An X'Pert PRO diffractometer with Cu Kα radiation (40 kV and 40 mA) was applied to record the powder X-ray diffraction (XRD) pattern of Cd-FAP. Owing to no ICDD reference code for cadmium fluorapatite exist, the mineral phase was identified by comparing with the ICDD reference code for cadmium hydroxyapatite (00-014-0302). The obtained Cd-FAP nanorods were morphologically observed using a Hitachi field emission scanning electron microscope (FE-SEM, S-4800) and characterized using a Nicolet Nexus Fourier transform infrared spectrophotometer (FT-IR, 470) in a KBr pellet within 4000–400 cm^{-1}.

2.2. Dissolution Experiments. The synthetic Cd-FAP solid (2 g) was put in a series of 0.1 L bottles which were then added with 0.1 L of HNO_3 solution (pH = 2), ultrapure water

(pH = 5.6), or NaOH solution (pH = 9). The capped bottles were placed in three temperature-controlled water baths (25°C, 35°C, and 45°C). The aqueous solution (5 mL) was sampled from each bottle for 20 times (1 h, 3 h, 6 h, 12 h, 24 h, 48 h, 72 h, 480 h, 720 h, 1080 h, 1440 h, 1800 h, 2160 h, 2880 h, 3600 h, 4320 h, 5040 h, 5760 h, 6480 h, and 7200 h). After each sampling, 5 mL of the corresponding initial solution was added to hold a constant solid/solution ratio. All solution samples were filtered through a 0.22 μm membrane filter and then stabilized in a 25 mL volumetric flask using 0.2% HNO_3 solution. The total cadmium, phosphor, and fluor contents were analyzed by a Dionex IC system (ICS-2100) and a PerkinElmer ICP-OES instrument (Optima 7000DV). After 7200 h dissolution, the Cd-FAP solids were taken out, rinsed with ultrapure water, dried, and analyzed with different instruments as described above.

2.3. Thermodynamic Calculations. All calculations were executed by the computer program PHREEQC using the minteq.v4.dat database (Version 3.1.2) [17]. The PHREEQC input files use order-independent keyword data blocks and are of free-format, which ease the model constructing to simulate numerous aqueous-based scenarios. The aqueous Cd^{2+}, PO_4^{3-}, and F^- activities were first computed using PHREEQC, and then, the ion activity product (IAP) for cadmium fluorapatite [$Cd_5(PO_4)_3F$] was calculated after the mass-action expression. The aqueous species Cd^{2+}, $CdOH^+$, $Cd(OH)_2^0$, $Cd(OH)_3^-$, $Cd(OH)_4^{2-}$, Cd_2OH^{3+}, CdF^+, and CdF_2^0 were included in the calculation for the total cadmium, PO_4^{3-}, HPO_4^{2-}, $H_2PO_4^-$, $H_3PO_4^0$, and $NaHPO_4^-$ for the total phosphate, and F^-, HF^0, HF_2^-, H_2F_2, NaF, CdF^+, and CdF_2^0 for the total fluoride.

3. Results and Discussion

3.1. Solid Characterization. The component of the synthesized cadmium fluorapatite [Cd-FAP, $Cd_5(PO_4)_3F$] is dependent on the initial Cd : P : F ratio molar ratio in the precursor solution. The crystal nanorods were confirmed to be the aimed composition of $Cd_5(PO_4)_3F$. The atomic Cd : P : F ratio was 5 : 3 : 1 that is the stoichiometric ratio of $Cd_5(PO_4)_3F$. No other components were detected in the white solid precipitate.

The Cd-FAP nanorods before and after dissolution were characterized using XRD, FT-IR, and FE-SEM. As showed in Figures 1–3, the synthetic Cd-FAP solids before and after dissolution were not distinguishable. No secondary minerals were evidenced after dissolution. The XRD patterns of the prepared solids are presented in Figure 1. The X-ray diffraction showed that all the solid samples were pure apatite with crystallizing in the hexagonal system P6$_3$/m, which was confirmed by comparing with the JCPDS reference for cadmium hydroxyapatite (00-014-0302) (Figure 1). The Cd-FAP solids were highly crystallized and showed the formation of apatite nanorods with the lattice parameters of $a = 9.3284$ and $c = 6.6378$.

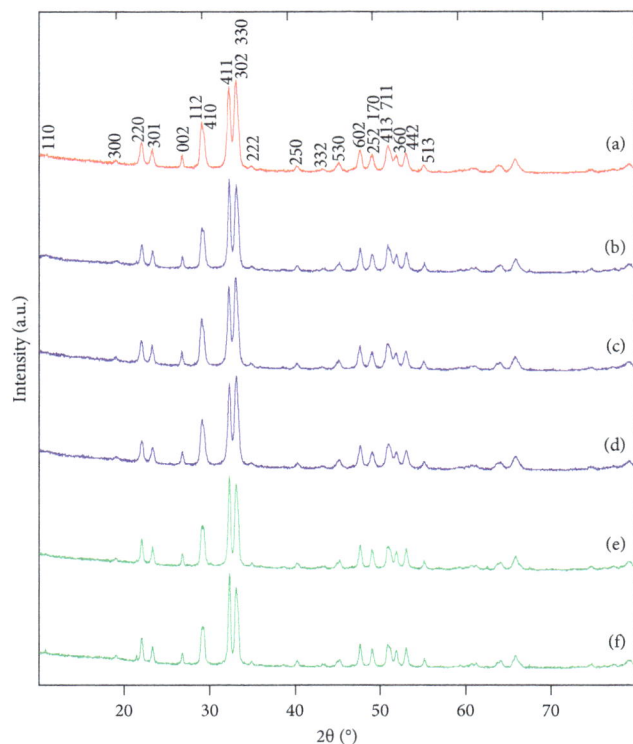

FIGURE 1: XRD analysis of the synthetic cadmium fluorapatite [$Cd_5(PO_4)_3F$] before (a) and after (b)–(f) dissolution at 25–45°C for 7200 h. (a) Synthetic Cd-FAP; (b) 25°C and pH = 2; (c) 25°C and pH = 5.6; (d) 25°C and pH = 9; (e) 35°C and pH = 2; (f) 45°C and pH = 2.

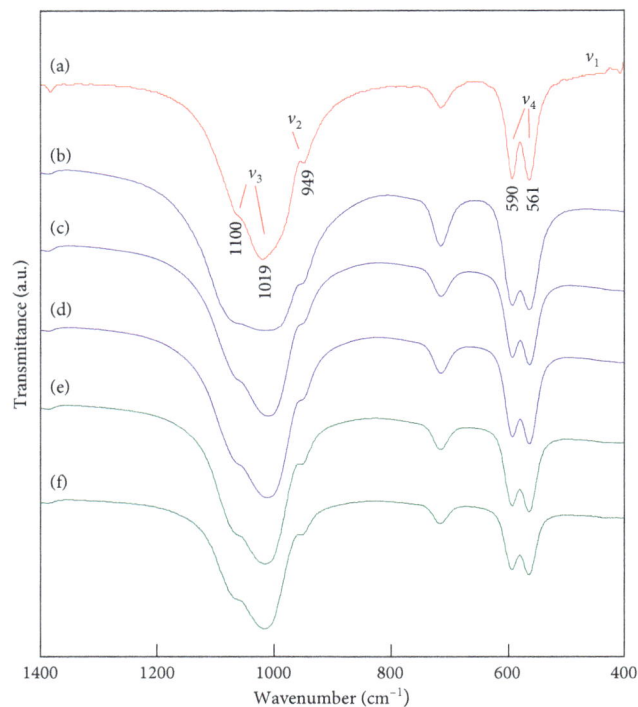

FIGURE 2: FT-IR analysis of the synthetic cadmium fluorapatite [$Cd_5(PO_4)_3F$] before (a) and after (b)–(f) dissolution at 25–45°C for 7200 h. (a) Synthetic Cd-FAP; (b) 25°C and pH = 2; (c) 25°C and pH = 5.6; (d) 25°C and pH = 9; (e) 35°C and pH = 2; (f) 45°C and pH = 2.

The regular mode of the tetrahedral PO_4^{3-} ions includes v_4 (the O-P-O bending), v_3 (the P-O stretching), v_2 (the O-P-O bending), and v_1 (the symmetric P-O stretching). Only the absorptions of the v_4 and v_3 vibrations can be observed in the undistorted situation. The v_2 and v_1 vibrations are infrared inactive as the symmetry of the tetrahedral PO_4^{3-} is reduced [18, 19]. The FT-IR spectra of the Cd-FAP solids before and after dissolution are plotted together in Figure 2. The regular mode of the tetrahedral PO_4^{3-} ions in the prepared cadmium fluorapatite could be observed in the region around 949 cm^{-1} (v_2), 1019 and 1100 cm^{-1} (v_3), and 561 and 590 cm^{-1} (v_4). The 712.60 cm^{-1} bands and the 3533~3537 cm^{-1} bands were assigned to the vibrational motion of OH^- ions and the stretching vibration of the bulk OH^- ions, respectively [4, 20]. The symmetric P-O stretching vibration (v_1) was not visible. The bands at 871 cm^{-1}, which were related to HPO_4^{2-} ions presenting in cation-deficient apatite, had not been detected [4]. The bands at 3650~3680 cm^{-1} representing the P-OH groups [4] and the bands at 1455 cm^{-1} representing the CO_3^{2-} vibration [21] were also not observed in the FT-IR spectra.

As shown in Figure 3, the FE-SEM examination indicated that all the synthesized cadmium fluorapatite nanorods were the typical hexagonal crystals that elongated along the c axis (diameter <50 nm) and with or without pinacoids as terminations. No obvious morphological variation was found after dissolution for 7200 h.

3.2. Dissolution Mechanism. The solution Cd^{2+}, PO_4^{3-}, and F^- concentrations and pHs during the Cd-FAP dissolution at 25°C, 35°C, and 45°C and different pHs as a function of time are shown in Figures 4(a)–4(f).

The early dissolution of the cadmium fluorapatite in water was nearly stoichiometric and then nonstoichiometric. During the Cd-FAP dissolution at an initial pH of 2 and 25°C (Figure 4(a)), the solution pHs increased from 2.00 to 3.80 within 1 h and after that, varied between 3.72 and 3.95. The release rates of cadmium and phosphate increased quickly until the highest cadmium and phosphate concentrations appeared after 480 h and 1080 h, respectively. Thereafter, the total dissolution rate of Cd-FAP slowly declined while the solution Cd : P molar ratios increased continually from 1.55~1.67, which are near to the stoichiometric Cd : P molar ratio for Cd-FAP, to 3.18~3.22 (Figure 4(f)). Generally, the aqueous fluoride concentrations reached the highest value in 1 h and decreased rapidly from 1 h to 2160 h, and then, it increased slowly and attained a steady state after 5760 h. For the Cd-FAP dissolution at different temperatures, the solution pHs and aqueous Cd^{2+}, PO_4^{3-}, and F^- concentrations became constant after 5760 h indicating an achievement of a steady state between the solid and the solution (Figures 4(a)–4(c)).

Moreover, the initial solution pH (Figures 4(a), 4(d), and 4(e)) and the solution temperature (Figures 4(a)–4(c)) appeared to have an obvious influence on the dissolution and thereafter, the solution concentrations of Cd^{2+}, PO_4^{3-},

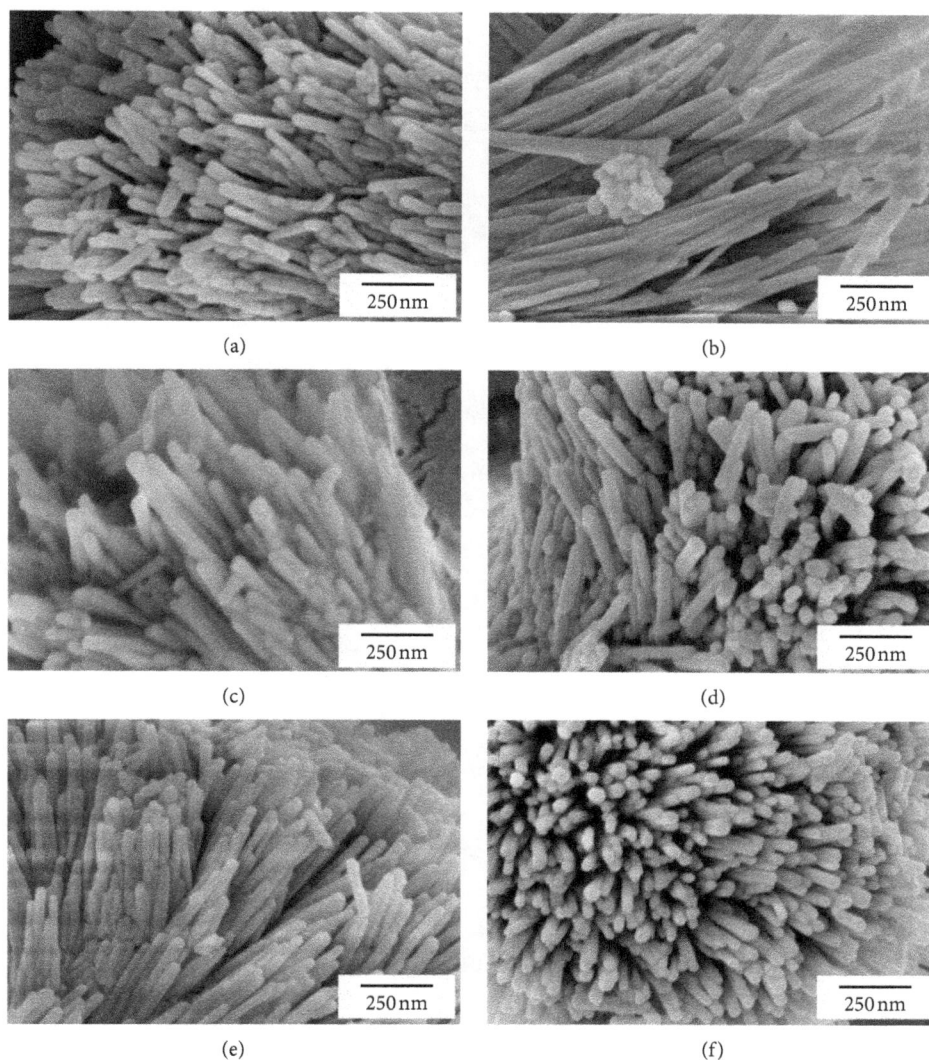

FIGURE 3: FE-SEM images of the synthetic cadmium fluorapatite [$Cd_5(PO_4)_3F$] before (a) and after (b)–(f) dissolution at 25–45°C for 7200 h. (a) Synthetic Cd-FAP; (b) 25°C and pH = 2; (c) 35°C and pH = 2; (d) 45°C and pH = 2; (e) 25°C and pH = 5.6; (f) 25°C and pH = 9.

and F^-. At the end of the dissolution, the solution concentrations of cadmium and phosphate at pH of 2.00 were higher than those at pH of 5.60 and 9.00, while the final pH and the solution concentration of fluoride at pH of 2.00 were lower than those at higher pHs. At the end of the dissolution, the solution concentrations of cadmium and fluoride at 35°C were a little lower than those at 25°C and 45°C, and the final pH and the solution concentration of phosphate at 35°C were a little higher than those at 25°C and 45°C (Figures 4(a)–4(c)). As a result, the aqueous Cd : P atomic ratios at 35°C were lower than those at 25°C and 45°C (Figure 4(f)). During the early dissolution (<120 h), the Cd^{2+} and PO_4^{3-} ions were released from Cd-FAP in the stoichiometric ratio with the dissolved Cd : P molar ratio being close to the stoichiometric ratio of 1.67 (Figure 4(f)). As the dissolution progressed, the solution Cd : P molar ratio rose and became larger than 1.67. This indicates that Cd^{2+} ions were preferentially released from the apatite structure in comparison with PO_4^{3-} ions during the

Cd-FAP dissolution. The solution Cd : P molar ratios at 45°C were significantly higher than those at 25°C and 35°C, which indicated that the Cd-FAP solubility and dissolution processes were related to the dissolution temperature.

The transient peak values in the aqueous component concentrations during the apatite dissolution were also reported in some earlier works [13], which were probably due to the grain size distribution. The smaller the apatite grains, the greater the dissolution rate of apatite, and the larger the apatite solubility. Hence, the peak values may be a consequence of the fast dissolution of the smaller grains, followed by reprecipitation to larger grains, which resulted in the achievement of the asymptotic solubility [13].

The decrease in solution pH indicated that protons were consumed during the Cd-FAP dissolution, that is, the negatively charged O ions of surface PO_4^{3-} groups adsorbed protons from solution and transformed into HPO_4^{2-}, which accelerated the Cd-FAP dissolution. Additionally, the

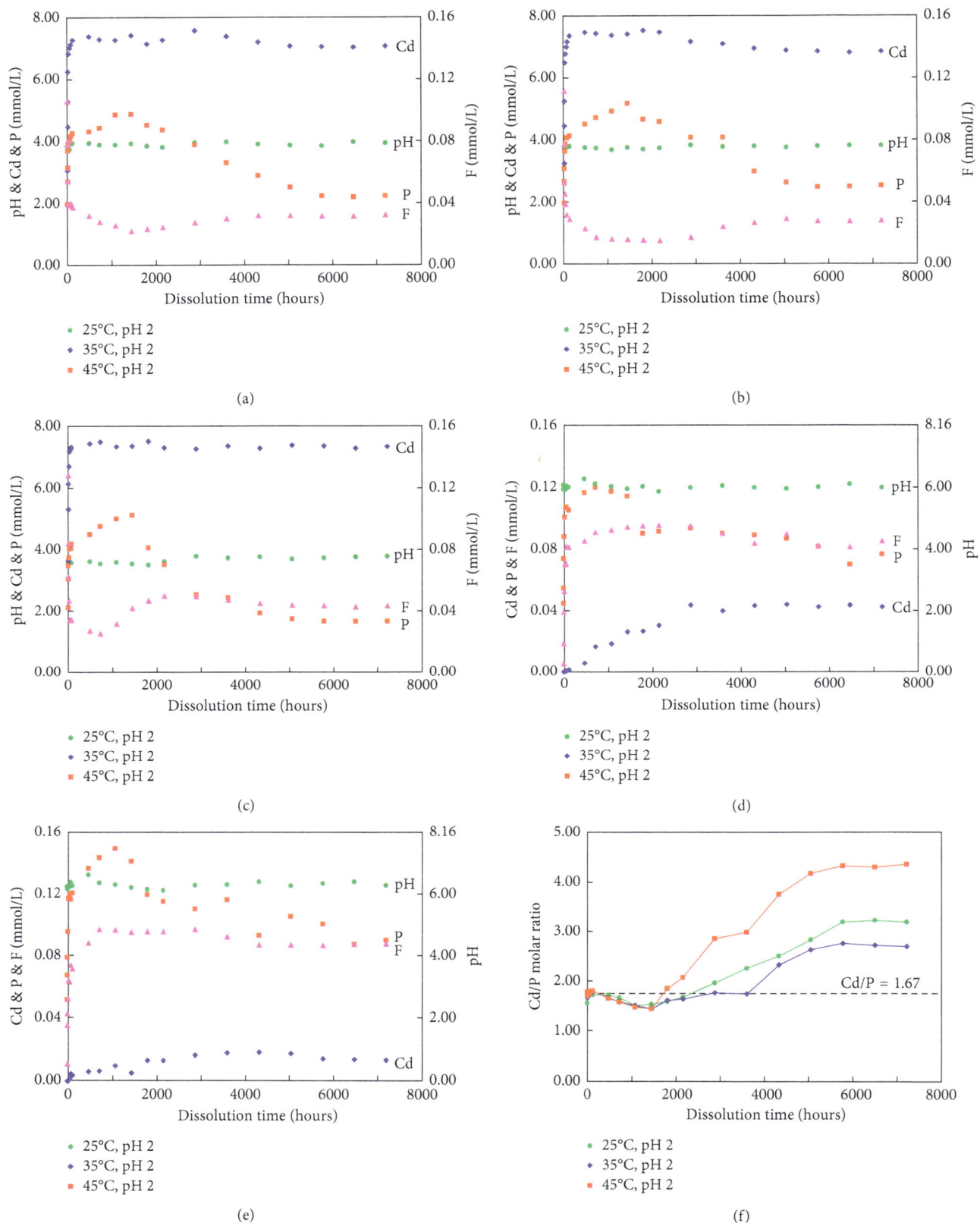

FIGURE 4: Aqueous evolution during the dissolution of the synthetic cadmium fluorapatite [Cd$_5$(PO$_4$)$_3$F] at 25–45°C for 7200 h. (a) 25°C and pH = 2; (b) 35°C and pH = 2; (c) 45°C and pH = 2; (d) 25°C and pH = 5.6; (e) 25°C and pH = 9; (f) Cd/P ratio.

stoichiometric ion exchange of $2H^+$ from solution for Cd^{2+} on the Cd-FAP surface and some other reactions also consumed protons [14, 21]. Besides, the dissolution mechanism of apatite is strongly dependent on the experimental condition such as temperature, solution composition, solid component, solid/solution ratio, and agitation [14]. Diverse models for the apatite dissolution have been proposed, but most of them take only some specific dissolution aspects into account and cannot describe the general dissolution mechanism of apatite [14].

Based on our experimental results and the data obtained for the apatite dissolution by different researchers, the cadmium fluorapatite (Cd-FAP) dissolution in the aqueous media is proposed to comprise the following simultaneous processes:

(i) Diffusion of protons from the aqueous solution to the Cd-FAP surface

(ii) H^+ adsorption onto the Cd-FAP surface coupled with protonation and chemical complexation/ transformation on the Cd-FAP surface

(iii) Stoichiometric dissolution with Cd^{2+}, PO_4^{3-}, and F^- release from the Cd-FAP surface

(iv) Nonstoichiometric dissolution with Cd^{2+} and PO_4^{3-} release from the Cd-FAP surface and F^- adsorption backwards onto the Cd-FAP surface

(v) Nonstoichiometric dissolution with F^- release from the Cd-FAP surface and Cd^{2+} and PO_4^{3-} adsorption backwards onto the Cd-FAP surface

For the dissolution of the synthetic cadmium fluorapatite at 25°C and initial pH 2, in process (i) and process (ii), H^+ ions were adsorbed from the aqueous acidic solution to F^- ions as well as to the negative charged O ions of PO_4^{3-} groups on the Cd-FAP surface [22], which resulted in an increase in the solution pHs from 2.00 to 3.80 in 1 h of dissolution. In comparison with the negative charged O ions of PO_4^{3-} groups on the Cd-FAP surface, H^+ ions should be adsorbed preferentially to the F^- ions owing to their higher electronegativity [14]. Meanwhile, the surface PO_4^{3-} groups were transformed into HPO_4^{2-}, which would catalyze the apatite dissolution [23].

In process (iii), Cd^{2+}, PO_4^{3-}, and F^- ions were dissociated stoichiometrically from the Cd-FAP solid surface (reaction (1)); that is, the aqueous Cd^{2+}, PO_4^{3-}, and F^- concentrations increased simultaneously with a Cd:P:F molar ratio of 5:3:1 in the short early dissolution period.

Apatites are complex minerals, and when investigating their dissolution, several possible reactions must be considered [3]. The dissolution of Cd-FAP according to reaction (1) is strongly dependent on the solution pH and coupled with protonation and complexation reactions (2)–(4), which could result in an increase in the aqueous pHs for the Cd-FAP dissolution in the acidic solution or a decrease in the aqueous pHs for the dissolution in the alkali solution:

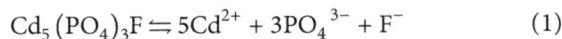

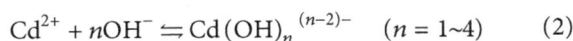

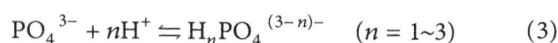

$$Cd_5(PO_4)_3F \rightleftharpoons 5Cd^{2+} + 3PO_4^{3-} + F^- \quad (1)$$

$$Cd^{2+} + nOH^- \rightleftharpoons Cd(OH)_n^{(n-2)-} \quad (n = 1\sim4) \quad (2)$$

$$PO_4^{3-} + nH^+ \rightleftharpoons H_nPO_4^{(3-n)-} \quad (n = 1\sim3) \quad (3)$$

$$H^+ + nF^- \rightleftharpoons HF_n^{(n-1)-} \quad (n = 1\sim2) \quad (4)$$

The speciation calculation using PHREEQC indicated that, for the Cd-FAP dissolution at 25°C and initial pH 2, the solution Cd species presented in the order of $Cd^{2+} > CdF^+$, $CdNO_3^+ > CdOH^+ > Cd_2OH^{3+} > Cd(OH)_2 > Cd(OH)_3^- > Cd(OH)_4^{2-}$; the solution phosphate species presented in the order of $H_2PO_4^- > H_3PO_4 > HPO_4^{2-} > PO_4^{3-}$; and the solution fluoride species presented in the order of $F^- > HF > CdF^+ \gg HF_2^-$.

In processes (iv) and (v), cadmium, phosphate, and fluoride ions were released from the Cd-FAP solid into solution nonstoichiometrically with the solution Cd:P molar ratios >1.67 and the solution Cd:F molar ratios >5.00, which might result in the formation of a surface layer with a component different from the bulk solid [14]. Some amount of cadmium, phosphate, and fluoride were adsorbed from the solution back onto the Cd-FAP surface after an initial portion of Cd-FAP had dissolved, the solution F concentrations had started to decrease gradually within 1 h of dissolution in the present experiment, the solution phosphate concentrations began to decrease gradually after 1440 h of dissolution, and the solution Cd concentrations began to decrease gradually after 2880 h of dissolution at 25°C and initial pH 2.

Fluorapatite dissolution might be started through a relatively fast and complete dissociation of F^- ions from the solid surface [14, 21]. In processes (iii)~(v), F^- ions were dissociated faster than cadmium and phosphate ions, and then the dissolved F^- ions were quickly adsorbed back to the Cd-FAP surface. As a result, the solution Cd:F ratio and P:F ratio became noticeably higher than the stoichiometric ratio of 5.00 and 3.00, respectively. The detachment of cadmium and phosphate ions followed the dissociation of F^- ions from the Cd-FAP surface. The F^- release was coupled to a quick hydrolysis of phosphate ions [21]. O ions of PO_4^{3-} groups cover 80~90% of the apatite surface [14]. $\equiv CdOH_2^+$ and $\equiv PO^-$ are thought to be the two distinct groups on the surface. According to the surface protonation model, the FAP surface protonation in the solution of pH 5~7 happened through the formation of $\equiv POH$ groups [24]. In the early dissolution of apatite (1–6 h), the solution Cd:P ratios (1.55~1.65) were a little smaller than its stoichiometric Cd:P ratio of 1.67, suggesting that PO_4^{3-} ions could be preferentially dissociated from the FAP surface in comparison to Cd^{2+} ions. The corners and edges of the apatite crystal were predominantly occupied by PO_4^{3-} ions, and the apatite dissolution could start through the complexation of H^+ ions from solution with these weakly bounded PO_4^{3-} groups on the solid surface [14].

Finally, the desorption-adsorption processes of cadmium, phosphate, and fluoride ions attained a steady state; that is, the solution cadmium, phosphate, and fluoride concentrations were constant for the Cd-FAP dissolution in the solution of pH 2 at 25°C from 5760 h to the end of experiment (7200 h).

3.3. *Determination of Solubility.* The aqueous Cd^{2+}, PO_4^{3-}, and F^- activities in the final equilibrated solution (5040 h, 5760 h, and 7200 h) were computed with the program PHREEQC to estimate the solubility product of the synthetic cadmium fluorapatite $[Cd_5(PO_4)_3F]$. The PHREEQC calculation also indicated that the final equilibrated solution was unsaturated with any possible secondary phases such as $Cd_3(PO_4)_2$, $Cd(OH)_2$, and monteponite (CdO).

The dissolution reaction of cadmium fluorapatite $[Cd_5(PO_4)_3F]$ and the dissociation of Cd^{2+}, PO_4^{3-}, and F^- are expressed by the dissolution equation (1). Assuming unit activity of $Cd_5(PO_4)_3F$:

$$K_{sp} = \left\{Cd^{2+}\right\}^5 \left\{PO_4^{3-}\right\}^3 \{F^-\}, \tag{5}$$

where {} is the thermodynamic activities of the Cd^{2+}, PO_4^{3-}, and F^- species and K_{sp} is the solubility product of cadmium fluorapatite $[Cd_5(PO_4)_3F]$ according to the dissolution equation (1).

The K_{sp} value under standard conditions (298.15 K and 0.101 MPa) is related to the standard free energy of reaction (ΔG_r^0) and can be described by

$$\Delta G_r^0 = -5.708 \log K_{sp}. \tag{6}$$

For Equation (1),

$$\Delta G_r^0 = 5\Delta G_f^0\left[Cd^{2+}\right] + 3\Delta G_f^0\left[PO_4^{3-}\right]$$
$$+ \Delta G_f^0[F^-] - \Delta G_f^0[Cd_5(PO_4)_3F]. \tag{7}$$

By rearranging,

$$\Delta G_f^0\left[Cd_5(PO_4)_3F\right] = 5\Delta G_f^0\left[Cd^{2+}\right] + 3\Delta G_f^0\left[PO_4^{3-}\right]$$
$$+ \Delta G_f^0[F^-] - \Delta G_r^0. \tag{8}$$

Apatite group minerals are sparingly soluble. Table 1 gives the dissolution temperature, the initial solution pH, the final solution pH, and Cd, P, and F analyses together with the calculated solubility product of Cd-FAP. The aqueous activities of Cd^{2+}, PO_4^{3-}, and F^- were firstly computed with PHREEQC. The $\log K_{sp}$ value for $Cd_5(PO_4)_3F$ was then calculated after Equation (5). The average K_{sp} values were determined for $Cd_5(PO_4)_3F$ of $10^{-60.03}$ ($10^{-59.74} \sim 10^{-60.46}$) at 25°C, $10^{-60.38}$ ($10^{-60.32} \sim 10^{-60.48}$) at 35°C, and $10^{-60.45}$ ($10^{-60.33} \sim 10^{-60.63}$) at 45°C, which were very close to the average K_{sp} values for $Ca_5(PO_4)_3F$ of $10^{-60.02}$ ($10^{-59.93} \sim 10^{-60.14}$) at 25°C, $10^{-60.84}$ ($10^{-60.83} \sim 10^{-60.85}$) at 25°C, and $10^{-61.13}$ ($10^{-61.02} \sim 10^{-61.28}$) at 45°C (Tables 1 and 2). Based on the calculated K_{sp} values at the initial pH of 2 and 25°C, the Gibbs free energies of formation (ΔG_f^0) were estimated according to Equations (6)–(8) to be −4065.76 kJ/mol (−4064.11~−4068.23 kJ/mol) for $Cd_5(PO_4)_3F$ and −6445.51 kJ/mol (−6445.01~−6446.15 kJ/mol) for $Ca_5(PO_4)_3F$ (Table 1).

No solubility data for cadmium fluorapatite [Cd-FAP, $Cd_5(PO_4)_3F$] have been reported in literatures. The average K_{sp} value for $Cd_5(PO_4)_3F$ of $10^{-60.03}$ at 25°C was about 4.81 log units smaller than $10^{-55.22}$ for $Cd_5(PO_4)_3OH$ at 37°C [25] and 4.59 log units higher than $10^{-64.62}$ reported for $Cd_5(PO_4)_3OH$ at 25°C [26], while the K_{sp} values for

$Ca_5(PO_4)_3F$ were reported to be $10^{-60.6}$ [27], 10^{-59} [28], 10^{-70} [12], and $10^{-55.71}$ [19]. This significant difference in the solubility product is thought to be related to the difference in the dissolution conditions and the hydroxyapatite materials used in the different experiments [3].

3.4. *Determination of Thermodynamic Data.* The solubility products (K_{sp}) for Cd-FAP and Ca-FAP can also be estimated using Equations (9)–(11) [25]:

$$-\log K_{sp} = \frac{A}{T} + B + CT. \tag{9}$$

For Cd-FAP,

$$-\log K_{sp} = -\frac{40857.3723}{T} + 319.3293 - 0.4100T. \tag{10}$$

For Ca-FAP,

$$-\log K_{sp} = -\frac{77482.5843}{T} + 547.2256 - 0.7624T. \tag{11}$$

The thermodynamic quantities ΔG^0, ΔH^0, ΔS^0, and ΔC_p^0 for the Cd-FAP and Ca-FAP dissolution at an initial pH of 2 are estimated after Equations (12)–(15) [25] and given in Table 2:

$$\Delta G^0 = 2.3026R\left(A + BT + CT^2\right), \tag{12}$$

$$\Delta H^0 = 2.3026R\left(A - CT^2\right), \tag{13}$$

$$\Delta S^0 = -2.3026R\left(B + 2CT\right), \tag{14}$$

$$\Delta C_p^0 = -2.3026R\left(CT\right), \tag{15}$$

where A, B, and C are the empirical constants of Equation (9). The negative values of ΔH^0 indicated that the Cd-FAP and Ca-FAP dissolution in the aqueous solution was an exothermic reaction, and their solubilities declined with the temperature rising. The larger negative value of ΔS^0 for the Cd-FAP dissolution showed that the order produced by the cadmium cations in the aqueous solution was a little higher than that by calcium cations [25], which was related to the ion size difference between Ca^{2+} (0.99 Å) and Cd^{2+} (0.97 Å). Generally, the smaller Cd^{2+} ions with a high charge can result in a lower entropy in aqueous solution, and consequently, Cd-FAP is less soluble than Ca-HAP. But the very similar ion sizes of Cd^{2+} and Ca^{2+} resulted in a close solubility product for Cd-FAP and Ca-HAP under the experimental conditions. The values of ΔG^0, ΔH^0, and ΔS^0 for the Cd-FAP and Ca-HAP dissolution increased with the increasing temperature and indicated that their dissolution was an energy-consuming process [25].

4. Conclusions

The synthetic cadmium fluorapatite $[Cd_5(PO_4)_3F]$ was the typical hexagonal columnar nanorod crystals (diameter <50 nm) with or without a pinacoid as a termination, which elongated along the c axis with the lattice cell parameters of

TABLE 1: Analytical data and solubility product of cadmium fluorapatite [$Cd_5(PO_4)_3F$].

Temperature (°C)	Initial pH	Reaction time (hour)	Analytical data				$\log K_{sp}$	ΔG_f^0 (kJ/mol)
			pH	Cd (mmol/L)	P (mmol/L)	F (mmol/L)		
25	2.00	5040	3.83	7.0385	2.2094	0.0311	−60.46	−4068.23
		5760	3.95	7.0154	2.1788	0.0312	−59.74	−4064.11
		7200	3.92	7.0474	2.2127	0.0319	−59.89	−4064.96
35	2.00	5040	3.78	6.8019	2.4679	0.0276	−60.48	—
		5760	3.80	6.7681	2.4834	0.0275	−60.35	—
		7200	3.80	6.8037	2.5215	0.0278	−60.32	—
45	2.00	5040	3.74	7.3463	1.6982	0.0437	−60.63	—
		5760	3.78	7.2840	1.6924	0.0432	−60.40	—
		7200	3.79	7.3356	1.6821	0.0439	−60.33	—
25	5.60	5040	6.01	0.0426	0.0814	0.0822	−61.20	−4072.43
		5760	6.10	0.0438	0.0698	0.0814	−60.82	−4070.27
		7200	5.98	0.0427	0.0768	0.0853	−61.43	−4073.74
25	9.00	5040	6.35	0.0143	0.1006	0.0869	−61.31	−4073.05
		5760	6.41	0.0139	0.0877	0.0876	−61.21	−4072.48
		7200	6.29	0.0137	0.0904	0.0882	−61.87	−4076.25

TABLE 2: Thermodynamic data of Cd-FAP than Ca-FAP in the aqueous acidic media (an initial pH of 2).

FAP	Temperature (°C)	Average $\log K_{sp}$	ΔG^0 (J/K·mol)	ΔH^0 (J/mol)	ΔS^0 (J/K·mol)	ΔC_p^0 (J/K·mol)
Ca-FAP	25	−60.02	342473.29	−187235.69	−1777.55	4349.85
	35	−60.84	358789.08	−98779.03	−1485.61	4495.82
	45	−61.13	372185.51	−7403.00	−1193.67	4641.79
Cd-FAP	25	−60.03	342515.78	−85088.80	−1434.91	2339.50
	35	−60.38	356079.86	−37513.71	−1277.90	2418.01
	45	−60.45	368073.80	11631.51	−1120.89	2496.51

$a = 9.3284$ and $c = 6.6378$. The essential vibrational modes of PO_4^{3-} tetrahedra appeared at 949 cm^{-1} (ν_2), 1019 and 1100 cm^{-1} (ν_3), and 561 and 590 cm^{-1} (ν_4).

For the Cd-FAP dissolution at an initial pH of 2 and 25°C, the dissociation rates of fluoride, phosphate, and cadmium increased quickly until their highest solution concentrations were attained after dissolution for <1 h, 1440 h, and 2880 h, respectively. After that, the Cd-FAP dissolution decreased slowly while the solution Cd/P molar ratios increased steadily from 1.65~1.67 to 3.18~3.22. The solution pH increased from 2.00 to 3.80 within 1 h and then varied between 3.72 and 3.95.

The average K_{sp} values were determined for $Cd_5(PO_4)_3F$ of $10^{-60.03}$ at 25°C, $10^{-60.38}$ at 35°C, and $10^{-60.45}$ at 45°C. Based on K_{sp} at an initial pH of 2 and 25°C, the Gibbs free energy of formation (ΔG_f^0) was calculated to be −4065.76 kJ/mol. The thermodynamic quantities, ΔG^0, ΔH^0, ΔS^0, and ΔC_p^0, for the Cd-FAP dissolution at an initial pH of 2 and 25°C were determined to be 342515.78 J/K·mol, −85088.80 J/mol, −1434.91 J/K·mol, and 2339.50 J/K·mol, respectively.

Conflicts of Interest

The authors declare that there are no conflicts of interest regarding the publication of this paper.

Acknowledgments

This work was financially supported by the National Natural Science Foundation of China (41763012, 41263009, and 21707024), the Guangxi Science and Technology Planning Project (GuiKe-AD18126018), the Special Fund for Guangxi Distinguished Experts, and the Provincial Natural Science Foundation of Guangxi (2014GXNSFBA118054).

References

[1] M. Masaoka, A. Kyono, T. Hatta, and M. Kimata, "Single crystal growth of Pb5(PxAs1−xO4)3Cl solid solution with apatite type structure," Journal of Crystal Growth, vol. 292, no. 1, pp. 129–135, 2006.

[2] M. C. F. Magalhães and P. A. Williams, "Chapter 18–apatite group minerals: solubility and environmental remediation," in Thermodynamics Solubility and Environmental Issues, pp. 327–340, Elsevier BV, New York City, NY, USA, 2007.

[3] A. Bengtsson, A. Shchukarev, P. Persson, and S. Sjöberg, "A solubility and surface complexation study of a non-stoichiometric hydroxyapatite," Geochimica et Cosmochimica Acta, vol. 73, no. 2, pp. 257–267, 2009.

[4] A. Yasukawa, M. Higashijima, K. Kandori, and T. Ishikawa, "Preparation and characterization of cadmium–calcium hydroxyapatite solid solution particles," Colloids and Surfaces A Physicochemical and Engineering Aspects, vol. 268, no. 1–3, pp. 111–117, 2005.

[5] M. S. Rosmi, S. Azhari, and R. Ahmad, "Adsorption of cadmium from aqueous solution by biomass: comparison of solid pineapple waste, sugarcane bagasse and activated carbon," *Advanced Materials Research*, vol. 832, no. 2014, pp. 810–815, 2013.

[6] M. Mahmood-Ul-Hassan, V. Suthar, E. Rafique, R. Ahmad, and M. Yasin, "Kinetics of cadmium, chromium, and lead sorption onto chemically modified sugarcane bagasse and wheat straw," *Environmental Monitoring and Assessment*, vol. 187, no. 7, pp. 1–11, 2015.

[7] N. Harouiya, C. Chaïrat, S. J. Köhler, R. Gout, and E. H. Oelkers, "The dissolution kinetics and apparent solubility of natural apatite in closed reactors at temperatures from 5 to 50°C and pH from 1 to 6," *Chemical Geology*, vol. 244, no. 3-4, pp. 554–568, 2007.

[8] R. Zhu, R. Yu, J. Yao, D. Mao, C. Xing, and D. Wang, "Removal of Cd^{2+} from aqueous solutions by hydroxyapatite," *Catalysis Today*, vol. 139, no. 1-2, pp. 94–99, 2008.

[9] F. Fernane, M. O. Mecherri, P. Sharrock, M. Hadioui, H. Lounici, and M. Fedoroff, "Sorption of cadmium and copper ions on natural and synthetic hydroxylapatite particles," *Materials Characterization*, vol. 59, no. 5, pp. 554–559, 2008.

[10] S. B. Chen, Y. B. Ma, L. Chen, and K. Xian, "Adsorption of aqueous Cd^{2+}, Pb^{2+}, Cu^{2+} ions by nano-hydroxyapatite: single- and multi-metal competitive adsorption study," *Geochemical Journal*, vol. 44, no. 3, pp. 233–239, 2010.

[11] Y. Zhu, Z. Zhu, X. Zhao, Y. Liang, L. Dai, and Y. Huang, "Characterization, dissolution and solubility of cadmium–calcium hydroxyapatite solid solutions at 25°C," *Chemical Geology*, vol. 423, pp. 34–48, 2016.

[12] E. Valsami-Jones, K. W. Ragnarsdottir, A. Putnis, D. Bosbach, A. J. Kemp, and G. Cressey, "The dissolution of apatite in the presence of aqueous metal cations at pH 2-7," *Chemical Geology*, vol. 151, no. 1-4, pp. 215–233, 1998.

[13] M. T. Fulmer, I. C. Ison, C. R. Hankermayer, B. R. Constantz, and J. Ross, "Measurements of the solubilities and dissolution rates of several hydroxyapatites," *Biomaterials*, vol. 23, no. 3, pp. 751–755, 2002.

[14] S. V. Dorozhkin, "A review on the dissolution models of calcium apatites," *Progress in Crystal Growth and Characterization of Materials*, vol. 44, no. 1, pp. 45–61, 2002.

[15] W. J. Tseng, C. C. Lin, P. W. Shen, and P. Shen, "Directional/acidic dissolution kinetics of (OH,F,Cl)-bearing apatite," *Journal of Biomedical Materials Research Part A*, vol. 76, no. 4, pp. 753–764, 2006.

[16] K. Skartsila and N. Spanos, "Surface characterization of hydroxyapatite: potentiometric titrations coupled with solubility measurements," *Journal of Colloid & Interface Science*, vol. 308, no. 2, pp. 405–412, 2007.

[17] D. L. Parkhurst and C. A. J. Appelo, "Description of input and examples for PHREEQC version 3—a computer program for speciation, batch-reaction, one-dimensional transport, and inverse geochemical calculations," in *Techniques and Methods, Book 6, Chapter A43*, pp. 1–497, U.S. Geological Survey, Reston, VA, United States, 2013.

[18] V. M. Bhatnagar, "X-Ray and infrared studies of lead apatites," *Canadian Journal of Chemistry*, vol. 49, no. 4, pp. 662-663, 1971.

[19] Y. Zhu, X. Zhang, Y. Chen et al., "A comparative study on the dissolution and solubility of hydroxylapatite and fluorapatite at 25°C and 45°C," *Chemical Geology*, vol. 268, no. 1-2, pp. 89–96, 2009.

[20] A. Yasukawa, T. Yokoyama, and T. Ishikawa, "Preparation of cadmium hydroxyapatite particles using acetamide," *Materials Research Bulletin*, vol. 36, no. 3-4, pp. 775–786, 2001.

[21] G. R. Qian, H.-M. Bai, F.-C. Sun, J.-Z. Zhou, W.-M. Sun, and X. Xu, "Preparation and stability of calcium cadmium hydroxyapatite," *Journal of Inorganic Materials*, vol. 23, no. 5, pp. 1016–1020, 2008.

[22] J. Christoffersen and M. R. Christoffersen, "Kinetics of dissolution of calcium hydroxyapatite: V. The acidity constant for the hydrogen phosphate surface complex," *Journal of Crystal Growth*, vol. 57, no. 1, pp. 21–26, 1982.

[23] J. Christoffersen, M. R. Christoffersen, and T. Johansen, "Some new aspects of surface nucleation applied to the growth and dissolution of fluorapatite and hydroxyapatite," *Journal of Crystal Growth*, vol. 163, no. 3, pp. 304–310, 1996.

[24] L. Wu, W. Forsling, and A. Holmgren, "Surface complexation of calcium minerals in aqueous solution," *Journal of Colloid and Interface Science*, vol. 147, no. 2, pp. 211–218, 2000.

[25] P. P. Mahapatra, H. Mishra, and N. S. Chickerur, "Solubility and thermodynamic data of cadmium hydroxyapatite in aqueous media," *Thermochimica Acta*, vol. 54, no. 1, pp. 1–8, 1982.

[26] Y. Zhu, Z. Zhu, X. Zhao, Y. Liang, L. Dai, and Y. Huang, "Characterization, dissolution and solubility of synthetic cadmium hydroxylapatite [$Cd_5(PO_4)_3OH$] at 25–45°C," *Geochemical Transactions*, vol. 16, no. 1, pp. 1–11, 2015.

[27] F. C. Driessens, "Mineral aspects of dentistry," *Monographs in Oral Science*, vol. 10, pp. 1–215, 1982.

[28] W. Stumm and J. J. Morgan, *Aquatic Chemistry, Chemical Equilibria and Rates in Natural Waters*, John Wiley & Sons, New York, NY, USA, 1996.

7

An Interaction of Anionic- and Cationic-Rich Mixed Surfactants in Aqueous Medium through Physicochemical Properties at Three Different Temperatures

K. M. Sachin,[1] Sameer A. Karpe,[1] Man Singh,[1] and Ajaya Bhattarai ⓘ[1,2]

[1]*School of Chemical Sciences, Central University of Gujarat, Gandhinagar, India*
[2]*Department of Chemistry, MMAMC, Tribhuvan University, Biratnagar 56613, Nepal*

Correspondence should be addressed to Ajaya Bhattarai; bkajaya@yahoo.com

Academic Editor: Tomokazu Yoshimura

The mixed micellization of aqueous binary mixtures of DTAB-rich and SDS-rich surfactants, comprising sodium dodecyl sulfate (SDS) and dodecyltrimethylammonium bromide (DTAB) is studied in aqueous solution by using the physicochemical properties (PCPs) at three different temperatures ($T = 293.15$, 298.15, and 303.15 K) and $P = 0.1$ MPa. The DTAB concentration is varied from 0.0001 to 0.03 M/mol·L^{-1} in the ~ 0.01 M/mol·L^{-1} SDS solution, while the concentration of SDS is varied from 0.001 to 0.015 M/mol·L^{-1} in the ~ 0.005 M/mol·L^{-1} DTAB. The stable formulations have been obtained by employing the DTAB-rich and SDS-rich surfactants solutions in $3:1$ ratio. Therefore, different phases and aggregated states formed in the ternary combinations of DTAB/SDS/H$_2$O have been identified and described. The calculated PCPs have been utilized for determining the nature of the solute-solvent interaction ($S_L S_0 I$). With increasing surfactants concentration, the polarisation of the solution also increases along with an increase in relative viscosity (η_r), viscous relaxation time (τ), and surface excess concentration (Γ_{max}). However, the surface area of the molecule (A_{min}), hydrodynamic volume (V_h), and hydrodynamic radius (R_h) decrease along with an increase in surfactants concentration.

1. Introduction

The role of mixed surfactants is very crucial in our daily life. It has widespread applications in the various households and industrial processes such as usages in the chemical purification, targeted drug delivery, synthesis of advanced nanomaterials [1–4], cosmetics, wastewater treatment, food industries, detergency, and oil recovery enhancement [5–8].

With the advantages of high biodegradability, greater surface activity, high biocompatibility, and application in various separation techniques, utilization in drug formulation and related biomedical applications makes the studies of the mixed surfactant system inevitable [9]. Due to the opposite charge, the surfactant induces several remarkable properties. However, cationic and anionic mixed surfactants in an aqueous medium show numerous noble features that arise from the strong electrostatic interactions between the oppositely charged head groups [10]. It has been already reported that several types of the binary surfactant systems,

cationic and anionic, show the strongest synergisms in the formation of mixed micelle and surface tension reduction of the solution [11].

The PCPs of surfactants, such as critical micellar concentration (CMC), the degree of ionization, and thermodynamics of micellization depend on the nature of the hydrophobic tail, hydrophilic head group, and the counterion species [12]. Mixed surfactants are also used in a personal cleaning product, laundry aids, shampoo, fabric softeners, and solubilizers for water-insoluble or sparingly soluble bioinspired molecules like polyphenolic compound, ionic liquid, and anticorrosive agents for steel and plastics and used as a catalyst for some industrially significant reactions, flotation collectors for mineral ores, and leveling agents for improving the dyeing processes [13–17]. Because it has an amphiphilic nature, the study of the interaction of mixed surfactants in an aqueous medium helps to decode functional and diverse information about the system and assist in harnessing their potential in technical applications [18–20].

Hence, the ternary system (DTAB/SDS/H$_2$O) can demonstrate arrays of self-assembled microstructures, viz, micelles, vesicles, planar bilayers, and bicontinuous structures. Earlier studies have been focused mostly on two critical facts which influence the interaction activities: (a) the type of the interactions involved during the formation of the micelles (b) and the resultant structure of the formed aggregates [21]. The SDS and DTAB surfactants (Figure 1) actively interact with each other due to opposite charge species. However, above the CMC, surfactants form aggregates into the micelle [22]. Maiti et al. [23] have been investigated on oppositely charged single-tailed surfactants that could associate through electrostatic, ion-dipole, and van der Waals force attraction under specific conditions. Thus, the various aggregated microstructures (micelles, vesicles, and lamellar phases) of catanionic surfactants have attracted the attention of researchers for their multifaceted potential application in the field of drug delivery and nanoparticle synthesis. The structure of the surfactants plays an essential role in their aggregation behavior. The critical packing parameters infer the type of possible assemblies in the solution. Due to these potentials, the mixed surfactants solution has remarkable properties such as lower surface tension with higher surface activities and critical aggregation concentrations (CACs) which are essential for detergency and pharmaceutical applications [24, 25]. The cationic surfactants can form many supramolecular structures, at the specific mole ratios and concentrations; they have formed a remarkable micelles structure [26, 27] and vesicles [28, 29]. Bakshi et al. studied single and mixed micellization of surfactants by using conductivity, turbidity, and NMR measurements [30, 31]. Therefore, anionic and cationic mixed surfactants can form a numerous type of aggregated microstructures like lamellar phases, vesicles, spheres, precipitates, and rod shape structures [32, 33]. Moreover, mixing of surfactants is also used in drug formulation, lowering the Krafft temperature, and with increasing the cloud point [34], and some studies have been reported on the electrical conductance of cationic and anionic mixed surfactants [35]. Recently, many researchers have been focused on the aggregation and micelles formation process in the aqueous and mixed solvent system [36, 37]. Earlier researchers have been focused mostly on spectroscopic and thermodynamic studies of single and mixed surfactants through UV-visible, CMC, CAC, entropy, enthalpy, Gibbs free energy, micelle ionization degree, Krafft temperature, dissociation constant, and the pre-slope and post-slope values of single and mixed surfactants in an aqueous medium and mixed solvent system at different temperatures [38–45].

There is a little work on PCPs of SDS-rich and DTAB-rich mixed surfactants in an aqueous medium at $T = 293.15$, 298.15, and 303.15 K [46]. In this research article, we are studying the various PCPs, which include relative viscosity, viscous relaxation time, acoustic impedance, hydrodynamic volume, hydrodynamic radius, intrinsic viscosity, friccohesity shift coefficient, surface excess concentration, and area of a molecule of the SDS-rich and DTAB-rich mixed surfactants in an aqueous medium at three different temperatures ($T = 293.15$, 298.15, and 303.15 K) at 0.1 MPa. This type of study on the mixed surfactant system could assist in

harnessing their potential in the household and industrial applications.

2. Materials and Methods

2.1. Materials. All chemicals were purchased from Sigma-Aldrich, and their details are given in Table 1. Dodecyltrimethylammonium bromide and sodium dodecyl sulfate surfactants were stored in the P$_2$O$_5$-filled vacuum desiccator due to their hygroscopic nature.

2.2. Solution Preparation. All solutions, water + SDS (aq-SDS) and water + DTAB (aq-DTAB), were prepared separately by dissolving 0.005 M/mol·L^{-1} and 0.01 M/mol·L^{-1} of DTAB and SDS surfactants separately into Milli-Q water and used as a stock solution. The 0.005 M/mol·L^{-1} DTAB and 0.01 M/mol·L^{-1} SDS solutions were used as a solvent for 0.000096 to 0.012 M/mol·L^{-1}SDS and 0.000864 to 0.00504 M/mol·L^{-1} DTAB, respectively. These solutions were kept for ~10 min sonication at 30 MHz for better homogenization. All solutions were prepared at the temperature 298.15 K and pressure 0.1 MPa using Milli-Q water at pH 7 and conductivity 0.71 μS·cm^{-1}. For weighing, Mettler Toledo NewClassic MS was used with <±0.1·10^{-6} kg repeatability. To avoid evaporation and contamination, all solutions were kept in an airtight volumetric flask at the temperature of 298.15 K.

Anton Paar DSA 5000M density meter was used for measurements of their densities (ρ) and sound velocity (u) data with ±5·10^{-6} g·cm^{-3} uncertainty, and the temperature was controlled by a built-in Peltier (PT100) device with ±1.10^{-3} K accuracy. Repeatability of the instrument corresponds to precision in ρ and u data with 1.10^{-3} kg·m^{-3} and 0.10 m·s^{-1}, respectively.

The instrument was calibrated with Milli-Q water at the temperature of 298.15 K, while aq-NaCl (1 M/mol·kg^{-1}) and 10% aq-DMSO were also used to check the performance of the instrument, and the values were in agreement with the literature within the experimental uncertainties (Table S1) [47, 48]. Reported densities were an average of three repeated measurements with ±3.10^{-6} g·cm^{-3} repeatability. The ρ and u at 3 MHz frequency of uncertainties were ±5 × 10^{-3} kg m^3 and ±0.5 m·s^{-1}, respectively. All experiments were carried out at the three different temperatures ($T = 293.15$, 298.15, and 303.15 K) with ±0.01 K accuracy [49]. Sound velocity work based on oscillation periods of quartz U-tube with air, solvent, and solutions [50]. After each measurement, the tube was cleaned with acetone and dried by passing dried through the U-tube by using an air pump. A process of drying continued till a constant oscillation period for air was obtained and noted as an initial calibration. Viscosity, surface tension, and friccohesity data were measured by Borosil Mansingh Survismeter [51] (Cal no. 06070582/1.01/C-0395, NPL, India) through viscous flow time (VFT) and pendant drop number (PDN) methods, respectively. Lauda Alpha RA 8 thermostat was used for controlling the temperature with ±0.05 K accuracy. After attaining a thermal equilibrium, the VFT was recorded by using an electronic timer with ±0.01 s accuracy, while the PDN counted with an electronic counter. The Survismeter was

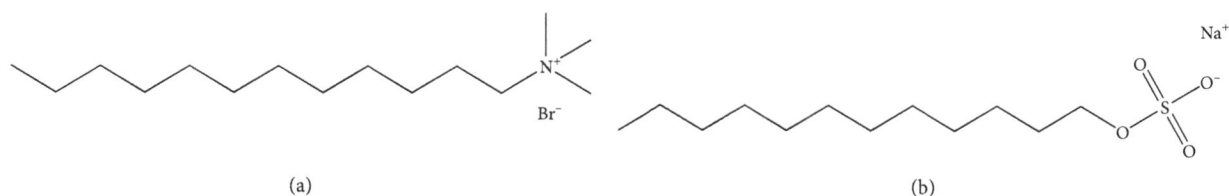

FIGURE 1: Molecular structure of dodecyltrimethylammonium bromide (DTAB) (a) and sodium dodecyl sulfate (SDS) (b) surfactant.

TABLE 1: Specification of chemicals used in this work.

Name of chemicals	Purity[a] (%)	Mw	Source	CAS no.
DTAB	~99	308.34	Sigma-Aldrich	1119-94-4
SDS	98	288.37	Sigma-Aldrich	151-21-3

[a]Purity as provided by suppliers; DTAB, dodecyltrimethylammonium bromide; SDS, sodium dodecyl sulfate.

TABLE 2: Relative viscosity (η_r) of SDS-rich and DTAB-rich surfactants in the aqueous medium at the three different temperatures $T = 293.15, 298.15$, and 303.15 K and at 0.1 MPa.

M (mol·L^{-1})	293.15 K	298.15 K	303.15 K
		SDS-rich	
0.005000	1.0245	1.0828	0.6281
0.000096	1.4596	0.9800	2.7704
0.000240	1.3026	1.0026	1.1631
0.000480	1.7872	0.9961	1.9515
0.000672	1.2815	0.8303	1.3448
0.000792	1.0601	1.2297	1.2073
0.000960	1.1046	1.0402	0.7981
0.006011	1.0234	1.1561	0.7698
0.007200	1.2651	1.1661	0.8053
0.007920	1.3315	1.4613	0.8415
0.009000	1.2510	1.4196	0.7796
0.010800	1.3326	1.1678	1.3417
0.012000	1.2202	2.4387	0.8836
		DTAB-rich	
0.010000	1.0858	0.4814	1.1978
0.000864	2.4619	1.6787	0.7013
0.000960	2.8325	2.2650	0.7551
0.001536	2.8749	1.8171	0.7718
0.002016	2.8246	2.8142	0.7288
0.002496	3.6614	4.6162	0.7161
0.002976	6.8520	5.8893	0.7620
0.003264	2.3537	8.9771	0.7392
0.003600	1.1194	1.0365	0.8286
0.005040	3.2060	1.7780	1.0040

M (mol·L^{-1}) is SDS and DTAB molarity in solvents ($\pm 3 \times 10^{-4}$ mol·L^{-1}) and standard uncertainties u are $u(m) = 0.00001$ mol·L^{-1}, $u(T) = \pm 0.01$ K, and $u(p) = \pm 0.01$ MPa.

washed with Milli-Q water, followed by acetone, and absolutely dried before measurements and 5, 10, 15, and 20% (w/w) aq-DMSO (AR grade, Rankem) solutions were used to check the performance of the Survismeter, and the values are in obedience to that of the literature values, given in Table S2 (supplementary material) [48, 52, 53]. The reported surface tension and viscosities are an average of three repeated measurements with $\pm 2 \times 10^{-6}$ kg·m^{-1}·s^{-1} and ± 0.03 mN·m^{-1} uncertainties, respectively.

3. Results and Discussion

3.1. Viscometric Study. Viscosity (η) values of SDS-rich and DTAB-rich mixed surfactants were measured at the three different temperatures ($T = 293.15, 298.15$, and 303.15 K) and at 0.1 MPa, and the same data are summarized in Table 2. Viscosity is a flowing, transporting property of the liquid mixture, and it is affected by molecular orientation and the nature of interaction ability of the solute and solvent interaction. And viscosity also gives the information about the interaction affinity of ionic species with the solvent system [54]. Table 2 shows that the aq-DTAB shows a higher η value than aq-SDS (Table S3). It indicates that the DTAB and SDS have the same hydrophobic part, except by only the head part (hydrophilic part). Due to the addition of DTAB into the aqueous system, the hydrophobic portion could be disrupted by the hydrogen bonding (HB) of the solvent system. Probably, it could also repel the solvent molecules to the surface site.

It could induce the weak CF with decreases in the surface tension (γ) value. DTAB has three methyl (-CH$_3$) groups in its head part which could also be developed by higher hydrophobicity; with stronger hydrophobic interaction, the γ value decreases with an increase in the η value. Generally, surfactants have a structure-breaking nature tendency of the solvent molecules which is present at the surface and strong electrostatic interaction with an increase in the η value. On increasing the concentration of surfactants, the η value increases with stronger IMF. SDS shows weaker hydrophobicity than DTAB because SDS has oxygen atoms in its head part. So, it could show weak hydrophobic interaction, and the η

values decrease. Thus, the aq-DTAB shows the highest η value with stronger van der Waals interactions and inducing stronger IMI affinities with solvent molecules. So, the DTAB shows lower γ values as the aq-DTAB could induce much solvent engagement. Addition of DTAB into the aq-SDS solution could form micelles at the air-liquid interfaces (ALIs). This study could be used for the preparation of drug formulation in the aqueous medium.

Table 2 shows increasing SDS and DTAB concentration due to stronger hydrophobic-hydrophobic interactions (HbHbI), stronger London dispersive force (LDF), and intermolecular force (IMF); then, the viscosity is increased. With increasing surfactants concentration, the population of the surface charges is increased in the solution, which could be induced by stronger interaction. The viscosity infers linkages of DTAB-rich and SDS-rich with a solvent system to determine fluid dynamics within the capillary with uniform water supply, contrary to static data like density. With increasing temperature, the kinetic energy increases as well as oscillation (rotational, vibrational, and transition) could be developed which shows weaker IMF and electrostatic interaction; then, the viscosity is decreased. The measurement of η data has been carried out in accordance with relative viscosity (η_r) as in [27]:

$$\eta_r = \frac{\eta}{\eta_0}, \tag{1}$$

where η_0 and η are the viscosity of the solvent and solution, respectively. The η_r value has been summarized in Table 2. The behavior of η_r versus M of SDS-rich and DTAB-rich is qualitatively the same as commonly observed in surfactant solutions [55, 56].

The η_r values of DTAB and SDS with the solvent systems follow the order: SDS > DTAB. This order inferred that the interaction affinity of the SDS molecule is stronger as compared to DTAB. However, SDS and DTAB both have the same tail part but different head groups. SDS contains oxygen atoms in its head part while $-CH_3$ groups in the DTAB could disrupt the HB of the solvent system, and DTAB could develop stronger ion-hydrophobic interaction (IH_bI). Due to the inclusion of 0.000864 to 0.00504 M/mol·L^{-1} DTAB into aq-SDS solution, the η_r value is more increased. It depicted that DTAB shows stronger hydrophobic interaction and maximum solvent molecules could repel with increase in the micelles formation rate. Similarly, 0.000096 to 0.012 M/mol·L^{-1} SDS was added into aq-DTAB. Hence, increasing rate of the η_r value decreases than the aq-DTAB system while decrement is higher compared to the DTAB-rich solution. Therefore, SDS shows the stronger ion-hydrophilic interaction (IHI) with a solvent system. On increasing the concentration of DTAB and SDS, the η_r value increases at a certain concentration, and after that, the η_r value decreases and further significantly increases. It indicates that, on increasing the concentration of surfactants, the micellization and aggregation processes could be occurred.

In our study, the trends of SDS-rich and DTAB-rich surfactants do not follow the regular trend. It means that the surfactant has a long alkyl chain (AC) which could trapped the air bubble, and so the graph trend of SDS-rich and DTAB-rich surfactants are obtained in the zic-zac order.

Chakraborty et al. [57] have reported that DTAB shows more interaction affinity towards the protein. The protein also has both hydrophilic and hydrophobic domains with the polar peptide bond in its molecular structure, due to stronger IH_bI dominant over IHI with increases in the η_r value. And the similar reason may be possible in the η_r value of the DTAB-rich mixed surfactant system. The η values have been further used to calculate viscous relaxation time (τ) using the following equation [58]:

$$\tau = \frac{4\eta}{3u^2\rho}, \tag{2}$$

where ρ is the density of the solution (Table S4), η is the viscosity of the solution (Table S3), and u is the sound velocity (Table S5) used for τ measurement.

The τ values are summarized in Table 3 and represented in Figures 2 and 3. The τ value is depending on the concentration, and interaction affinity of the solute with the

TABLE 3: Viscous relaxation time (τ/ps) of SDS-rich and DTAB-rich surfactants in the aqueous medium at three different temperatures $T = 293.15$, 298.15, and 303.15 K and at 0.1 MPa.

M (mol·L^{-1})	293.15 K	298.15 K	303.15 K
		SDS-rich	
0.005000	6.22E − 07	5.74E − 07	2.95E − 07
0.000096	9.08E − 07	5.53E − 07	8.14E − 07
0.000240	8.10E − 07	5.66E − 07	3.42E − 07
0.000480	1.11E − 06	5.62E − 07	5.73E − 07
0.000672	7.97E − 07	4.76E − 07	3.95E − 07
0.000792	6.59E − 07	7.05E − 07	3.55E − 07
0.000960	6.87E − 07	5.97E − 07	2.35E − 07
0.006011	6.37E − 07	6.64E − 07	2.26E − 07
0.007200	7.87E − 07	6.70E − 07	2.37E − 07
0.007920	8.29E − 07	8.40E − 07	2.48E − 07
0.009000	7.78E − 07	8.15E − 07	2.29E − 07
0.010800	8.29E − 07	6.71E − 07	3.94E − 07
0.012000	7.58E − 07	1.40E − 06	2.60E − 07
		DTAB-rich	
0.010000	6.58E − 07	6.34E − 07	2.25E − 07
0.000864	4.54E − 07	3.79E − 07	1.62E − 06
0.000960	4.79E − 07	5.09E − 07	1.87E − 06
0.001536	4.90E − 07	4.09E − 07	1.89E − 06
0.002016	4.63E − 07	6.34E − 07	1.86E − 06
0.002496	4.55E − 07	1.04E − 06	2.41E − 06
0.002976	4.84E − 07	1.33E − 06	4.52E − 06
0.003264	4.70E − 07	2.02E − 06	1.55E − 06
0.003600	5.26E − 07	2.34E − 07	7.38E − 07
0.005040	6.38E − 07	4.01E − 07	2.11E − 06

M (mol·L^{-1}) is SDS and DTAB molarity in solvents ($\pm 3 \times 10^{-4}$ mol·L^{-1}) and standard uncertainties u are $u(m) = 0.00001$ mol·L^{-1}, $u(T) = \pm 0.01$ K, and $u(p) = \pm 0.01$ MPa, and the expanded uncertainties, U_c (0.95 confidence level), is $U_c(\tau) = \pm 0.003$ ps (0.95 level of confidence).

solvent systems and temperature may be related to the structural relaxation processes occurring due to the rearrangement and reorientation of the molecules [59].

With an increase in the temperature, the τ value decreases with the increasing KE and weakening of electrostatic and binding forces. The τ value order of solvent is SDS > DTAB. This τ value order is also supported for η_r and ρ data. It infers that the SDS strongly interacts with the solvent medium by multiple intermolecular interactions (MIMI), and due to the strong interaction between solute and solvent, the solution could slowly pass through the capillary with an increase in the τ value. By increasing the concentrations of DTAB and SDS, the τ value increases with the weakening of CF and stronger electrostatic interaction, IMF, van der Waal forces. An inclusion of SDS into the aqueous system, the τ value is increased, while with DTAB, the τ value slightly decreases due to the stronger IHI domination over IH_bI. On increasing 0.000096 to 0.0012 M/mol·L^{-1} SDS, the τ value drastically increased with higher polarization, strong compactness, and the mobility of the micelles could be decreased. Similarly, with DTAB 0.000864 to 0.00504 M/mol·L^{-1} into aq-SDS, the τ value is less increased compared to the DTAB-rich surfactant solution. It infers that, due to the stronger IHI, the flow rate of the solution is decreased with increase in the τ value, while with DTAB-rich surfactant solution, because of stronger IH_bI and with the weakening of CF, the solution quickly passes and the τ value is decreased.

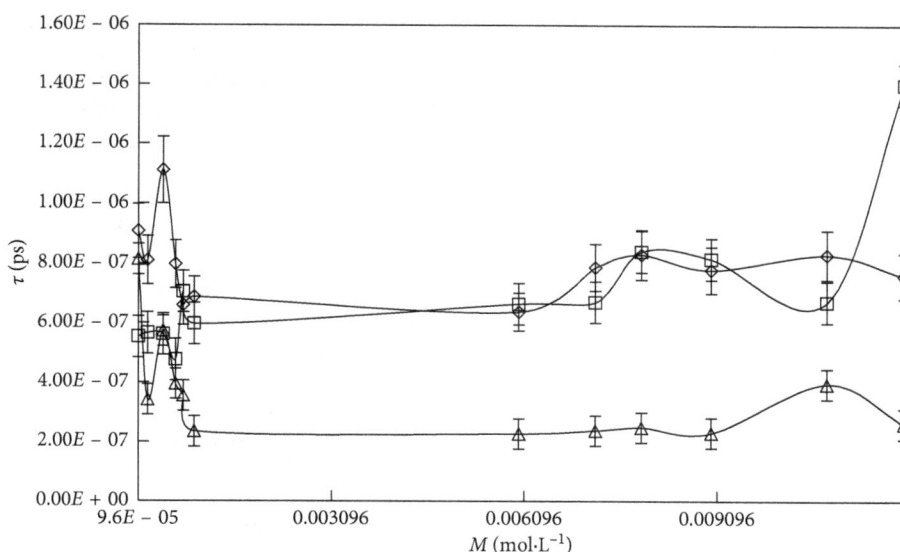

FIGURE 2: The τ value of the SDS-rich surfactant at the three different temperatures = 293.15 (◊), 298.15 (□), and 303.15 K (△), respectively.

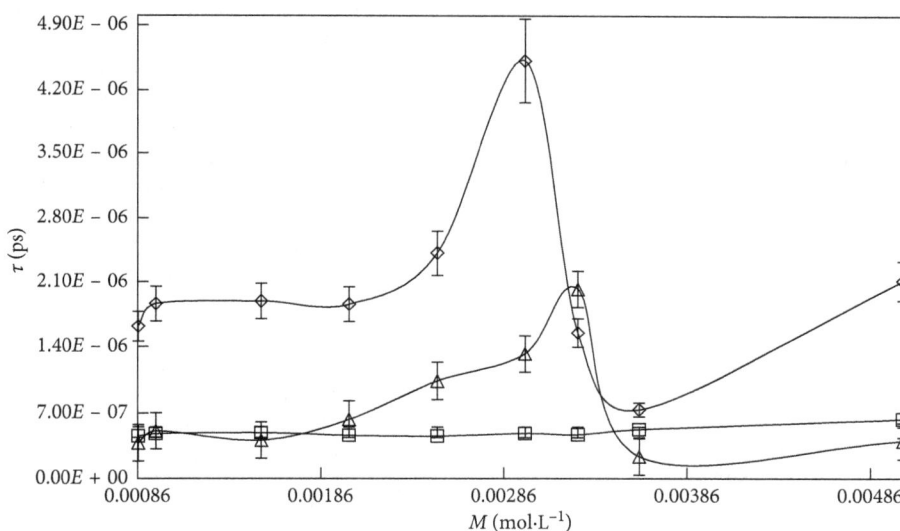

FIGURE 3: The τ value of the DTAB-rich surfactant at the three different temperatures $T = 293.15$ (◊), 298.15 (□), and 303.15 K (△), respectively.

3.2. Acoustic Impedance (Z). Initially, an inclusion of DTAB into the aqueous system, the acoustic impedance (Z) value decreases, while with SDS, the Z value is increased. Furthermore, on increasing the concentration of SDS and DTAB, the Z value (Table 4) increases. To measure a sound velocity, which is generated by the vibration due to $S_L S_O I$, the Z value was calculated by using the following equation:

$$Z = \rho \cdot u, \tag{3}$$

where ρ is the density and u is the sound velocity (u) of the solution.

The Z value infers an increase in the u value at a fixed composition and temperature. However, on increasing the temperature, the Z value is increased. It indicates that the Z property is directly proportional to the u value (Table S5) because of the heat which is a kind of KE.

On increasing the temperature, the molecules could gain energy which could induce rotational, electronic, transformational, and vibrational transitions and because of these transitions, the sound waves could travel quickly and the Z value is increased. The Z value (Figures 4 and 5) of the solvent systems follows the order: SDS > DTAB. The Z value also supported the ρ, η_r, and τ data. This order reflected that aq-SDS shows the higher Z value than aq-DTAB. SDS has a higher hydrophilic nature and stronger interaction abilities with an increase in the compactness of the solution. Thus, the aq-DTAB shows the higher hydrophobic nature which could induce stronger $IH_b I$ repelling the solvent molecules to the surface site and weakening the CFs with decreases in the γ value and with stronger IMI, the compactness and internal pressure (IP) increases, so the Z value is increased.

TABLE 4: Acoustic impedance (Z/g·cm^{-2}·s^{-1}) of SDS-rich and DTAB-rich surfactants in the aqueous medium at the three different temperatures $T = 293.15$, 298.15, and 303.15 K and at 0.1 MPa.

M (mol·L^{-1})	293.15 K	298.15 K	303.15 K
		SDS-rich	
0.005000	1481.77	1493.79	1498.87
0.000096	1481.90	1507.06	1504.04
0.000240	1482.03	1507.14	1504.12
0.000480	1482.02	1507.23	1504.21
0.000672	1481.93	1495.07	1504.28
0.000792	1481.94	1495.15	1504.31
0.000960	1481.89	1494.30	1503.99
0.006011	1481.31	1494.34	1503.56
0.007200	1482.11	1493.99	1502.86
0.007920	1480.60	1493.54	1503.57
0.009000	1481.86	1494.32	1503.72
0.010800	1481.76	1494.17	1504.16
0.012000	1482.45	1494.74	1504.81
		DTAB-rich	
0.010000	1483.00	1495.30	1504.82
0.000864	1482.79	1480.63	1502.52
0.000960	1482.62	1494.48	1506.18
0.001536	1482.50	1494.53	1504.13
0.002016	1482.53	1494.60	1504.22
0.002496	1482.13	1494.52	1504.29
0.002976	1482.16	1494.48	1504.26
0.003264	1480.20	1492.06	1503.21
0.003600	1482.23	1494.31	1503.82
0.005040	1481.90	1493.87	1503.56

M (mol·L^{-1}) is SDS and DTAB molarity in solvents ($\pm 3 \times 10^{-4}$ mol·L^{-1}) and standard uncertainties u are u(m) = 0.00001 mol·L^{-1}, $u(T) = \pm 0.01$ K, and $u(p) = \pm 0.01$ MPa.

All parameters are supporting each other on the basis of these interlink (coordinative) properties.

On increasing 0.000096 to 0.012 M/mol·L^{-1} SDS and 0.000864 to 0.00504 M/mol·L^{-1} DTAB concentration with aq-DTAB and aq-SDS, respectively, the Z value increases. However, in the case of DTAB-rich surfactant, the increasing rate of the Z value is higher than the SDS-rich surfactant at the three different temperatures ($T = 293.15$, 298.15, and 303.15 K).

Both surfactants have the same hydrophobicity spacer (tail region) except the hydrophilic spacer (head region). The higher Z value of aq-SDS infers that the aq-SDS could strongly interact by stronger IHI and ion-dipole interaction (IDI) forms small size micelles of the aq-SDS solution while with aq-DTAB, by stronger IH$_b$I forms large size micelles with weaker compactness in the solution, and the Z value is decreased. However, with increasing SDS concentration, the Z value increases with stronger IHI dominant over IH$_b$I and stronger electrostatic, van der Waals interaction with higher compactness occurring in the solution. For the DTAB-rich system, the Z value decreases with stronger IH$_b$I dominant over IHI.

3.3. Surface Property.
The DTAB-rich and SDS-rich systems have been applied in several technological applications because of the formation of micelles during the aggregation method under certain functioning conditions. Several

physical properties of surfactants have been reported in the literature because of their ability of characterizing different physical properties that have been analysed in the literature and due to their ability of describing the aggregation processes by using electrical conductivity(κ) and surface tension (γ) values [30, 31].

Table S6 shows that the γ value of SDS-rich and DTAB-rich decreases with increases in surfactants concentration in an aqueous system at the three different temperatures ($T = 293.15$, 298.15, and 303.15 K). It is evident from Table S6 that the γ value initially decreases with increasing concentration of SDS and then reaches a minimum. It indicates that micelles could form and the concentration of the break point is CMC, whereas for DTAB-rich, the surface tension reduced by adsorption of the surfactant at the interface, and a sigmoidal curve between surface tension (γ) and log (surfactant) is produced by the distinct break after which the γ value remains almost unchanged. Due to the presence of DTAB and SDS, surfactants produce a decrease in the γ value. Nevertheless, this decrease in surface tension reaches a constant γ value at a certain surfactant concentration. It depicted that the surface tension is a physical property influenced by the aggregation phenomenon due to a change in the surface concentration of the surfactant. Due to this reason, the surface tension has been used to determine the colloidal dynamics of numerous systems [60]. Thus, the aggregation process creates the concentration of SDS and DTAB remains constant due to the addition of different surfactants that are engaged in the formation of micelles. However, it could not effect on the surfactant concentration in the free liquid surface. Hence, the surface tensions remain with a constant value.

3.4. Interfacial Behavior.
The packing symmetry of the solvent spread monolayer of the ion-pair amphiphiles at the air-water interface depends on the stoichiometry and the magnitude of the charged head groups and the symmetry and the dissymmetry in the precursors' hydrophobic spear (the alkyl chain). The alkyl chains packed them in a way to maximize their van der Waals interaction, LDF, and electrostatic interaction in the bulk site. However, molecular packing at the air-water interfaces (AWI) to be more compact results in the lower molecular lift-off area [61].

The surface excess concentration (Γ_{max}) and the minimum surface area of the molecule (A_{min}) are two important parameters which determine the adsorption behavior and packing density of the micelles at the air/water interface [60, 62]. Γ_{max} is the concentration difference between the interface and a virtual interface in the interior of the volume phase, while A_{min} describes the minimum area of the amphiphile molecules at the surfactant-saturated monolayer at the air/solution interface [58, 60].

A reverse result is observed with A_{min}. The solvent system follows the order: aq-SDS > aq-DTAB and aq-DTAB > aq-SDS, the surface excess concentration and area of molecules, respectively. The very low A_{min} and the high Γ_{max} values for pure aq-SDS suggest that it is a poor self-assembly behavior presumably owing to the planar head group which could not provide an appropriate packing at the interface.

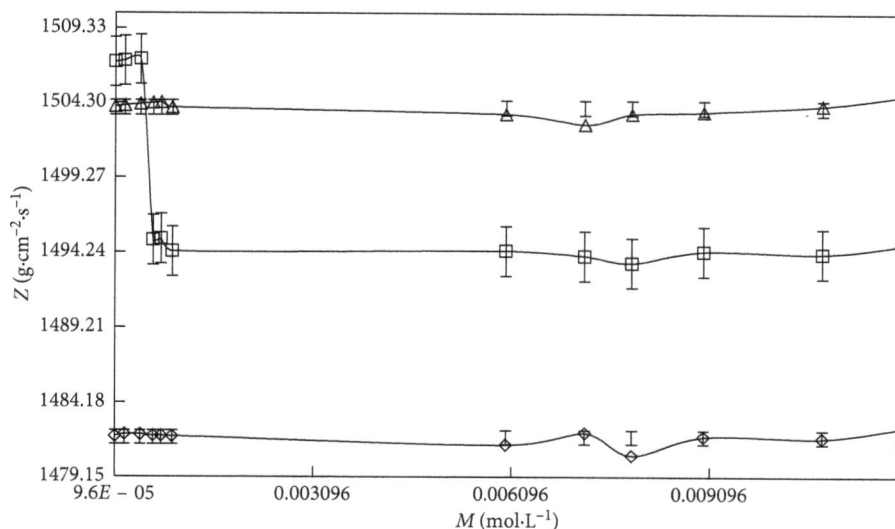

FIGURE 4: The Z value of the SDS-rich surfactant at the three different temperatures $T = 293.15$ ($\Diamond$), 298.15 ($\Box$), and 303.15 K ($\triangle$), respectively.

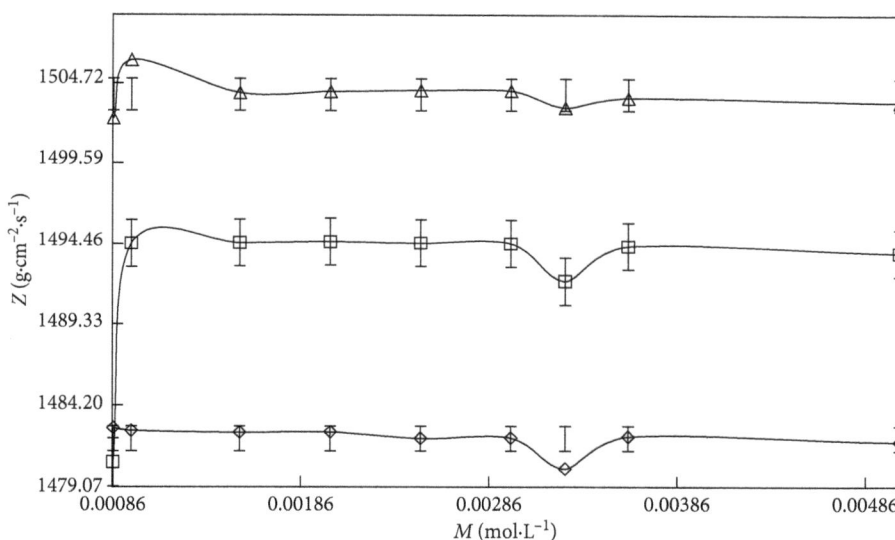

FIGURE 5: The Z value of the DTAB-rich surfactant at the three different temperatures $T = 293.15$ ($\Diamond$), 298.15 ($\Box$), and 303.15 K ($\triangle$), respectively.

The sudden change in the interfacial parameters by adding DTAB may presumably be connected to the efficient self-assembly behavior of DTAB, which favors the self-assembly process even at this low doping. With further increases in the concentration of DTAB and SDS in the aq-SDS and aq-DTAB, respectively, the total Γ_{max} increased indicating an antagonistic effect by the doping of aq-SDS in the DTAB system in this concentration range. A reverse effect is observed for the A_{min} value (in this concentration range) which is decreased with the increase in the concentration of DTAB. The increase in Γ_{max} and decrease in A_{min} with an increase in the DTAB content indicate that the packing density of surfactant molecules at the interface decreases with an increase in the SDS content. The surface excess concentration (Γ_{max}) value for 0.000096 to 0.012 M/mol·L^{-1} SDS and 0.000864 to 0.00504 M/mol·L^{-1} DTAB in aq-DTAB and aq-SDS solution is summarized in Table 5, and the area of molecules (A_{min}) is

calculated according to the following Gibbs adsorption equation [63], given in Table 6:

$$\Gamma_{max} = -\frac{c}{2RT} \cdot \frac{d\gamma}{dc}, \qquad (4)$$

$$A_{min} = \frac{1 \times 10^{18}}{\Gamma_{max} N_A}, \qquad (5)$$

where N_A is the Avogadro number, Γ_{max} is the surface excess concentration, A_{min} is area of molecules, R is the gas constant, T is the temperature in Kelvin, $d\gamma$ is the difference in the surface tension value, and c is the surfactant concentration. For Γ_{max} calculation, the 0.012 M/mol·L^{-1} and 0.00504 M/mol·L^{-1} as the limiting SDS and DTAB concentration is written in equation (4) contrary to CMC reported [64]. Furthermore, Γ_{max} is calculated, and surface pressure (π) is noted as follows:

TABLE 5: Surface excess concentration (Γ_{max}/mol·m^{-2}) of SDS-rich and DTAB-rich surfactants in the aqueous medium at the three different temperatures $T = 293.15$, 298.15, and 303.15 K and at 0.1 MPa.

M (mol·L^{-1})	293.15 K	298.15 K	303.15 K
		SDS-rich	
0.000096	104.57	32.32	40.87
0.000240	8903.09	6826.05	9244.40
0.000480	1983.06	3906.60	2792.54
0.000672	−4771.73	−4278.36	−3228.07
0.000792	14396.98	15220.63	7270.84
0.000960	−4249.40	10801.67	−8857.06
0.006011	−7020.87	−6358.77	−5015.42
0.007200	−35875.12	36095.19	30640.08
0.007920	−6680.42	−7485.56	11272.62
0.009000	−4131.37	−1995.31	−4163.18
0.010800	−2974.58	−1587.85	−1767.75
0.012000	−1611.23	−5292.82	−1280.98
		DTAB-rich	
0.000864	96550.00	40830.30	54441.63
0.000960	2148.31	14651.86	16499.02
0.001536	−1561.66	2858.12	−3389.47
0.002016	21654.96	15465.11	13188.97
0.002496	−11603.35	27424.35	18028.51
0.002976	125576.98	129682.4	105979.7
0.003264	−60686.80	42126.50	52703.16
0.003600	−6723.80	14731.67	15019.49
0.005040	96550.00	40830.30	54441.63

M (mol·L^{-1}) is SDS and DTAB molarity in solvents ($\pm 3 \times 10^{-4}$ mol·L^{-1}) and standard uncertainties u are $u(m) = 0.00001$ mol·L^{-1}, $u(T) = \pm 0.01$ K, and $u(p) = \pm 0.01$ MPa.

TABLE 6: Area of the molecule (A_{min}/nm^2·mol^{-1}) of SDS-rich and DTAB-rich surfactants in the aqueous medium at the three different temperatures $T = 293.15$, 298.15, and 303.15 K and at 0.1 MPa.

M (mol·L^{-1})	293.15 K	298.15 K	303.15 K
		SDS-rich	
0.000096	$1.59E - 08$	$5.14E - 08$	$4.06E - 08$
0.000240	$1.86E - 10$	$2.43E - 10$	$1.80E - 10$
0.000480	$8.37E - 10$	$4.25E - 10$	$5.95E - 10$
0.000672	$-3.48E - 10$	$-3.88E - 10$	$-5.14E - 10$
0.000792	$1.15E - 10$	$1.09E - 10$	$2.28E - 10$
0.000960	$-3.91E - 10$	$-1.54E - 10$	$-1.87E - 10$
0.006011	$-2.36E - 10$	$-2.61E - 10$	$-3.31E - 10$
0.007200	$-4.63E - 11$	$-4.60E - 11$	$-5.42E - 11$
0.007920	$-2.49E - 10$	$-2.22E - 10$	$-1.47E - 10$
0.009000	$-4.02E - 10$	$-8.32E - 10$	$-3.99E - 10$
0.010800	$-5.58E - 10$	$-1.05E - 09$	$-9.39E - 10$
0.012000	$-1.03E - 09$	$-3.14E - 10$	$-1.30E - 09$
		DTAB-rich	
0.000864	$2.68E - 09$	$1.02E - 08$	$1.08E - 08$
0.000960	$1.72E - 11$	$4.07E - 11$	$3.05E - 11$
0.001536	$7.73E - 10$	$1.13E - 10$	$1.01E - 10$
0.002016	$-1.06E - 09$	$5.81E - 10$	$-4.90E - 10$
0.002496	$7.67E - 11$	$1.07E - 10$	$1.26E - 10$
0.002976	$-1.43E - 10$	$-6.05E - 11$	$-9.21E - 11$
0.003264	$-1.32E - 11$	$-1.28E - 11$	$-1.57E - 11$
0.003600	$-2.74E - 11$	$-3.94E - 11$	$-3.15E - 11$
0.005040	$-2.47E - 10$	$-1.13E - 10$	$-1.11E - 10$

M (mol·L^{-1}) is SDS and DTAB molarity in solvents ($\pm 3 \times 10^{-4}$ mol·L^{-1}) and standard uncertainties u are $u(m) = 0.00001$ mol·L^{-1}, $u(T) = \pm 0.01$ K, and $u(p) = \pm 0.01$ MPa.

$$\pi_{binary} = \gamma_W - \gamma_{DTAB} \, (aq - DTAB),$$
$$\gamma_{binary} = \gamma_{water} - \gamma_{SDS} \, (aq - SDS),$$
$$\pi_{ter} = \gamma_{W+DTAB} - \gamma_{SDS} \, (anionic\ rich), \qquad (6)$$
$$\pi_{ter} = \gamma_{W+SDS} - \gamma_{DTAB} \, (cationic\ rich).$$

An inclusion of SDS into the water, the Γ_{max} value (Figures 6 and 7) is more increased. It indicates that the SDS has oxygen atoms in the head part which is small in size. So, the maximum number of SDS molecules could go to the surface site. Similarly, an addition of DTAB into the aqueous system, the Γ_{max} value is decreased than the aq-SDS system. It indicates that the larger size of DTAB has three -CH$_3$ groups in its structure which could induce a hindrance for a move to the surface site. So, the less number of DTAB molecules could move to the surface, and the Γ_{max} value decreases. On increasing the concentration from 0.000096 to 0.012 M/mol·L^{-1} SDS and 0.000864 to 0.00504 M/mol·L^{-1} DTAB, the maximum surfactant molecules move to the surface site with surfactant molecules occupying a less area with stronger H$_b$H$_b$I and stronger LDF due to a more considerable difference in the chemical potential of the surface and in the bulk phase. Due to the stronger LDF occurrence with stronger BF and stronger IMF, A_{min} is decreased. On increasing the temperature, A_{min} expands due to increased KE and weakening of BF. Due to the increase in the temperature, Γ_{max} value decreases with increasing area of molecules with the weakening of BF and IMI, and the least number of surfactant molecules could go to the surface with the increased A_{min} value (Figures 8 and 9).

3.5. Friccohesity Shift Coefficient (FSC). Friccohesity predicts working or functional ability of solution where the residual molecular forces remain in a reversible mode. Fundamentally, the ability of the medium or the solvent and the constituent molecules to promote the S_LS_0I rather than self-binding individually is a fundamental need for sparing the molecular surface area. The disruption of the self-binding state could be attained by the weakening of the CF on increasing friccohesity attracting other molecules like drugs or others for binding. The self-binding state could have stronger homomolecular potential noted as an anti-dispersion activity. Hence, the potentializing homomolecular intramolecular potential to trap other molecules is an essential need to weaken CF and to develop intermolecular or the heteromolecular forces to get stuck to the solution. The shear stress and strains lead to velocity gradients and interlayer distance. The interlayer distance directly reflects the strength of the IMF when the solute molecules could align along with line subjected to the interlayer thickness. The intermolecular strength is determined with HB and also the weakening of the solvent structures and tends to form a structure with the solute. It becomes an urgent need that the status of CF and IMF is measured simultaneously which is rightly and logically determined by friccohesity data.

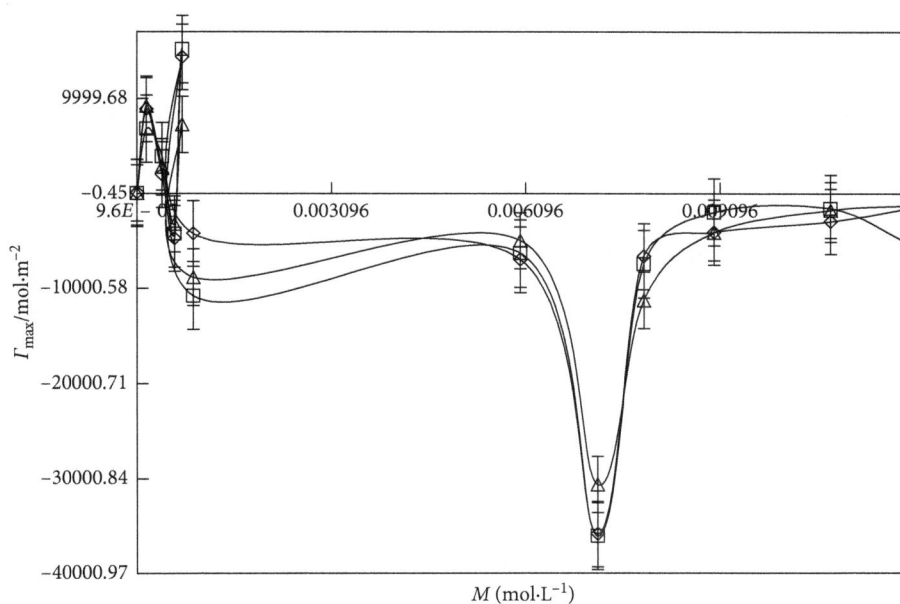

FIGURE 6: The Γ_{max} value of the SDS-rich surfactant at the three different temperatures $T = 293.15$ ($\Diamond$), 298.15 ($\Box$), and 303.15 K ($\triangle$), respectively.

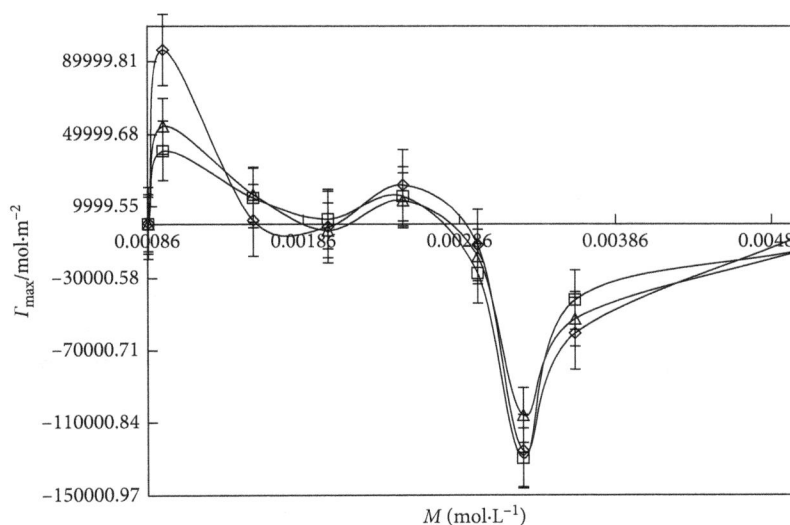

FIGURE 7: The Γ_{max} value of the DTAB-rich surfactant at the three different temperatures $T = 293.15$ ($\Diamond$), 298.15 ($\Box$), and 303.15 K ($\triangle$), respectively.

The η measurements deal with intra- and intermolecular networking of electronic forces materialized through electrostatic forces, and γ (Table S6) tracks damages of IMF or the molecular forces working within the similar molecules through HB and another interaction mechanism. The molecular forces have two separate domains where one of them remains operational at the surface, causing a continuous thin film where even air could not enter. Therefore, an aqueous electrolyte or surfactants in aqueous solutions even on shaking do not develop bubble. So, such engineering is confined to the surface force which is tracked by the surface tension. Another interaction between the two forces remains defunct because the force factors counterbalance

the linear elements of molecular interactions. The σ data have higher resolution and reproducibility and illustrate the interfaces of CFs and frictional forces (FFs) where these forces are the core theories of γ and η measurements, respectively. Therefore, σ of DTAB-rich and SDS-rich is given Table S7 and is calculated by using the following Man singh equation [51]:

$$\sigma = \frac{\eta_0}{\gamma_0} \left[\left(\frac{t}{t_0} \right) \left(\frac{n}{n_0} \right) \right], \tag{7}$$

where η_0, γ_0, t_0, and n_0 and η, γ, t, and n are viscosity, surface tension, viscous flow time, and pendant drop numbers of the solvent and solution, respectively.

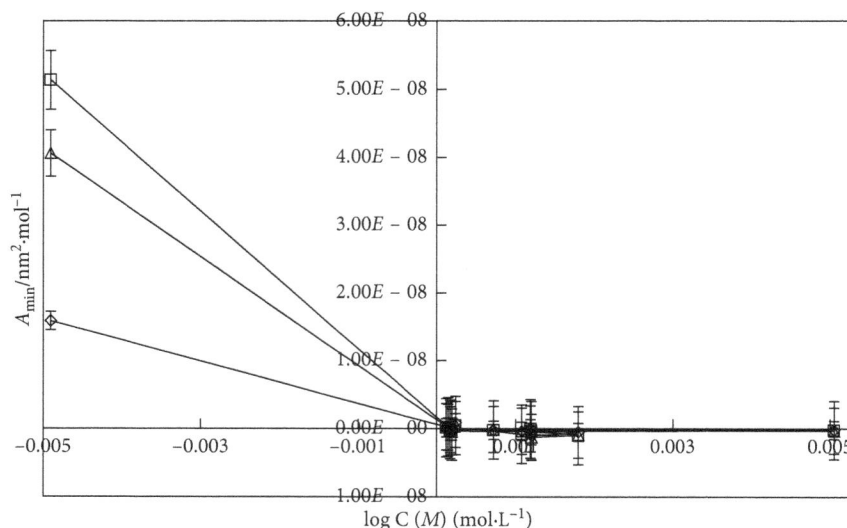

FIGURE 8: The A_{min} value of the SDS-rich surfactant at the three different temperatures $T = 293.15$ ($\Diamond$), 298.15 ($\Box$), and 303.15 K ($\triangle$), respectively.

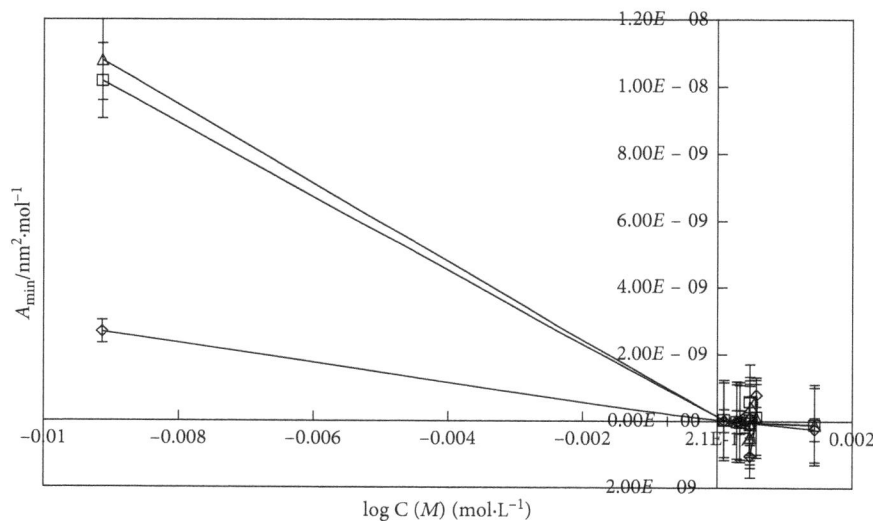

FIGURE 9: The A_{min} value of the DTAB-rich surfactant at the three different temperatures $T = 293.15$ ($\Diamond$), 298.15 ($\Box$), and 303.15 K ($\triangle$), respectively.

The friccohesity shift coefficient (σ_C) is calculated by using the following equation:.

$$\text{Friccohesity shift coefficient } (\sigma_C) = \frac{1}{\sigma \cdot \gamma}, \qquad (8)$$

where σ_C is the friccohesity shift coefficient, σ is the friccohesity, and γ is the surface tension of the solution.

The solvent systems follow the order in the aqueous medium: SDS > DTAB. This order infers that the σ_C value increased the aq-SDS (Table 7) than aq-DTAB due to SDS which could develop weak CFs with stronger FFs and IMF; the γ value decreases with the higher ρ value. While with aq-DTAB, the σ_C value is decreased. The σ_C value of aq-DTAB is more decreased than aq-SDS solutions because both surfactants have the same tail part, except the only head part, and so the stronger IH$_b$I and weak CFs with

stronger FFs. On increasing 0.000096 to 0.012 M/mol·L^{-1} SDS and 0.000864 to 0.00504 M/mol·L^{-1} DTAB concentration, the σ_C value increases due to stronger IMI with the weakening of CFs.

This parameter reveals the mechanism of $S_L S_0 I$ and $S_L S_L I$ of surfactants [65]. Such parameters determined a critical and comparative study of γ (Figures 10 and 11) and friccohesity of the SDS-rich and DTAB-rich surfactant solution summarized in Table S7. It also infers the efficacy of interacting activity of SDS and DTAB with the solvent system, its fluidity and absorptivity. We obtained a conversion relation between γ and η, and the aq-DTAB shows higher η and lower γ compared to the aq-SDS solution due to stronger hydrophobic interaction. The η value is increased because of the interaction with dissimilar molecules. Due to the inclusion of SDS and DTAB in aq-DTAB and aq-SDS

TABLE 7: Friccohesity shift coefficient (FSC) of SDS-rich and DTAB-rich surfactants in the aqueous medium at the three different temperatures $T = 293.15$, 298.15, and 303.15 K and at 0.1 MPa.

M (mol·L^{-1})	293.15 K	298.15 K	303.15 K
		SDS-rich	
0.005000	1.997570	0.988696	1.126186
0.000096	0.472963	0.992845	1.524808
0.000240	1.229149	1.181705	0.751272
0.000480	0.722956	0.870212	1.363012
0.000672	1.056841	1.252171	1.983979
0.000792	1.226570	1.526528	0.629227
0.000960	1.478152	1.467933	1.274735
0.006011	0.795106	1.402486	0.608560
0.007200	0.878679	1.052150	0.702083
0.007920	0.999447	0.772808	0.630889
0.009000	1.021779	0.884746	0.853029
0.010800	0.780298	0.823337	0.546062
0.012000	0.625146	0.758690	0.391575
		DTAB-rich	
0.010000	2.606188	1.205797	1.135148
0.000864	2.752721	1.536592	0.719237
0.000960	2.556441	1.138905	0.625152
0.001536	3.491674	1.419604	0.615893
0.002016	2.648554	0.916497	0.626744
0.002496	3.554227	0.558781	0.483651
0.002976	2.533495	0.437935	0.258442
0.003264	2.444864	0.251481	0.772786
0.003600	2.074703	2.167979	1.686891
0.005040	1.490789	1.431129	0.440847

M (mol·L^{-1}) is SDS and DTAB molarity in solvents ($\pm 3 \times 10^{-4}$ mol·L^{-1}) and standard uncertainties u are $u(m) = 0.00001$ mol·L^{-1}, $u(T) = \pm 0.01$ K, and $u(p) = \pm 0.01$ MPa.

solution, the friccohesity shift coefficient decreases with stronger FFs and weak CFs. On increasing the concentration of surfactants, the σ_C value is decreased. On increasing the temperature, the σ_C value is decreased due to the weakening of FF, electrostatic interaction, IDI, and binding forces.

3.6. Hydrodynamic Volume (V_h).

Hydrodynamic values are a significant factor in determining a magnitude to the volume change of the hydrated molecules with increasing solute concentration. On increasing the temperature, the SDS-rich and DTAB-rich have negative V_h values which decrease as the size of the KE increases. The V_h values in this study (Table 8) shows negative at the temperatures $T = 298.15$ and 303.15 K. Thus, the sign of the V_h values reflects the nature of the $S_L S_0 I$; then, we can conclude that, at different temperatures ($T = 298.15$ and 303.15 K), the DTAB-rich and SDS-rich mixed surfactants have structure-making effects on water, whereas temperature 293.15 K in this study shows structure-breaking effects [66].

The hydrodynamic volume (V_h) reflected the $S_L S_0 I$ and solute-solute interaction ($S_L S_L I$). V_h is calculated with the following equation and summarized in Table 9:

$$V_h = \frac{\phi M}{N_A c}, \qquad (9)$$

where ϕ is the fractional volume (Table 8), M is a molar mass of the solute, N_A is the Avogadro number, and c is the concentration.

Fractional volume (ϕ) is calculated by the following equation:

$$\phi = \frac{4}{3\pi r^3 N_A c}, \qquad (10)$$

where r is the particles size, N_A is Avogadro's number, and c is the solute concentration.

The V_h values for solvent follow the order: SDS > DTAB. This order indicates that the interaction activity of SDS with H$^+$ ions of solvent molecules is stronger than DTAB because SDS could be strongly towards H$^+$ ions of water by the O$^-$ ion which is present at the head region in the SDS. So, the interaction affinity of SDS with H$^+$ ions is higher, while with DTAB is the lower because DTAB has -CH$_3$ groups in its head region, which could repel the water molecules. Thus, the V_h value of DTAB is lesser than SDS in an aqueous medium. An inclusion of SDS into aq-DTAB, the V_h value drastically increased due to a higher concentration of SDS, it has more O$^-$ ions which could show the stronger interaction affinity with the IHI domain over IH$_b$I, and SDS could form a more hydrogen sphere compared to the DTAB, while with DTAB into aq-SDS solution, the V_h value is decreased as compared to SDS-rich surfactants. It depicted that the stronger hydrophobic interaction and DTAB could show weak interaction ability with water molecules with decreases the V_h value. On increasing the surfactants concentration, the V_h values decrease with stronger $S_L S_L I$ and weaker $S_L S_0 I$. On increasing the temperature, the V_h value increases due to the weakening of electrostatic interaction and binding forces.

3.7. Hydrodynamic Radius (R_h).

Hydrodynamic radius (R_h) depicts the basic activities of solute and solvent interaction. So, micelles of SDS and DTAB with solvent systems could change in R_h along with other amphiphilic solutes which could reflect various modes of interactions. Hydrophobicity and structural constituents of surfactants could develop stronger molecular networking with an effect of the solvent cage, and R_h is calculated using the following equation (Table 10):

$$R_h = \frac{kT}{6\pi\eta D}, \qquad (11)$$

where κ is the Boltzmann constant, D is the diffusion coefficient of the medium, and the R_h value is as DTAB > SDS in the aqueous medium. Due to the inclusion of DTAB in the aqueous system, the R_h value is increased while with SDS, the R_h value decreases. It indicates that the DTAB having -CH$_3$ groups could be repelled by the solvent molecules, so the size of the radius is increased, while SDS contains hydrophilic atoms in its head part which could strongly interact, so the value of hydrodynamic radius is decreased. Due to the addition of SDS into the aq-DTAB solution, the R_h value is decreased and with DTAB in the aq-SDS, the R_h value is also decreased. It depicted that the dominance of IHI over IH$_b$I. On increasing the concentration of 0.000096 to 0.012 M/mol·L^{-1} SDS and 0.000864 to 0.00504 M/mol·L^{-1} DTAB into aq-DTAB and aq-SDS solution, the R_h value decreases due to the stronger IMI, electrostatic interaction, van der Waals interactions, and IDI. On increasing the

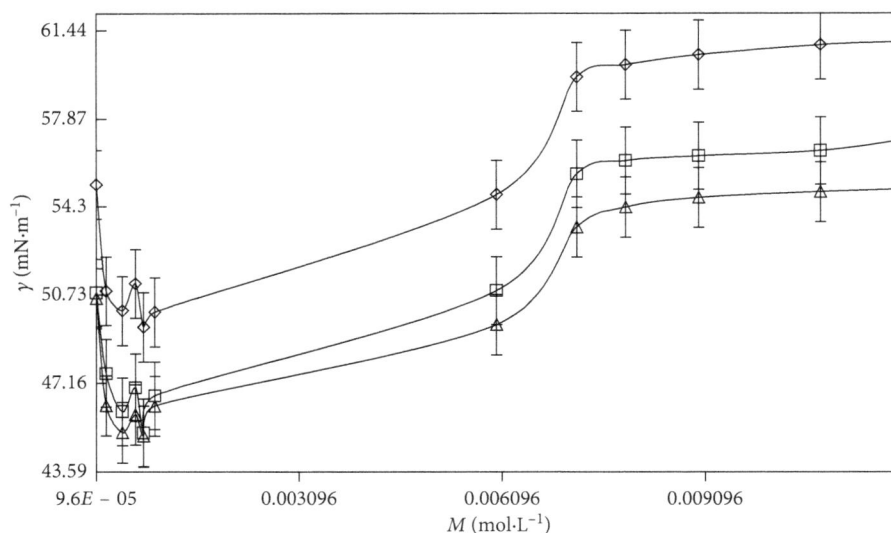

FIGURE 10: The γ value of the SDS-rich surfactant at the three different temperatures $T = 293.15$ ($\Diamond$), 298.15 ($\Box$), and 303.15 K ($\triangle$), respectively.

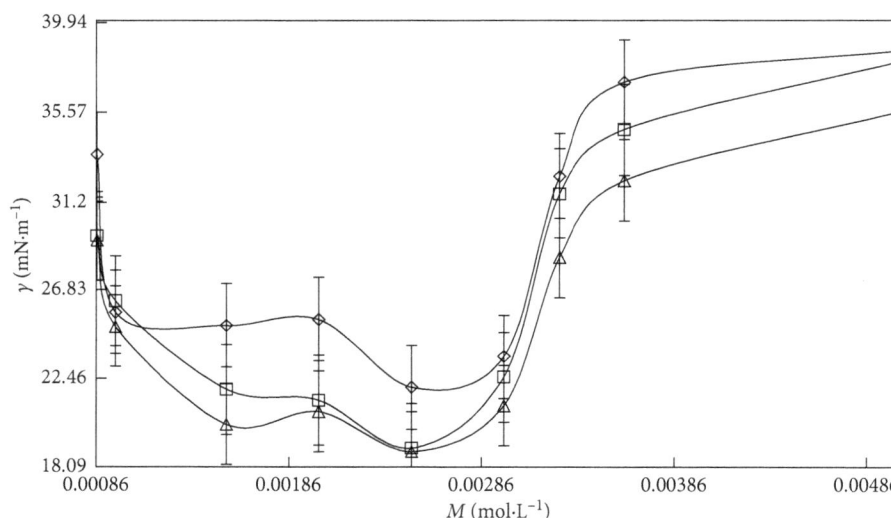

FIGURE 11: The γ value of the DTAB-rich surfactant at the three different temperatures $T = 293.15$ ($\Diamond$), 298.15 ($\Box$), and 303.15 K ($\triangle$), respectively.

temperature, the R_h values are increased due to the weakening of BFs and IMF with increased kinetic expansion. Their R_h values depict a solvent entangling around the surfactants that affect a mutual contact of solvent molecules.

3.8. *Viscosity B-Coefficient (B)*. Viscosity B-coefficient (B) of SDS-rich and DTAB-rich is calculated by using the following Jones-Dole equation:

$$\left(\frac{\eta_r - 1}{M}\right) = B + Dm + D'm^2. \quad (12)$$

$[\eta]$ is obtained from $(\eta_r - 1)/M$ versus M.

$$\left(\frac{\eta_r - 1}{M}\right) = [\eta], \quad (13)$$

where $[\eta]$ is the intrinsic viscosity, η_r is the relative viscosity; M is the molarity; B is the viscosity B-coefficient, and D and D' are Falkenhagen's coefficients. D illustrates $S_L S_L I$, while B illustrates $S_L S_0 I$ [54, 66] at the three different temperatures ($T = 293.15$, 298.15, and 303.15 K), respectively. The positive B values depict stronger $S_L S_0 I$ with stronger IMF (Table 11). The higher and positive B values for SDS-rich and DTAB-rich describe stronger IHI, IDI, and IMI. The B value predicts solute solvation and their effect on the structure of solvent in the vicinity of the solute molecule having either negative or positive magnitude. The B coefficient measures structural modifications induced by $S_L S_0 I$. Thus, Table 11 reveals that DTAB has higher positive B values at $T = 293.15 K$ compared to SDS. Initially, surfactants in water could

TABLE 8: Fractional volume (ϕ) of SDS-rich and DTAB-rich surfactants in the aqueous medium at the three different temperatures $T = 293.15$, 298.15, and 303.15 K and at 0.1 MPa.

M (mol·L^{-1})	293.15 K	298.15 K	303.15 K
		SDS-rich	
0.005000	0.0098	0.0331	−0.1488
0.000096	0.1838	−0.0080	0.7081
0.000240	0.1210	0.0010	0.0653
0.000480	0.3149	−0.0016	0.3806
0.000672	0.1126	−0.0679	0.1379
0.000792	0.0241	0.0919	0.0829
0.000960	0.0418	0.0161	−0.0808
0.006011	0.0094	0.0624	−0.0921
0.007200	0.1060	0.0664	−0.0779
0.007920	0.1326	0.1845	−0.0634
0.009000	0.1004	0.1678	−0.0882
0.010800	0.1330	0.0671	0.1367
0.012000	0.0881	0.5755	−0.0466
		DTAB-rich	
0.010000	0.0343	0.0791	−0.2075
0.000864	0.5848	−0.1195	0.2715
0.000960	0.7330	−0.0980	0.5060
0.001536	0.7499	−0.0913	0.3268
0.002016	0.7299	−0.1085	0.7257
0.002496	1.0646	−0.1136	1.4465
0.002976	2.3408	−0.0952	1.9557
0.003264	0.5415	−0.1043	3.1908
0.003600	0.0477	−0.0686	0.0146
0.005040	0.8824	0.0016	0.3112

M (mol·L^{-1}) is SDS and DTAB molarity in solvents ($\pm 3 \times 10^{-4}$ mol·L^{-1}) and standard uncertainties u are $u(m) = 0.00001$ mol·L^{-1}, $u(T) = \pm 0.01$ K, and $u(p) = \pm 0.01$ MPa.

TABLE 9: Hydrodynamic volume (V_h/nm^3) of SDS-rich and DTAB-rich surfactants in the aqueous medium at the three different temperatures $T = 293.15$, 298.15, and 303.15 K and at 0.1 MPa.

M (mol·L^{-1})	293.15 K	298.15 K	303.15 K
		SDS-rich	
0.005000	1.00	3.39	−15.23
0.000096	916.87	−39.94	3531.75
0.000240	241.47	2.07	130.18
0.000480	314.06	−1.57	379.62
0.000672	80.21	−48.37	98.28
0.000792	14.54	55.54	50.12
0.000960	20.86	8.03	−40.27
0.006011	0.75	4.97	−7.34
0.007200	7.05	4.42	−5.18
0.007920	8.02	11.15	−3.83
0.009000	5.34	8.93	−4.69
0.010800	5.90	2.98	6.06
0.012000	3.51	22.96	−1.86
		DTAB-rich	
0.010000	1.64	3.79	−10.62
0.000864	346.48	−70.79	160.87
0.000960	390.88	−52.25	269.83
0.001536	249.95	−30.43	108.93
0.002016	185.34	−27.55	184.28
0.002496	218.35	−23.30	296.68
0.002976	402.67	−16.38	336.42
0.003264	84.93	−16.36	500.46
0.003600	6.79	−9.75	2.07
0.005040	89.63	0.16	31.61

M (mol·L^{-1}) is SDS and DTAB molarity in solvents ($\pm 3 \times 10^{-4}$ mol·L^{-1}) and standard uncertainties u are $u(m) = 0.00001$ mol·L^{-1}, $u(T) = \pm 0.01$ K, and $u(p) = \pm 0.01$ MPa.

repel out the hydrophobic part of the surfactants to surface which results in a decrease of γ at the surface. Furthermore, the inclusion of SDS and DTAB to aq-DTAB and aq-SDS, the hydrophilic part accommodates in the bulk solution instead of the surface because surfactants being hydrophobic could not go to the surface site which is already occupied by the hydrophobic region of the SDS and DTAB. So, hydrophobicity increases in the bulk solution. Also, this mechanism leads to a stable formulation out of such solution mixtures. Thus, the DTAB-rich at $T = 298.15$ K could be induced hydrophobicity to a maximum extent and behaves as a structure maker at this temperature because -CH$_3$ could be heat sensitive. Positive B value supports the structure, making tendency of SDS-rich and DTAB- rich at the three different temperatures ($T = 293.15$, 298.15, and 303.15 K).

The stronger H$_b$HI is decreased the B value with a tendency to behave as a structural breaker [54]. The surfactants induced stronger hydrophilic and hydrophobic interactions with the water system. The B values reflect the structure, making or breaking effects noted as $(\eta_{r-1}/m) > 1$. It indicates an ability of a solute to interact with the medium via the IMF and HB.

4. Conclusion

In this study, the relative viscosity, viscous relaxation time, and acoustic impedance values increase with increasing of concentration of the surfactants due to stronger ion-hydrophobic interaction with the weakening of cohesive forces with stronger frictional forces. By the addition of SDS and DTAB into water, the surface tension value decreases while the viscosity and friccohesity value increase due to weakening of cohesive forces and stronger intermolecular forces. These properties are correlated to each other. Mixed surfactants form self-assembly which could be applicable in the industry, pharmaceuticals, and drug formulation. Therefore, friccohesity determined the surface and bulk properties of the solution. With increasing concentration of the surfactant, surface excess concentration values are increased with stronger hydrophobicity pushing larger DTAB and SDS amount to the surface with more Brownian motion and stronger LDF. A less volume because of stronger H$_b$H$_b$I, bringing together the stronger LDF, and stronger LDF causes stronger binding forces have produced a greater internal pressure and lower surface area. On increasing the temperature, the area of the molecule increases because of weak intermolecular interaction and bond forces. SDS and DTAB mixed surfactant could be applicable in industrial and pharmaceutical for the formation of the drug, drug delivery, drug loading, enhanced solubility, and dispersion of drug. We have calculated the surface and bulk properties of the mixed surfactant which can be used in these applications.

TABLE 10: Hydrodynamic radius (R_h/nm) of SDS-rich and DTAB-rich surfactants in the aqueous medium at the three different temperatures T = 293.15, 298.15, and 303.15 K and at 0.1 MPa.

M (mol·L^{-1})	293.15 K	298.15 K	303.15 K
		SDS-rich	
0.005000	59.38	60.97	76.27
0.000096	52.35	61.38	54.30
0.000240	54.37	60.92	72.52
0.000480	48.93	61.05	61.03
0.000672	54.67	64.87	69.09
0.000792	58.24	56.91	71.63
0.000960	57.44	60.17	82.22
0.006011	58.92	58.09	83.22
0.007200	54.90	57.93	81.98
0.007920	53.97	53.73	80.78
0.009000	55.11	54.25	82.87
0.010800	53.96	57.90	69.15
0.012000	55.57	45.30	79.48
		DTAB-rich	
0.010000	58.24	58.95	83.34
0.000864	43.13	66.35	70.12
0.000960	41.16	64.74	63.46
0.001536	40.96	64.27	68.30
0.002016	41.20	65.51	59.03
0.002496	37.79	65.90	50.05
0.002976	30.66	64.54	46.15
0.003264	43.78	65.20	40.10
0.003600	56.09	62.77	82.35
0.005040	39.50	58.87	68.79

M (mol·L^{-1}) is SDS and DTAB molarity in solvents ($\pm 3 \times 10^{-4}$ mol·L^{-1}) and standard uncertainties u are u(m) = 0.00001 mol·L^{-1}, $u(T)$ = ± 0.01 K, and $u(p)$ = ± 0.01 MPa.

TABLE 11: Intrinsic viscosity (η) of SDS-rich and DTAB-rich surfactants in the aqueous medium at the three different temperatures T = 293.15, 298.15, and 303.15 K and at 0.1 MPa.

T/K	SDS-rich
293.15	1.3689
298.15	1.0474
303.15	1.7095
T/K	DTAB-rich
293.15	2.5019
298.15	0.6706
303.15	1.2427

M (mol·L^{-1}) is SDS and DTAB molarity in solvents ($\pm 3 \times 10^{-4}$ mol·L^{-1}) and standard uncertainties u are u(m) = 0.00001 mol·L^{-1}, $u(T)$ = ± 0.01 K, and $u(p)$ = ± 0.01 MPa.

Conflicts of Interest

The authors declare that there are no conflicts of interest regarding the publication of this paper.

Acknowledgments

Ajaya Bhattarai is thankful to The World Academy of Sciences (TWAS), Italy, for providing funds to work in the Department of Chemical Sciences, Central University of Gujarat, Gandhinagar (India).

Supplementary Materials

Table S1 compares the measured densities values at $\pm 5 \cdot 10^{-6}$ g·cm^{-3} uncertainty by controlling the temperature by the help of the Peltier (PT100) device in $\pm 1 \cdot 10^{-3}$ K accuracy obtained from the Anton Paar DSA 5000M density meter with the literature values. Repeatability of the instrument corresponds to precision in ρ and u with $1 \cdot 10^{-3}$ kg·m^{-3} and 0.10 m·s^{-1}, respectively, in 1.0 M/mol·kg^{-1} sodium chloride and 10% (w/w) DMSO in aqueous solutions and were used for instrument calibration at T = 298.15 K. Table S2 compares the experimental data of surface tension and viscosity which were measured by Borosil Mansingh Survismeter through the viscous flow time (VFT) and pendant drop number (PDN) methods, respectively, with the literature values in 5, 10, 15, and 20% (w/w) aq-DMSO. Viscosity and surface tension both have been an average of three replicate measurements with $\pm 2 \times 10^{-6}$ kg·m^{-1}·s^{-1} and ± 0.03 mN·m^{-1} uncertainties, respectively. There is also the difference in density measured with the Anton Paar DSA 5000M density meter with the literature values for 5, 10, 15, and 20% (w/w) aq-DMSO. Table S3 compares the experimental data of viscosity at three different temperatures (293.15, 298.15, and 303.15 K) which were measured by Borosil Mansingh Survismeter. In all investigated concentrations of SDS-rich and DTAB-rich surfactants in the aqueous medium, there is an increase of viscosity from 293.15 K to 298.15 K, whereas in the case of DTAB-rich, there is a decrease in viscosity in all investigated concentrations from 298.15 K to 303.15 K. But there is not a regular pattern of viscosity change for SDS-rich in the increment of the temperature from 298.15 K to 303.15 K. Table S4 confers the density value is higher in the aq-DTAB solution which is used as a stock solution for SDS-rich surfactant solution. However, due to addition of SDS in the aqueous DTAB solution, the ρ value decreases. It depicted that the SDS and DTAB both have a same hydrophobic tail part, except the head part (counterpart). DTAB has three methyl (-CH$_3$) which could develop stronger hydrophobic interaction with the weakening of CF and stronger intermolecular interaction with the increase in the ρ value. Due to addition of SDS into aq-DTAB, the ρ value is decreased. It indicates that the SDS is less hydrophobic compared to DTAB, so it could induce less weak CFs than DTAB. With increasing the concentration of SDS, the ρ value increases due to stronger van der Waals interaction, electrostatic interaction, and intermolecular interaction. But at the particular concentration, the ρ value drastically decreased. However, at the particular concentration, the density value drastically decreased; it means that the concentration leading to CMC generates maximum assembly in a particular shape which is influenced by the nature of the surfactant and surrounding environment. The similar kind of trend is observed in the case of DTAB-rich surfactant. With increasing the temperature, the ρ value decreases due to increase in KE with weakening of binding forces (BF) and electrostatic interaction. Table S5 compares the sound velocity of the DTAB-rich surfactant and SDS-rich surfactant solutions. The sound velocities of the solvent follow the order: aq-SDS > aq-DTAB. The order indicates that the number of interacting molecules per unit volume increases, and the molecules become tightly packed in the presence of SDS, resulting in faster sound wave propagation. The u values were observed to increase with increasing temperature. This suggests that the molecules upon gaining the KE oscillate very strongly, weakening the S_0S_0I. The sound velocity is increased with increasing DTAB and SDS concentration and temperature. With increasing the surfactant concentration, the IMF strengthens, and the hydrophilic sites of DTAB and SDS become closer to greater KE transfer, thereby increasing u with higher density. On increasing the temperature, the interacting groups of DTAB and SDS with a solvent system obtain more energy with greater vibration, causing faster sound wave circulation. This subsequently increases the u while decreasing the ρ values (Table 2). The slopes for ρ is steeper than those for u (Tables S4 and S5) which mutually supports the first order of interaction with increasing surfactants concentration. Table S6 compares the surface tension data and the γ value of aq-DTAB is lower than the SDS-rich surfactant because of the weakening of CFs with stronger ion-hydrophobic interaction. The surface tension (γ) or surface activities define the involvement of solvent with surfactants activities where the CFs or surface energy of the solvent decreases to interact with SDS and DTAB. Stronger surfactants-solvent interactions reflect weaker CFs with disruption of the HB network with lower γ values and vice versa. The hydrophobic alkyl chain of the surfactants accumulates on the solvent surface, thereby decreasing the γ value. However, the aq-SDS γ value is lower than aq-DTAB due to stronger hydrophobic-hydrophobic interaction. With increasing the concentration of the surfactants, the γ value decreases with disruption of HB with weakening of CFs of the solution. Table S7 compares the data of friccohesity of SDS-rich and DTAB-rich surfactant solution at three different temperatures. In all investigated concentrations of DTAB-rich surfactants in the aqueous medium, there is increase of friccohesity from 293.15 K to 298.15 K, whereas in the case of $0.003264\,mol\cdot L^{-1}$, there is decrease in friccohesity. It is found that there is decrease in friccohesity from 298.15 K to 303.15 K in all investigated concentrations of DTAB-rich surfactants in the aqueous medium, while in the concentration $0.01\,mol\cdot L^{-1}$, there is an opposite trend. But there is not a regular pattern of friccohesity change for SDS-rich in the increment of temperature from 293.15 K to 303.15 K. (*Supplementary Materials*)

References

[1] X. Xu, P. Chow, C. Quek, H. Hng, and L. Gan, "Nanoparticles of polystyrene latexes by semicontinuous microemulsion polymerization using mixed surfactants," *Journal of Nanoscience and Nanotechnology*, vol. 3, no. 3, pp. 235–240, 2003.

[2] P. Li, K. Ma, R. K. Thomas, and J. Penfold, "Analysis of the asymmetric synergy in the adsorption of zwitterionic–ionic surfactant mixtures at the air–water interface below and above the critical micelle concentration," *Journal of Physical Chemistry B*, vol. 120, no. 15, pp. 3677–369, 2016.

[3] Y. Moroi, *Micelles: Theoretical and Applied Aspects*, Springer Science+Business Media, New York, NY, USA, 1992.

[4] A. Pal and A. Pillania, "Thermodynamic and aggregation properties of aqueous dodecyltrimethylammonium bromide in the presence of hydrophilic ionic liquid 1,2-dimethyl-3-octylimidazolium chloride," *Journal of Molecular Liquids*, vol. 212, pp. 818–824, 2015.

[5] R. Wang, Y. Li, and Y. Li, "Interaction between cationic and anionic surfactants: detergency and foaming properties of mixed systems," *Journal of Surfactants and Detergents*, vol. 17, no. 5, pp. 881–888, 2014.

[6] K. Sharma and S. Chauhan, "Effect of biologically active amino acids on the surface activity and micellar properties of industrially important ionic surfactants," *Colloids and Surfaces A: Physicochemical and Engineering Aspects*, vol. 453, pp. 78–85, 2014.

[7] M. J. Rosen and J. T. Kunjappu, *Surfactants and Interfacial Phenomenon*, Wiley, Hoboken, NJ, USA, 2012.

[8] A. K. Tiwari and S. K. Saha, "Study on mixed micelles cationic gemini surfactants having hydroxyl groups in the spacers with conventional cationic surfactants: effects of spacer and hydrocarbon tail length," *Industrial and Engineering Chemistry Research*, vol. 52, no. 17, pp. 5895–5905, 2013.

[9] U. Masafumi, A. Hiroshi, K. Nana, Y. Takumi, T. Junzo, and K. Tsuyoshi, "Synthesis of high surface area hydroxyapatite nanoparticles by mixed surfactant-mediated approach," *Langmuir*, vol. 21, no. 10, pp. 4724–4728, 2005.

[10] N. E. Kadi, F. Martins, D. Clausse, and P. C. Schulz, "Critical micelle concentrations of aqueous hexadecytrimethylammonium bromide-sodium oleate mixtures," *Colloid and Polymer Science*, vol. 281, no. 4, pp. 353–362, 2003.

[11] B. Sohrabi, H. Gharibi, B. Tajik, S. Javadian, and M. Hashemianzadeh, "Molecular interactions of cationic and anionic surfactants in mixed monolayers and aggregates," *Journal of Physical Chemistry B*, vol. 112, no. 47, pp. 14869–14876, 2008.

[12] J. Mata, D. Varade, and P. Bahadur, "Aggregation behavior of quaternary salt based cationic surfactants," *Thermochemica Acta*, vol. 428, no. 1-2, pp. 147–155, 2005.

[13] T. F. Tadros, *Applied Surfactants: Principles and Applications*, Wiley-VCH, Weinheim, Germany, 2005.

[14] S. B. Sulthana, P. V. C. Rao, S. G. T. Bhat, T. Y. Nakano, G. Sugihara, and A. K. Rakshit, "Solution properties of nonionic surfactants and their mixtures: polyoxyethylene (10) alkyl ether CnE10 and MEGA-10," *Langmiur*, vol. 16, no. 3, pp. 980–987, 2000.

[15] H. Fauser, M. Uhlig, R. Miller, and R. von Klitzing, "Surface adsorption of oppositely charged SDS:C12TAB mixtures and the relation to foam film formation and stability," *Journal of Physical Chemistry B*, vol. 119, no. 40, pp. 12877–12886, 2015.

[16] H. Akba, A. Elimenli, and M. Boz, "Aggregation and thermodynamic properties of some cationic gemini surfactants," *Journal of Surfactants and Detergents*, vol. 15, no. 1, pp. 33–40, 2012.

[17] L. Arriaga, D. Varade, D. Carriere, W. Drenckhan, and D. Langevin, "Adsorption, organization, and rheology of catanionic layers at the air/water interface," *Langmuir*, vol. 29, no. 10, pp. 3214–3222, 2013.

[18] P. Norvaisas, V. Petrauskas, and D. Matulis, "Thermodynamics of cationic and anionic surfactant interaction," *Journal of Physical Chemistry B*, vol. 116, no. 7, pp. 2138–2144, 2012.

[19] O. Cudina, K. Karljikivic-Rajic, and I. Ruvarac-Bugarcic, "Interaction of hydrochlorothiazide with cationic surfactant micelles of cetyltrimethylammonium bromide," *Colloids and Surfaces A: Physicochemical and Engineering Aspects*, vol. 256, no. 2-3, pp. 225–232, 2005.

[20] A. Ali, S. Uzair, N. A. Malik, and M. Ali, "Study of interaction between cationic surfactants and cresol red dye by electrical conductivity and spectroscopic methods," *Journal of Molecular Liquids*, vol. 196, pp. 395–403, 2014.

[21] S. K. Mahta, S. Chaudhary, and K. K. Bhasin, "Self-assembly of cetylpyridinium chloride in water-DMF Binary mixtures. a spectroscopic and physicochemical approach," *Journal of Colloid Interfaces Science*, vol. 321, no. 2, pp. 426–433, 2008.

[22] A. Bhattarai, K. Pathak, and B. Dev, "Cationic and anionic surfactants interaction in pure water and methanol-water mixed solvent media," *Journal of Molecular Liquids*, vol. 229, pp. 153–160, 2017.

[23] K. Maiti, S. C. Bhattacharya, S. P. Moulik, and A. K. Panda, "Physicochemical studies on ion-pair amphiphiles: solution and interfacial behaviour of systems derived from sodium dodecylsulfate and n-alkyltrimethylammonium bromide homologues," *Journal of Chemical Sciences*, vol. 122, no. 6, pp. 867–879, 2010.

[24] A. Stocco, D. Carriere, M. Cottat, and D. Langevin, "Interfacial behavior of catanionic surfactants," *Langmuir*, vol. 26, no. 13, pp. 10663–10669, 2010.

[25] Z. G. Cui and J. P. Canselier, "Interfacial and aggregation properties of some anionic/cationic surfactant binary systems II. Mixed micelle formation and surface tension reduction effectiveness," *Colloid and Polymer Science*, vol. 279, no. 3, pp. 259–267, 2001.

[26] G. Kume, M. Gallotti, and G. Nunes, "Review on anionic/cationic surfactant mixtures," *Journal of Surfactants and Detergents*, vol. 11, no. 1, pp. 1–11, 2008.

[27] K. Sharma and S. Chauhan, "Apparent molar volume, compressibility and viscometric studies of sodium dodecyl benzene sulfonate (SDBS) and dodecyltrimethylammonium bromide (DTAB) in aqueous amino acid solutions: a thermoacoustic approach," *Thermochimica Acta*, vol. 578, pp. 15–27, 2014.

[28] S. Segota and D. Tezak, "Spontaneous formation of vesicles," *Advances in Colloid and Interface Science*, vol. 121, no. 1–3, pp. 51–75, 2006.

[29] A. Bahramian, R. K. Thomas, and J. Penfold, "The adsorption behavior of ionic surfactants and their mixtures with nonionic polymers and with polyelectrolytes of opposite charge at the air–water interface," *Journal of Physical Chemistry B*, vol. 118, no. 10, pp. 2769–2783, 2014.

[30] M. S. Bakshi and I. Kaur, "Benzylic and pyridinium head groups controlled surfactant-polymer aggregates of mixed cationic micelles and anionic polyelectrolytes," *Colloid and Polymer Science*, vol. 282, no. 5, pp. 476–485, 2004.

[31] M. S. Bakshi and S. Sachar, "Surfactant polymer interactions between strongly interacting cationic surfactants and anionic polyelectrolytes from conductivity and turbidity measurements," *Colloid and Polymer Science*, vol. 282, no. 9, pp. 993–999, 2004.

[32] A. K. Panda, F. Possmayer, N. O. Petersen, K. Nag, and S. P. Moulik, "Physico-chemical studies on mixed oppositely charged surfactants: their uses in the preparation of surfactant ion selective membrane and monolayer behavior at the air water interface," *Colloids and Surfaces A: Physicochemical and Engineering Aspects*, vol. 264, no. 1–3, pp. 106–113, 2005.

[33] H. Xu, P. X. Li, K. Ma, R. K. Thomas, J. Penfold, and J. R. Lu, "Limitations in the application of the Gibbs equation to anionic surfactants at the air–water surface: sodium dodecylsulfate and sodium dodecylmonooxyethylenesulfate above and below the CMC," *Langmuir*, vol. 29, no. 30, pp. 9324–9334, 2013.

[34] H. Guo, Z. Liu, S. Yang, and C. Sun, "The feasibility of enhanced soil washing of p-nitrochlorobenzene (pNCB) with SDBS/Tween80 mixed surfactants," *Journal of Hazardous Materials*, vol. 170, no. 2-3, pp. 1236–1241, 2009.

[35] P. M. Devinsky and F. I. Lacko, "Critical micelle concentration, ionization degree and micellisation energy of cationic dimeric (gemini) surfactants in aqueous solution and in mixed micelles with anionic surfactant," *Acta Facultatis Pharmaceuticae Universitatis Comenianae*, vol. 50, pp. 119–131, 2003.

[36] D. G. mez-Dıaz, J. M. Navaza, and B. Sanjurjo, "Density, kinematic viscosity, speed of sound, and surface tension of hexyl, octyl, and decyl trimethyl ammonium bromide aqueous solutions," *Journal of Chemical Engineering Data*, vol. 52, no. 3, pp. 889–891, 2007.

[37] T. P. Niraula, S. K. Chatterjee, and A. Bhattarai, "Micellization of sodium dodecyl sulphate in presence and absence of alkali metal halides at different temperatures in water and methanol-water mixtures," *Journal of Molecular Liquids*, vol. 250, pp. 287–294, 2018.

[38] J. X. Xiao and Y. X. Bao, "An unusual variation of surface tension with concentration of mixed cationic–anionic surfactants," *Chinese Journal of Chemistry*, vol. 19, no. 1, pp. 73–75, 2001.

[39] J. Rodriguez, E. Clavero, and D. Laria, "Computer simulations of catanionic surfactants adsorbed at air/water interfaces," *Journal of Physical Chemistry B*, vol. 109, no. 51, pp. 24427–24433, 2005.

[40] T. P. Niraula, S. K. Shah, S. K. Chatterjee, and A. Bhattarai, "Effect of methanol on the surface tension and viscosity of sodiumdodecyl sulfate (SDS) in aqueous medium at 298.15–323.15 K," *Karbala International Journal of Modern Science*, vol. 4, no. 1, pp. 26–34, 2017.

[41] A. Bhattarai, "Studies of the micellization of cationic–anionic surfactant systems in water and methanol–water mixed solvents," *Journal of Solution Chemistry*, vol. 44, no. 10, pp. 2090–2105, 2015.

[42] S. K. Shah, S. K. Chatterjee, and A. Bhattarai, "The effect of methanol on the micellar properties of dodecyltrimethylammonium bromide (DTAB) in aqueous medium at different temperatures," *Journal of Surfactants and Detergents*, vol. 19, no. 1, pp. 201–207, 2016.

[43] S. K. Shah, S. K. Chatterjee, and A. Bhattarai, "Micellization of cationic surfactants in alcohol—water mixed solvent media," *Journal of Molecular Liquids*, vol. 222, pp. 906–914, 2016.

[44] A. Bhattarai, S. K. Chatterjee, and T. P. Niraula, "Effects of concentration, temperature and solvent composition on density and apparent molar volume of the binary mixtures of cationic–anionic surfactants in methanol–water mixed solvent media," *SpringerPlus*, vol. 2, no. 1, p. 280, 2013.

[45] A. Bhattarai, A. K. Yadav, S. K. Sah, and A. Deo, "Influence of methanol and dimethyl sulfoxide and temperature on the micellization of cetylpyridinium chloride," *Journal of Molecular Liquids*, vol. 242, pp. 831–837, 2017.

[46] K. M. Sachin, S. Karpe, M. Singh, and A. Bhattarai, "Physicochemical properties of dodecyltrimethylammounium bromide (DTAB) and sodiumdodecyl sulphate (SDS) rich surfactants in aqueous medium, at T = 293.15, 298.15, and

303.15 K," *Macromolecular Symposia*, vol. 379, no. 1, article 1700034, 2018.

[47] R. K. Ameta, M. Singh, and R. K. Kale, "Comparative study of density, sound velocity and refractive index for (water + alkali metal) phosphates aqueous systems at T = (298.15, 303.15, and 308.15) K," *Journal of Chemical Thermodynamics*, vol. 60, pp. 159–168, 2013.

[48] R. G. Lebel and D. A. I. Goring, "Density, viscosity, refractive index, and hygroscopicity of mixtures of water and dimethyl sulfoxide," *Journal of Chemical Engineering Data*, vol. 7, no. 1, pp. 100-101, 1962.

[49] S. Ryshetti, A. Gupta, S. J. Tangeda, and R. L. Gardas, "Acoustic and volumetric properties of betaine hydrochloride drug in aqueous D (+)-glucose and sucrose solutions," *Journal of Chemical Thermodynamics*, vol. 77, pp. 123–130, 2014.

[50] A. Pal, H. Kumar, S. Sharma, R. Maan, and H. K. Sharma, "Characterization and adsorption studies of *Cocos nucifera* L. activated carbon for the removal of methylene blue from aqueous solutions," *Journal of Chemical Engineering Data*, vol. 55, no. 8, pp. 1424–1429, 2010.

[51] M. Singh, "Survismeter—Type I and II for surface tension, viscosity measurements of liquids for academic, and research and development studies," *Journal of Biochemistry Biophysics*, vol. 67, no. 2, pp. 151–161, 2006.

[52] B. Naseem, A. Jamal, and A. Jamal, "Influence of sodium acetate on the volumetric behavior of binary mixtures of DMSO and water at 298.15 to 313.15 K," *Journal of Molecular Liquids*, vol. 181, pp. 68–76, 2013.

[53] W. J. Cheong and P. W. Carr, "The surface tension of mixtures of methanol, acetonitrile, tetrahydrofuran, isopropanol, tertiary butanol and dimethyl-sulfoxide with water at 25°C," *Journal Liquid Chromatography*, vol. 10, no. 4, pp. 561–581, 1987.

[54] K. M. Sachin, A. Chandra, and M. Singh, "Nanodispersion of flavonoids in aqueous DMSO-BSA catalysed by cationic surfactants of variable alkyl chain at T = 298.15 to 308.15 K," *Journal of Molecular Liquids*, vol. 246, pp. 379–395, 2017.

[55] R. Singh, S. Chauhan, and K. Sharma, "Surface tension, viscosity, and refractive index of sodium dodecyl sulfate (SDS) in aqueous solution containing poly(ethylene glycol) (PEG), poly(vinyl pyrrolidone) (PVP), and their blends," *Journal of Chemical Engineering Data*, vol. 62, no. 7, pp. 1955–1964, 2017.

[56] J. George, S. M. Nair, and L. Sreejith, "Interactions of sodium dodecyl benzene sulfonate and sodium dodecyl sulfate with gelatin: a comparison," *Journal of Surfactants and Detergents*, vol. 11, no. 1, pp. 29–32, 2008.

[57] T. Chakraborty, I. Chakraborty, S. P. Moulik, and S. Ghosh, "Physicochemical and conformational studies on BSA-surfactant interaction in aqueous medium," *Langmuir*, vol. 25, no. 5, pp. 3062–3074, 2009.

[58] S. Chauhan, V. Sharma, and K. Sharma, "Maltodextrin-SDS interactions: volumetric, viscometric and surface tension study," *Fluid Phase Equilib*, vol. 354, pp. 236–244, 2013.

[59] S. S. Aswale, S. R. Aswale, and R. S. Hajare, "Adiabatic compressibility, intermolecular free length and acoustic relaxation time of aqueous antibiotic cefotaxime sodium," *Journal of Chemical and Pharmaceutical Research*, vol. 4, pp. 2671–2677, 2012.

[60] M. Gutie´rrez-Pichel, S. Barbosa, P. Taboada, and V. Mosquera, "Surface properties of some amphiphilic antidepressant drugs in different aqueous media," *Progress in Colloid and Polymer Science*, vol. 281, no. 6, pp. 575–579, 2003.

[61] E. Alami, G. Beinert, P. Marie, and R. Zana, "Alkanediyl-.alpha, omega. bis(dimethylalkylammonium bromide) surfactants behavior at the air-water interface," *Langmiur*, vol. 9, no. 6, pp. 1465–1467, 1993.

[62] T. Chakraborty, S. Ghosh, and S. P. Moulik, "Micellization and related behavior of binary and ternary surfactant mixtures in aqueous medium: cetylpyridiniumchloride(CPC), cetyltrimethyl ammonium bromide (CTAB), and polyoxyethylene (10) cetyl ether (Brij-56) derived system," *Journal of Physical Chemistry B*, vol. 109, no. 31, pp. 14813–14823, 2005.

[63] D. K. Chattoraj and K. S. Birdi,, "Adsorption and the Gibbs Surface Excess," Plenum Press, New York, NY, USA, 1984.

[64] R. K. Ameta, M. Singh, and R. K. Kale, "Synthesis and structure-activity relationship of benzylamine supported platinum (IV) complexes," *New Journal of Chemistry*, vol. 37, no. 5, pp. 1501–1508, 2013.

[65] A. L. Chavez and G. G. Birch, "The hydrostatic and hydrodynamic volumes of polyols in aqueous solutions and their sweet taste," *Chemical Senses*, vol. 22, no. 2, pp. 149–161, 1997.

[66] C. Bai and G. B. Yan, "Viscosity B-coefficients and activation parameters for viscous flow of a solution of heptanedioic acid in aqueous sucrose solution," *Carbohydrate Research*, vol. 338, no. 23, pp. 2921–2927, 2003.

8

Modification of Nafion® Membrane via a Sol-Gel Route for Vanadium Redox Flow Energy Storage Battery Applications

Shu-Ling Huang, Hsin-Fu Yu, and Yung-Sheng Lin

Department of Chemical Engineering, National United University, Miaoli 36003, Taiwan

Correspondence should be addressed to Shu-Ling Huang; simone@nuu.edu.tw and Yung-Sheng Lin; linys@nuu.edu.tw

Academic Editor: José M. G. Martinho

Nafion 117 (N-117)/SiO_2-SO_3H modified membranes were prepared using the 3-Mercaptopropyltrimethoxysilane (MPTMS) to react with H_2O_2 via in situ sol-gel route. Basic properties including water uptake, contact angle, ion exchange capacity (IEC), vanadium ion permeability, impedance, and conductivity were measured to investigate how they affect the charge-discharge characteristics of a cell. Furthermore, we also set a vanadium redox flow energy battery (VRFB) single cell by the unmodified/modified N-117 membranes as a separated membrane to test its charge/discharge performance and compare the relations among the impedance and efficiency. The results show that the appropriate amount of SiO_2-SO_3H led into the N-117 membrane contributive to the improvement of proton conductivity and vanadium ion selectivity. The permeability was effectively decreased from original 3.13×10^{-6} cm²/min for unmodified N-117 to 0.13×10^{-6} cm²/min for modified membrane. The IEC was raised from original 0.99 mmol/g to 1.24 mmol/g. The modified membrane showed a good cell performance in the VRFB charge/discharge experiment, and the maximum coulombic efficiency was up to 94%, and energy efficiency was 82%. In comparison with unmodified N-117, the energy efficiency of modified membrane had increased more than around 10%.

1. Introduction

Search of a higher efficiency, lower pollution, and greener alternative energy has become an important trend for rapid growth of nowadays global economy. Recently, scientists are actively involved in the exploitation of renewable, sustainable, and clean energy, such as wind turbine and photovoltaic, to produce clean and sustainable energy [1, 2]. However, power produced from those devices is fluctuating, and it is easily affected by the climate change. Consequently, electrical energy storage is needed to buffer the peak power on electrical grid. There are several available storage technologies, namely, hydropump, compressed air energy storage, and secondary batteries [3, 4]. Great accomplishment has been made to develop new types of redox flow storage battery (RFB) [5–7]. RFB is a promising energy storage technology due to its low cost and long cycle life, which could be up to 13,000 cycles.

Vanadium redox flow battery (VRFB) is one of the most promising technologies for mid-to-large scale (KW-MW)

energy storage, which was first put forward by Sum and Skyllas-Kazacos in 1985 [8]. High cycle life, low cost, reasonable efficiency, and safe operation of VRFB make it very attractive in many energy related applications, such as load leveling, peak shaving, and voltage stabilizing. VRFB is the most mature energy storage technology among others [6, 9, 10]. A constant supply of V^{2+}/V^{3+} ions and VO^{2+}/VO_2^+ ions, dissolved in sulfuric acid, is provided to the negative and positive electrodes, respectively, through two pumps connected to external storage tanks, as illustrated in Figure 1.

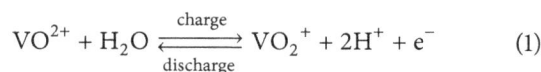

$$VO^{2+} + H_2O \underset{\text{discharge}}{\overset{\text{charge}}{\rightleftharpoons}} VO_2^+ + 2H^+ + e^- \qquad (1)$$

VRFB is a flow battery where electrolytes are circulating between electrolytic cells and storage tank. During charge-discharge cycle, reaction of (4) took place on the electrodes.

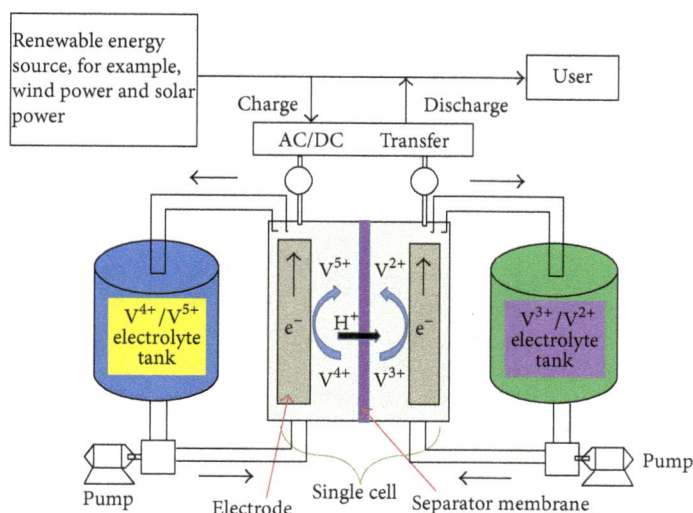

FIGURE 1: Schematic of a vanadium redox flow battery.

Positive half-cell reaction:

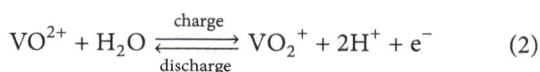

$$VO^{2+} + H_2O \underset{\text{discharge}}{\overset{\text{charge}}{\rightleftharpoons}} VO_2^+ + 2H^+ + e^- \quad (2)$$

Negative half-cell reaction:

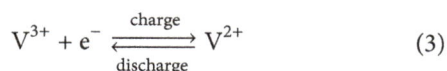

$$V^{3+} + e^- \underset{\text{discharge}}{\overset{\text{charge}}{\rightleftharpoons}} V^{2+} \quad (3)$$

Overall reaction:

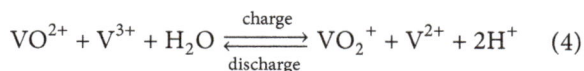

$$VO^{2+} + V^{3+} + H_2O \underset{\text{discharge}}{\overset{\text{charge}}{\rightleftharpoons}} VO_2^+ + V^{2+} + 2H^+ \quad (4)$$

The direction of above reactions is reversed, during charge-discharge cycle. Two electrodes are separated by a separation membrane. This membrane is electrical insulated and is highly ionic conductive. The H^+ ion is the major charge carrier inside the membrane. However, due to the electrical field and concentration gradient across the membrane, vanadium ions (VO^{2+}, VO_2^+, V^{+2}, V^{+3}) are also migrated/diffused through the membrane. The crossover of vanadium ions not only reduces the discharge cell voltage but also reduces its Faraday efficiency. During battery operation, the water is transported from one half-cell to the other half-cell by osmosis dragging and diffusion. The phenomena of water transfer cause dilution of one half-cell electrolytes and electrolyte concentration of the other half-cell. The water transport properties of Selemion CMV, AMV, and DMV (Asahi Glass, Japan) had been studied by Mohammadi et al. [11]. Many other composite membranes were also fabricated and tested, including cationic exchange membranes [12–14], anionic exchange membranes [15–17], and amphoteric ion exchange membranes [18, 19]. Nowadays commercial available cationic separate membranes of Nafion based membranes have been

developed by DuPont, USA. Although its film has high ionic conductivity, chemical stability, and thermal stability, it cannot overcome the penetration of vanadium ions, which caused the decrease of energy density of battery.

Separation membrane plays an important role in VRFB system. An ideal separated membrane should exhibit low vanadium ions diffusion (crossover) and high proton conductivity. Many modifications were made on Nafion membranes, either to decrease their vanadium permeability or to develop a new membrane with low cost and low vanadium permeability [20–22]. Some investigations [23, 24] have reported that the inorganic-organic hybrid route, sol-gel method, enables inorganic silica or TiO_2 particles to be led into the channel structure of Nafion membrane and then improve the cycle performance of the VRFB. In fact, it has been verified that both Nafion/SiO_2 and Nafion/TiO_2 hybrid membranes show nearly the same IEC and proton conductivity as that of pristine Nafion 117 (N-117) membrane. This could be a promising strategy to overcome the vanadium ions cross-mixed.

Nevertheless, Vijayakumar et al. [25] had proposed explanation with many spectroscopic analyses, showing that SiO_2 is still present in the structure of Nafion-SiO_2 composite membrane even after 30 cycles of charge/discharge operation, but its ion diffusivity was similar as that of pristine N-117 membrane. Note that under the highly acidic condition porous SiO_2 material condenses, shrinks, and forms agglomeration, which in turn decreases the amount of interactions area between SiO_2 and Nafion, causing unbinding V^{4+} ion transport via its usual pathway of reversible binding to the sulfonic acid groups of a side of Nafion channel walls.

Based on above deduction, we used the silica bonding sulfonic acid groups (SiO_2-SO_3H) with the sulfonated 3-mercaptopropyl trimethoxysilane (MPTMS) to react with an oxidized reagent H_2O_2 via in situ sol-gel route to modify Nafion membrane, expecting that the unbinding V^{4+} ion can be bound with either SiO_2 or -SO_3H group and avoid binding with the side -SO_3H of Nafion membrane. It could improve the ion crossover and raise the proton conductivity to obtain better cell performances for the VRFB charge/discharge experiment.

2. Experimental

2.1. Materials. The preparation of N-117/SiO_2-SO_3H hybrid membrane was carried out in our laboratory. 3-Mercaptopropyl trimethoxysilane (Acros Organics, USA), peroxide hydrogen (H_2O_2) (SHIMAKYU, Japan), vanadyl sulfate ($VOSO_4$) (Alfa Aesar, USA), trimethylamine solution 35% (SHIMAKYU, Japan), $MgSO_4 \cdot 7H_2O$ (SHOWA, Japan), H_2SO_4 (Scharlau, Australia), and ethanol, 99.5% (up) anhydrous (ECHO Chemical, Taiwan), were used without further purification. N-117 membranes were purchased from DuPont Inc., USA.

2.2. The Synthesis of Sulfonated MPTMS. MPTMS was mixed with EtOH, in MPTMS/EtOH volume ratio of 1 : 5, in a flask equipped with mechanical stirring, followed by adding H_2O_2, as an oxidizing agent at room temperature. Under vigorous stirring, a premixed MPTMS/EtOH solution was added into H_2O_2 (35 wt%) solution. The volume ratio of MPTMS/EtOH/H_2O_2 was ranged from 1 : 5 : 2 to 1 : 5 : 18. In order to understand the oxidative stability of MPTMS, the pH value of the reaction solution was recorded instantaneously under the different H_2O_2 adding amount. The thiol group (-SH) and alkoxysilane groups (-SiOCH_3) of MPTMS with an oxidizing agent (H_2O_2) were oxidized and hydrolyzed to form a sulfonated MPTMS within the Si-OH and SO_3H groups.

2.3. Preparation of N-117/SiO_2-SO_3H Modified Membranes. The N-117 membranes were pretreated by heating them in a 3% H_2O_2 solution at 80°C for one hour, followed by washing with deionized H_2O for 30 minutes at 80°C, then immersed in 1 M H_2SO_4 solution for 1 hour at 80°C, and lastly rinsed repeatedly in deionized H_2O to ensure that all membranes were fully protonated before being chemically modified [23, 24]. The pretreated N-117 film was dried for 3 hours at 110°C and then soaked in a sulfonated MPTMS solution (MPTMS/EtOH/H_2O_2 = 1 mL : 5 mL : 10 mL) in a two-neck reaction vessel with a mechanical stirrer. The N-117/SiO_2-SO_3H membrane was fabricated, and the hydrolysis/polycondensation reaction was allowed to proceed for 0.5 hours (NM-0.5H), 1 hour (NM-1H), and 24 hours (NM-24H), respectively. The formation of SiO_2-SO_3H via the sol-gel reaction was embedded to inside channel network of the Nafion membrane.

The structural characteristics of membranes were analyzed using a Fourier transform infrared spectrometer FT-IR (U-2001, HITACHI, Japan) in transmission mode, wavelength ranging from 400 to 4000 cm^{-1}, with a 4 cm^{-1} resolution. Thermophysical properties of N-117 and N-117/SiO_2-SO_3H hybrid membranes were performed by a differential

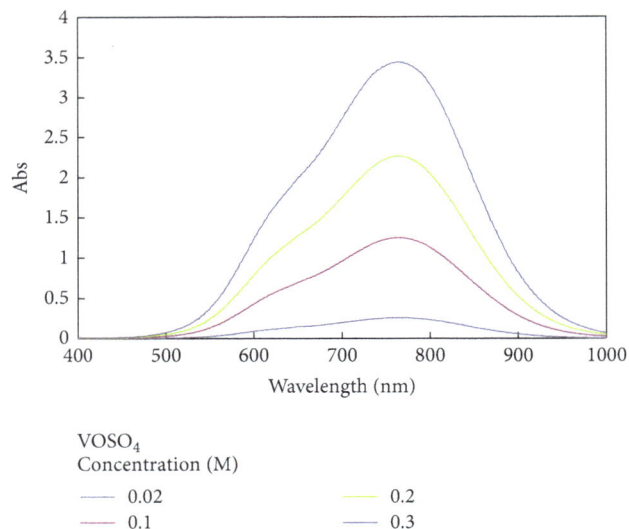

FIGURE 2: Absorption spectra of $VOSO_4$ solution for four different concentrations at λ_{max} = 766 nm.

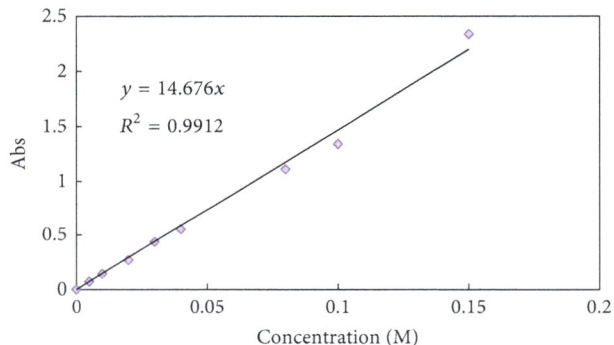

FIGURE 3: Calibration curve of $VOSO_4$ solution with the concentration range between 0.005 M and 0.15 M.

scanning calorimeter, DSC-Q10 (TA Instruments, USA), in a nitrogen atmosphere, at a heating rate of 10°C/min.

2.4. Membrane Properties

2.4.1. Vanadium Ion Permeability. The rate change of the vanadium ion in $VOSO_4$ solution absorbance is used to calculate the diffusion coefficient, that is, permeability, according to Fick's First Law and Beer's Law [26]. The diffusion cell has two compartments, compartment A and compartment B. The former was filled with 50 mL of 2.0 M $VOSO_4$/2.0 M H_2SO_4 solution, and the latter was filled with 50 mL of 2.0 M $MgSO_4$/2.0 M H_2SO_4 solution.

Figure 2 is the spectra of $VOSO_4$ solution at four different concentrations, and the maximum absorption peak for VO^{2+} ions is located at 766 nm. A calibration curve based on the results of Figure 2 was carried out in the concentration range from 0.005 to 0.15 M $VOSO_4$ solution, at wavelength λ_{max} = 766 nm, as shown in Figure 3.

2.4.2. Water Uptake and Contact Angle.

The water uptake is defined as the ratio of the weight of absorbed water to that of the dry membrane. The water uptake was calculated by (5).

$$\text{Water uptake} = \frac{W_W - W_d}{W_d} \times 100\%, \qquad (5)$$

where W_W is the weight of the wetted membrane after the membrane is immersed in H_2O for 24 hours and W_d is the weight of the dry membrane. The contact angle between the water and the membranes was directly measured by a contact angle measuring instrument (FTA-125, APPR, Germany) for evaluation of their hydrophilic/hydrophobic properties.

2.4.3. Ion Exchange Capacity.

Ion exchange capacity was measured by the typical acid-base titration (inverse-titrated method). Membranes in acidic form were first immersed in excessive 0.1 M NaOH solution for 24 hours to exchange the fixed H^+ ions by Na^+ ions. The unreacted NaOH solution with membrane was inverse-titrated by 0.1 M HCl solution, and the IEC could be calculated by (6):

$$\text{IEC} = \frac{N_{\text{NaOH}}V_{\text{NaOH}} - N_{\text{HCl}}V_{\text{HCl}}}{W_{\text{sample}}}, \qquad (6)$$

where $N_{\text{NaOH}} \times V_{\text{NaOH}}$ is the total mmoles of NaOH and $N_{\text{HCl}} \times V_{\text{HCl}}$ is consumed moles by HCl solution inverse-titrated.

2.4.4. Resistance and Ionic Conductivity.

The membrane ionic conductivity was measured with a single cell. The membrane was sandwiched between two composite graphite plates. Flow channels on the carbon plate were filled with 2.0 M $VOSO_4$/2.0 M H_2SO_4 solution. The electrochemical impedance spectroscopy (EIS) of this single cell with different membranes was measured with a Frequency Response Detector & Potentiostats system (Princeton Applied Research, FRD100&VersaSTAT™, USA). The sinusoidal excitation voltage applied to the cells was 10 mV with a frequency range between 0.1 Hz and 100 kHz. Proton conductivity (κ) and area resistance were calculated by (7) and (8).

$$\kappa = \frac{L}{R \times A}, \qquad (7)$$

$$R_A = (R_1 - R_2) \times A, \qquad (8)$$

where A is area, L is thickness, R is electric resistance, κ is conductivity, and R_1 and R_2 are the electric resistance of the cell with and without a membrane, respectively.

2.5. Single Cell Performance of VRFB.

An H-type single cell for charge-discharge experiments was designed by Green Energy and Environmental Lab. of Industrial Technology Research Institute, Taiwan. The single cell of VRFB is consisted of two pieces of carbon paper (Shenhe Carbon Fiber Materials Co., Ltd.) and two current collectors. The VRFB for charge-discharge test was performed by sandwiching the membrane between two pieces of graphite carbon paper electrodes. A 2.0 M $VOSO_4$/2.0 M H_2SO_4 solution was employed

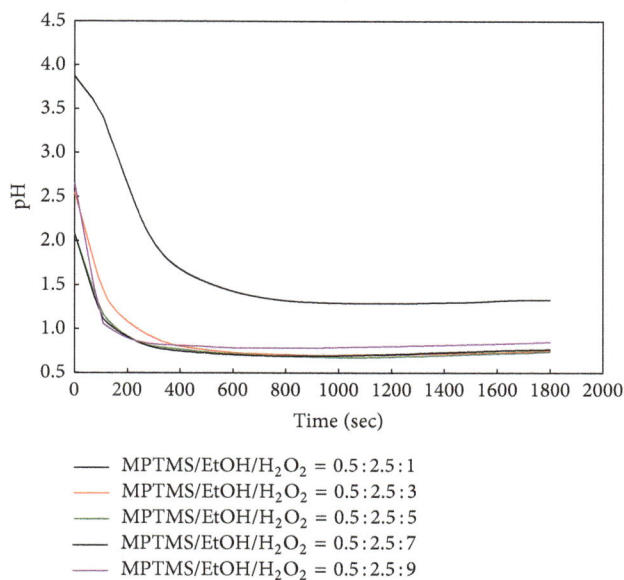

FIGURE 4: The reaction time dependence of pH value for the sulfonated MPTMS solution.

as negative and positive electrolytes. The effective area of electrode was 5×5 cm^2 and the volume of the electrolytes solution in each half-cell was 30 mL and cyclically pumped into the corresponding half-cell. The charge and discharge test was carried out using a Battery Cycler Test System (WBCS3000S, WonATech, Korea) and CT2001C-10 V/2 A (Wuhan Land Co., China) between 0.8 and 1.8 V at a current density of 20 mA cm^{-2}.

3. Results

3.1. Preparation of N-117/SiO$_2$-SO$_3$H Hybrid Membranes

3.1.1. The Syntheses of Sulfonated MPTMS and N-117/SiO$_2$-SO$_3$H Hybrid Membranes.

MPTMS was oxidized by an oxidizing agent, H_2O_2 in EtOH solution to form a strong sulfuric acid (-SO$_3$H) group. The pH value of the sulfonated solution was reduced with increase of reaction time. Figure 4 shows that the pH of the solution decreases strongly for all H_2O_2 contents in the first 400 s. The more complete the sulfonated reaction of MPTMS, the lower the pH value it reaches. This is due to the formation of sulfonic acid groups. The solution has the lowest pH value at a MPTMS/EtOH/H_2O_2 volume ratio of 0.5 : 2.5 : 5.

The N-117/SiO$_2$-SO$_3$H hybrid membranes were prepared using the sulfonated MPTMS to react with an oxidized reagent H_2O_2 via sol-gel method. The content of SiO$_2$-SO$_3$H increases with the increasing of sol-gel reaction time, as shown in Table 1. The SiO$_2$-SO$_3$H content of samples NM-0.5H, NM-1H, and NM-24H was 1.51 wt%, 1.91 wt%, and 1.99 wt%, respectively.

3.1.2. FT-IR Spectra of Membranes.

The structural comparison of N-117/SiO$_2$-SO$_3$H and pretreated N-117 membranes

TABLE 1: Comparison of basic properties of N-117 and N-117/SiO$_2$-SO$_3$H membranes.

Samples	SiO$_2$-SO$_3$H (wt%)	Thickness (μm)	Contact angle Angle ($°$)	Water uptake (wt%)	IEC (mmol/g)	Permeability ($\times 10^{-6}$ cm^2/min)
N-117	0.00	195 ± 5	86 ± 2	21	0.99	3.13
NM-0.5H	1.51	195 ± 5	86 ± 2	17	1.23	0.20
NM-1H	1.91	195 ± 5	86 ± 2	17	1.24	0.13
NM-24H	1.99	195 ± 5	86 ± 2	18	1.24	0.12

The reaction times for NM-0.5H, NM-1H, and NM-24H were 0.5 h, 1 h, and 24 h, respectively.

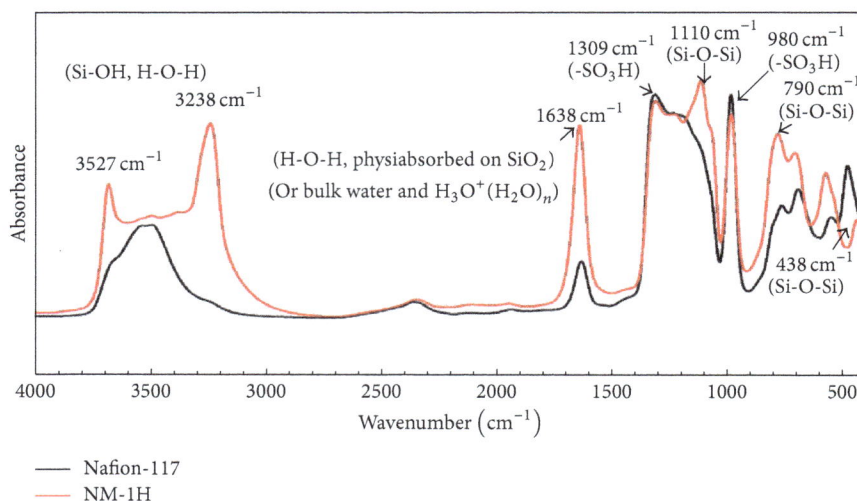

FIGURE 5: FT-IR spectra and for the pretreated N-117 and NM-1H modified membranes.

was confirmed by FT-IR spectra, as shown in Figure 5. The spectrum of the N-117 membrane has been reported in previous research [27]. In comparing spectrum of N-117 with NM-1H, a new peak at 1110 cm^{-1} can be observed, which is attributed to the vibration of Si-O-Si groups [24, 28, 29]. However, there are not any new peaks of N-117 membrane found in that region. In addition, Figure 5 (red line) shows there is one specific peak appearing at 3238 cm^{-1}. This peak is corresponding to the O-H stretching vibration of SO$_3$H groups [28, 29].

Furthermore, for both spectrums of N-117 and NM-1H, the band at 1309 cm^{-1} is from the antisymmetric CF$_3$ stretch, which appears overlapped with the antisymmetric and symmetric CF$_2$ stretching modes, around 1230 and 1150 cm^{-1}, and S=O antisymmetric and symmetric stretching bands from SO$_3$H groups would appear at 1435 and 1320 cm^{-1}, respectively [24]. There are no peaks found in 2500~2600 cm^{-1} (stretching vibration of -SH group) for the NM-1H, which indicates the -SH groups of MPTS have been oxidized to sulfonic acid groups [27]. Moreover, there is a stronger absorption peak at 1638 cm^{-1} for NM-1H membrane, showing that some physical-absorbed water may be present in the NM-1H membrane. That is associated with SiO$_2$ particles and either bulk water, (H$_2$O)$_n$, or highly hydrated oxonium ions, H$_3$O$^+$(H$_2$O)$_n$ [25, 27].

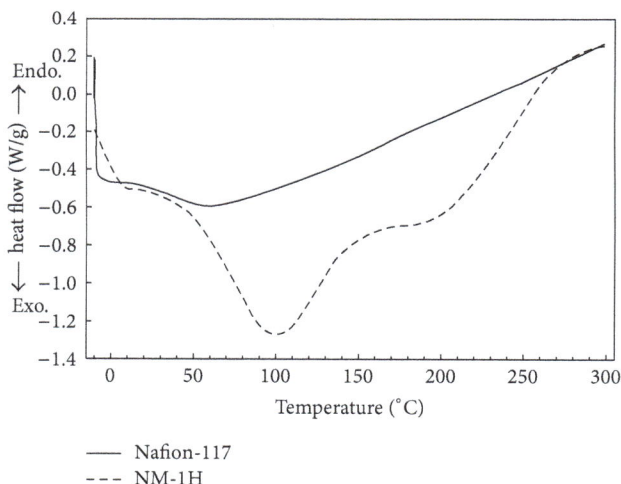

FIGURE 6: DSC thermogram of N-117 and NM-1H modified membranes, in a nitrogen atmosphere at a heating rate of 10°C/min.

3.1.3. DSC. The DSC thermograms (Figure 6) also imply the formation of the Si-O-Si groups. The exothermic peaks appear at 100°C and around 178°C corresponding to the release of H$_2$O molecules and formation of the Si-O-Si

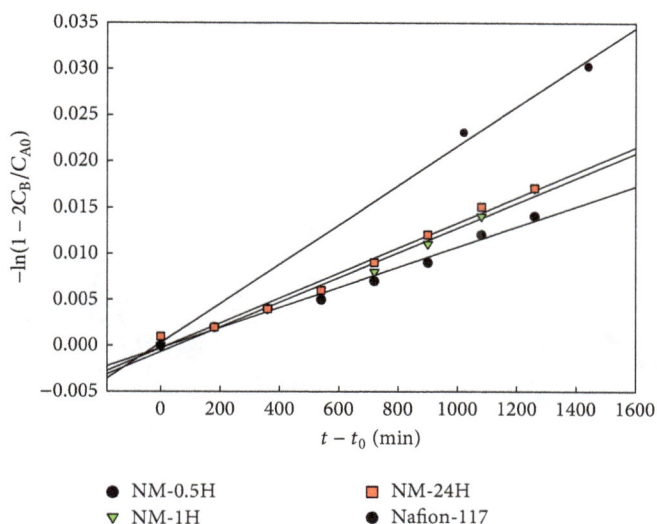

FIGURE 7: The vanadium concentration is plotted as a function of time according to (9).

groups which come from the condensation of Si-OH [30] and formation of crystalline SiO_2 particles and release of H_2O for sample NM-1H. The unmodified N-117 has not shown any peaks in these regions.

3.2. Basic Properties of Membranes

3.2.1. Vanadium Ion (V^{4+}) Permeability.

The membrane was placed between two compartments, compartment A and compartment B, and vanadium concentration in compartment B (C_B) was measured after experimental time t from the calibration curve in Figure 3.

According to the previous literature [31], V^{4+} permeability can be obtained by (9). t_0 is the initial time of experiment, and t_0 is equal to zero in this case. C_{A0} is the initial vanadium concentration and P is permeability. A and L are the membrane area and thickness, respectively. V_B is the solution volume in compartment B.

$$-\ln\left(1 - \frac{2C_B(t)}{C_{A0}}\right) = \frac{PA}{V_B L}(t - t_0). \tag{9}$$

The logarithmic function of vanadium concentration is calculated according to (9) and is plotted as a function of time in Figure 7. As shown in Figure 7, a linear behavior of $-\ln[1 - (2C_B/C_{A0})]$ versus $(t - t_0)$ was obtained. The permeability was calculated from the slope of Figure 7, and detailed data were summarized on Table 1. As exhibited in Figure 7 and Table 1, the permeability of membranes N-117/SiO_2-SO_3H, NM-0.5H, NM-1H, and NM-24H is 3.13×10^{-6} cm²/min, 0.2×10^{-6} cm²/min, 0.13×10^{-6} cm²/min, and 0.12×10^{-6} cm²/min, respectively.

3.2.2. Water Uptake, Contact Angle, IEC, and Other Properties of Membranes.

Table 1 lists basic properties of four membranes, N-117, NM-0.5H, NM-1H, and NM-24H, including water uptake, contact angle, IEC, and permeability. Water uptake is one of basic factors of ion exchange membrane.

An optimal amount of water uptake enables the membrane to achieve good ion conductivity. The membrane contains excessive water uptake which results in the vanadium ion cross-over mixing and even reduces its mechanical properties. The water uptakes of the modified N-117 membranes (NM-0.5H, NM-1H, and NM-24H) are in the range of 17 to 18 wt%, which are lower than that of N-117 which is 21 wt%.

Contact angles were measured to evaluate the changes in the hydrophilic/hydrophobic properties. In general case, Nafion series membranes are water-swellable. When water droplets come into contact with membrane, the membrane was swollen by water, and a projection phenomenon could be observed on the surface of membrane as exhibited in Figure 8. The contact angles of the N-117/SiO_2-SO_3H membranes were in the range of $82 \pm 2°$ to $86 \pm 2°$, which are slightly lower than that of N-117($88° \pm 2$). These hybrid membranes were slightly hydrophilic.

IEC is defined as the ability of H^+ proton to exchange between positive and negative electrolytes. The separate membrane with a high IEC will exhibit great proton conductivity for VRFB system. As listed in Table 1, the IEC for an unmodified N-117 and three modified membranes, NM-0.5H, NM-1H, and NM-24H, are 0.99 mmol/g, 1.23 mmol/g, 1.24 mmol/g, and 1.24 mmol/g, respectively. The ion exchange capacity increases with the reaction time and SiO_2-SO_3H from NM-0.5H to NM-1H. There is no further increase when the reaction time is extended to 24 hours. It can be found that the modified membranes have improved its IEC by using the sulfonated MPTMS.

3.3. Electrochemical Impedance Spectroscopy.

The membrane conductivity was measured with a single cell. A cationic exchange membrane (N-117) was used as both separator and reference for comparison with the modified membranes. The resistance, area resistance, and conductivity of membranes can be estimated from the electrochemical impedance spectroscopy data. The Nyquist plots and Equivalent circuit are

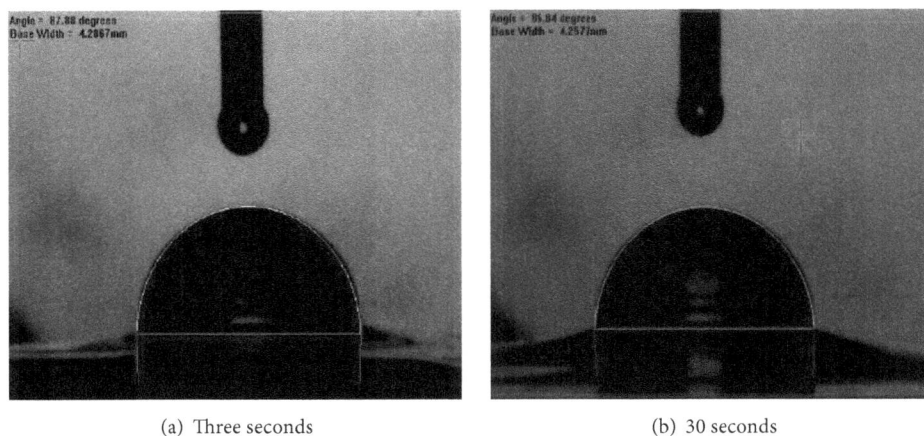

(a) Three seconds

(b) 30 seconds

FIGURE 8: Water drops into the NM-1H membrane; after 30 secs, the membrane was swollen and caused a projection phenomenon.

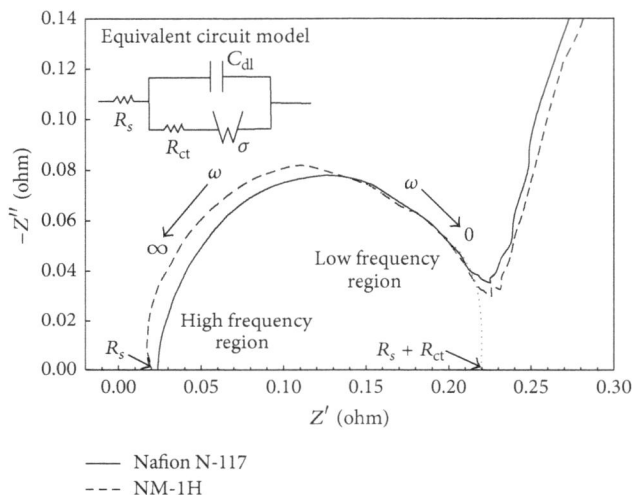

FIGURE 9: Nyquist plots of the impedance data for N-117 and NM-1H membranes.

TABLE 2: Comparison of resistance and conductivity for N-117 as well as modified membranes.

Samples	R_S (mΩ)	Area resistance (R_A) ($\Omega\times$cm^2)	Conductivity (mS/cm)
Nafion N-117	23.70	0.59	33.05
NM-1H	22.32	0.56	34.82
NM-24H	36.40	0.91	21.43

$2 \times C_{dl}\sigma^2 - R_s - R_{ct}$, where C_{dl} is electric double layer capacity and σ is Warburg constant. The resistance (R_s and R_A) and conductivity of N-117 and modified membranes were summarized in Table 2.

The R_s value of membranes is in the order as NM-24H (36.40 mΩ), N-117 (23.70 mΩ), NM-1H (22.32 mΩ). The NM-1H modified membrane has a slightly lower area resistance ($R_A = 0.56$ Ωcm^2) than that of N-117 ($R_A = 0.59$ Ωcm^2). In general, a membrane with high area resistance exhibits low conductivity.

3.4. Single Cell Performance for VRFB System. A single cell for VRFB was charged to 1.8 V and discharged to 0.8 V at a constant current density of 20 mA/cm^2. The columbic efficiency (CE), voltage efficiency (VE), and energy efficiency (EE) are defined according to (10):

$$CE = \frac{C_{discharge}}{C_{charge}} \times 100\%,$$

$$VE = \frac{V_{discharge}}{V_{charge}} \times 100\%, \tag{10}$$

$$EE = CE \times VE,$$

where C_{charge} and $C_{discharge}$ are charge/discharge capacity and V_{charge} and $V_{discharge}$ are the middle point voltage of charge/discharge, respectively. The higher CE, meaning lower capacity loss, is mainly due to the lower crossover diffusion rate for vanadium ions and the side reactions. Higher VE

often used to model the electrochemical system. Figure 9 is the Nyquist plots of the impedance data with the simulating Equivalent circuit model for N-117 and NM-1H sample.

The semicircle in high frequency region is concerned with the charge transfer process. The membrane resistance R_s representing the resistance of electrolyte solution can be obtained from the intercept of the impedance curve on the x-axis at frequency close to infinity, that is, at $Z'' = 0$. The charge transfer resistance $R_s + R_{ct}$ is related to across electrode/solution interface charge transfer process. The semicircle R_{ct} is the charge transfer resistance of the electrode, as similar to R_s, and can be obtained from the semicircle diameter at frequency close to zero and at $Z'' = 0$ [32, 33].

The semicircle in low frequency region represents the mass diffusion control process. This resistance is called Warburg resistance when the electrochemical system achieves in mass transfer control process. The Warburg resistance equals

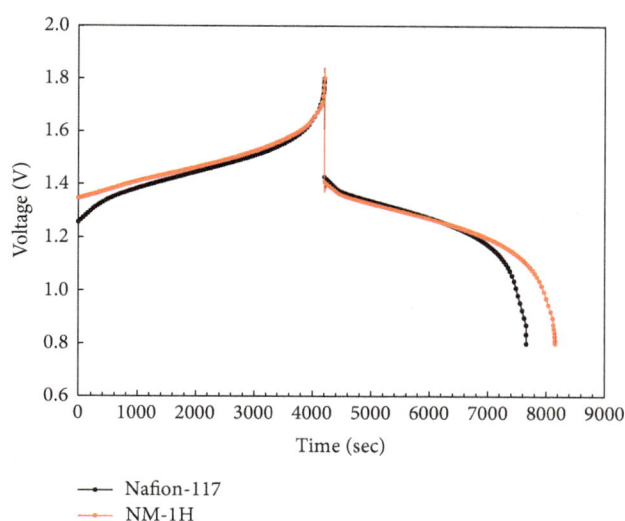

FIGURE 10: Charge-discharge curves of single cell performance for VRFB by N-117 and NM-1H membrane, respectively, as a separator; a current density is 20 mA cm^{-2}.

TABLE 3: The single cell efficiency for VRFB system based on the pristine N-117 membrane and modified NM-1H membrane.

Samples	Coulombic efficiency (%)	Voltage efficiency (%)	Energy efficiency (%)
N-117	82	87	71
NM-1H	94	87	82

indicates lower battery resistances, which are referred to the electrodes, separator, flow path, and so on.

The charge-discharge experiments were carried out by a single cell. Figure 10 displays the charge-discharge curves of the VRFB with a 2.0 M vanadium electrolyte and using N-117 and NM-1H modified membrane as a separator, respectively. Comparing the charge-discharge curves of VRFB between N-117 and NM-1H modified membranes, we can find that the discharge capacity of the VRFB with NM-1H modified membrane is higher than N-117 membrane. The relationships among the CE, VE, and EE are listed in Table 3. The CE and EE of the VRFB with NM-1H membrane are 94% and 82%, and the CE and EE of the VRFB for N-117 membrane are 82% and 71%, respectively. The VE of the VRFB is close to around 87% for both N-117 and NM-1H membrane. The VRFB with NM-1H modified membrane displays higher CE and EE than that of N-117 membrane.

4. Discussion

The investigation of the pH change for sulfonated MPTMS solution and the FTIR spectra in Figures 4 and 5 shows that the present condition of synthesis, at a MPTMS/EtOH/H$_2$O$_2$ volume ratio of 1 : 5 : 10, enables the MPTMS to be oxidized completely.

By full hydrolysis/polycondensation and in situ sol-gel route, the inorganic SiO$_2$-SO$_3$H groups are introduced into the channel network structure of N-117 membrane. The deposition of SiO$_2$ particle on the N-117 membrane can be confirmed by the presence of a new absorption band and a specific band, at 1110 cm^{-1} and 1638 cm^{-1} of the FTIR spectra, which are referred to Si-O-Si groups and physiabsorbed water associated with SiO$_2$ particles.

Furthermore, the stretching vibration absorption peak of -SH group in the MPTST disappeared in the range of 2500~2600 cm^{-1}, and the specific surfonic acid bands in the range 900~1000 cm^{-1} and 1300~1400 cm^{-1} can be observed for the NM-1H modified membrane. It can be corroborated that the -SH groups of MPTS have been oxidized to the sulfonic acid groups. Summarizing these FT-IR spectra and DSC data, it can be established that the N-117/SiO$_2$-SO$_3$H modified membranes could be successfully synthesized via the sol-gel reaction.

By combining the results from the basic properties of membranes in Table 1, the contact angle of NM-1H membrane, 86° ± 2, is similar to that of N-117, 88° ± 2. This can be explained as follows: the strong hydrophilicity of sulfonated group can reduce the contact angle, but the hydrophobicity of silica group will cause the contact angles to increase, under the cross effect between the hydrophilic and hydrophobic groups; the contact angles of hybrid membranes are similar to that of N-117. The water uptake of the hybrid membranes is based on above reasons, too.

In addition, Table 1 shows that the permeability of modified membrane reduces obviously with the reaction time, and the SiO$_2$-SO$_3$H amounts are increased with reaction time. However, the formation of SiO$_2$-SO$_3$H via the sol-gel route grafting on the channel network of the Nafion membrane reaches equilibrium after reacting around one hour; hence the permeability of NM-1H sample, 0.13 × 10^{-6} cm^2/min, is similar to that of NM-24H sample, 0.12 × 10^{-6} cm^2/min. It is clear that the permeability was effectively decreased for modified membrane, because SiO$_2$-SO$_3$H particles result in the polar clusters inside the Nafion membrane and hinder the vanadium ion diffusion. In fact, these results verified that both Nafion/SiO$_2$ and Nafion/TiO$_2$ hybrid membrane show nearly the same IEC and proton conductivity as that of pristine N-117 membrane. The results of this research suggest that this approach is a promising strategy to overcome the vanadium ions cross-mixed problem.

From Table 1, it can be proved that the IEC of N-117/SiO$_2$-SO$_3$H modified membrane was raised more, and the permeability was reduced stronger than those of other references [23–25]. The outcomes can be attributed to the introduction of silica and surfonic acid groups into the pore cluster of N-117.

In the same case, resistance and conductivity data of N-117/SiO$_2$-SO$_3$H modified membrane for the EIS test of VRFB shown in Table 2 indicate the proton conductivity of NM-1H modified membrane is higher than that of N-117, signifying that the increasing proton conductivity is referred to the SiO$_2$-SO$_3$H immobilized in the network channel of N-117. At the same time, the NH-24H modified membrane has the lowest proton conductivity. As shown in Table 1, the IEC of NH-24H membrane is the same as that of NM-1H membrane, implying that the equivalent of sulfonic acid for both NH-1H and NH-24H is equal. However, the SiO_2-SO_3H amount of NH-24H membrane, 1.99 wt%, is larger than that of NM-1H, 1.91 wt%, inferring that the SiO_2 content of NH-24H is more than that of NM-1H, but its -SO_3H groups are similar to that of NM-1H. It can be deduced that the higher resistance for NH-24H sample was ascribed to the SiO_2 particles.

Furthermore, as shown in Figure 10 and Table 3, test data of a vanadium redox single cell performance, it can be seen that the columbic efficiency of the single cell using NM-1H as a separator was increased from 82% to 94%; fortunately, the voltage efficiency was not reduced, holding at 87%. Related references [23–25] revealed that the voltage efficiency of modified membrane was reduced because of the higher area resistance of SiO_2 or TiO_2 inorganic particles.

The EE of the NM-1H modified membrane is higher than that of unmodified N-117. It could be attributed to the -SO_3H groups of NM-1H membrane having improved the proton conductivity and the SiO_2-SO_3H groups having restrained vanadium ion cross-mixed, especially shown in the increase of CE.

5. Conclusions

The present condition of synthesis, at a MPTMS/EtOH/H_2O_2 volume ratio of 1 : 5 : 10, enables the MPTMS to be oxidized completely to form a sulfonated SiO_2-SO_3H functional group. The N-117/SiO_2-SO_3H modified membranes could be synthesized using an oxidation reaction and a simple sol-gel route. It can be confirmed by FT-IR spectra and DSC that the SiO_2-SO_3H groups can be successfully grafted on the N-117 network structure to form modified membranes. The permeability was effectively decreased and the ion exchange capacity was raised obviously for modified membrane. It is mainly attributed to its SiO_2-SO_3H particles which hinder the diffusion of vanadium ion and then improve the transfer of H^+ proton.

In the EIS test of VRFB, the improvement of proton conductivity can be attributed to the introduction of SiO_2-SO_3H into the Nafion channel network; however, excessive SiO_2 in the modified membrane would cause the resistance to increase. The NM-1H modified membrane has a lower R_s, lower area resistance, and higher proton conductivity than those of unmodified N-117 membrane. Hence, the NM-1H sample exhibits a superior cell performance than the pristine N-117 in the VRFB charge/discharge experiment. The maximum coulombic efficiency is up to 94%, and energy efficiency is 82% for the NM-1H modified membrane.

Competing Interests

The authors declare that there is no conflict of interests regarding the publication of this paper.

Acknowledgments

This work was supported by the Ministry of Science and Technology, Taiwan, under Grant no. MOST 105-2221-E-239-031.

References

[1] H. Lund and B. V. Mathiesen, "Energy system analysis of 100% renewable energy systems—the case of Denmark in years 2030 and 2050," *Energy*, vol. 34, no. 5, pp. 524–531, 2009.

[2] S. Vazquez, S. M. Lukic, E. Galvan, L. G. Franquelo, and J. M. Carrasco, "Energy storage systems for transport and grid applications," *IEEE Transactions on Industrial Electronics*, vol. 57, no. 12, pp. 3881–3895, 2010.

[3] H. Lund and G. Salgi, "The role of compressed air energy storage (CAES) in future sustainable energy systems," *Energy Conversion and Management*, vol. 50, no. 5, pp. 1172–1179, 2009.

[4] Á. J. Duque, E. D. Castronuovo, I. Sánchez, and J. Usaola, "Optimal operation of a pumped-storage hydro plant that compensates the imbalances of a wind power producer," *Electric Power Systems Research*, vol. 81, no. 9, pp. 1767–1777, 2011.

[5] S. Ha and K. G. Gallagher, "Estimating the system price of redox flow batteries for grid storage," *Journal of Power Sources*, vol. 296, pp. 122–132, 2015.

[6] A. Lucas and S. Chondrogiannis, "Smart grid energy storage controller for frequency regulation and peak shaving, using a vanadium redox flow battery," *International Journal of Electrical Power & Energy Systems*, vol. 80, pp. 26–36, 2016.

[7] Y. Zeng, T. Zhao, X. Zhou, L. Wei, and H. Jiang, "A low-cost iron-cadmium redox flow battery for large-scale energy storage," *Journal of Power Sources*, vol. 330, pp. 55–60, 2016.

[8] E. Sum and M. Skyllas-Kazacos, "A study of the V(II)/V(III) redox couple for redox flow cell applications," *Journal of Power Sources*, vol. 15, no. 2-3, pp. 179–190, 1985.

[9] C. Fabjan, J. Garche, B. Harrer et al., "The vanadium redox-battery: an efficient storage unit for photovoltaic systems," *Electrochimica Acta*, vol. 47, no. 5, pp. 825–831, 2001.

[10] C. J. Rydh and B. A. Sandén, "Energy analysis of batteries in photovoltaic systems. Part I: performance and energy requirements," *Energy Conversion and Management*, vol. 46, no. 11-12, pp. 1957–1979, 2005.

[11] T. Mohammadi, S. C. Chieng, and M. S. Kazacos, "Water transport study across commercial ion exchange membranes in the vanadium redox flow battery," *Journal of Membrane Science*, vol. 133, no. 2, pp. 151–159, 1997.

[12] T. Mohammadi and M. Skyllas-Kazacos, "Preparation of sulfonated composite membrane for vanadium redox flow battery applications," *Journal of Membrane Science*, vol. 107, no. 1-2, pp. 35–45, 1995.

[13] W. Wei, H. Zhang, X. Li, Z. Mai, and H. Zhang, "Poly(tetrafluoroethylene) reinforced sulfonated poly(ether ether ketone) membranes for vanadium redox flow battery application," *Journal of Power Sources*, vol. 208, pp. 421–425, 2012.

[14] J. G. Austing, C. N. Kirchner, L. Komsiyska, and G. Wittstock, "Layer-by-layer modification of Nafion membranes for increased life-time and efficiency of vanadium/air redox flow batteries," *Journal of Membrane Science*, vol. 510, pp. 259–269, 2016.

[15] T. Mohammadi and M. Skyllas Kazacos, "Modification of anion-exchange membranes for vanadium redox flow battery

applications," *Journal of Power Sources*, vol. 63, no. 2, pp. 179–186, 1996.

[16] B. Zhang, S. Zhang, D. Xing, R. Han, C. Yin, and X. Jian, "Quaternized poly(phthalazinone ether ketone ketone) anion exchange membrane with low permeability of vanadium ions for vanadium redox flow battery application," *Journal of Power Sources*, vol. 217, pp. 296–302, 2012.

[17] L. Zeng, T. Zhao, L. Wei, Y. Zeng, and Z. Zhang, "Highly stable pyridinium-functionalized cross-linked anion exchange membranes for all vanadium redox flow batteries," *Journal of Power Sources*, vol. 331, pp. 452–461, 2016.

[18] G. Hu, Y. Wang, J. Ma et al., "A novel amphoteric ion exchange membrane synthesized by radiation-induced grafting α-methylstyrene and N,N-dimethylaminoethyl methacrylate for vanadium redox flow battery application," *Journal of Membrane Science*, vol. 407-408, pp. 184–192, 2012.

[19] J. B. Liao, M. Z. Lu, Y. Q. Chu, and J. L. Wang, "Ultra-low vanadium ion diffusion amphoteric ion-exchange membranes for all-vanadium redox flow batteries," *Journal of Power Sources*, vol. 282, pp. 241–247, 2015.

[20] Q. Luo, H. Zhang, J. Chen, D. You, C. Sun, and Y. Zhang, "Preparation and characterization of Nafion/SPEEK layered composite membrane and its application in vanadium redox flow battery," *Journal of Membrane Science*, vol. 325, no. 2, pp. 553–558, 2008.

[21] J. Zeng, C. Jiang, Y. Wang et al., "Studies on polypyrrole modified nafion membrane for vanadium redox flow battery," *Electrochemistry Communications*, vol. 10, no. 3, pp. 372–375, 2008.

[22] B. Schwenzer, S. Kim, M. Vijayakumar, Z. Yang, and J. Liu, "Correlation of structural differences between Nafion/polyaniline and Nafion/polypyrrole composite membranes and observed transport properties," *Journal of Membrane Science*, vol. 372, no. 1-2, pp. 11–19, 2011.

[23] J. Xi, Z. Wu, X. Qiu, and L. Chen, "Nafion/SiO_2 hybrid membrane for vanadium redox flow battery," *Journal of Power Sources*, vol. 166, no. 2, pp. 531–536, 2007.

[24] X. Teng, Y. Zhao, J. Xi, Z. Wu, X. Qiu, and L. Chen, "Nafion/organic silica modified TiO_2 composite membrane for vanadium redox flow battery via in situ sol-gel reactions," *Journal of Membrane Science*, vol. 341, no. 1-2, pp. 149–154, 2009.

[25] M. Vijayakumar, B. Schwenzer, S. Kim et al., "Investigation of local environments in NafionSiO_2 composite membranes used in vanadium redox flow batteries," *Solid State Nuclear Magnetic Resonance*, vol. 42, pp. 71–80, 2012.

[26] X. Teng, Y. Zhao, J. Xi, Z. Wu, X. Qiu, and L. Chen, "Nafion/organic silica modified TiO_2 composite membrane for vanadium redox flow battery via in situ sol-gel reactions," *Journal of Membrane Science*, vol. 341, pp. 149–154, 2009.

[27] D. Chen, S. Wang, M. Xiao, D. Han, and Y. Meng, "Sulfonated poly (fluorenyl ether ketone) membrane with embedded silica rich layer and enhanced proton selectivity for vanadium redox flow battery," *Journal of Power Sources*, vol. 195, no. 22, pp. 7701–7708, 2010.

[28] A. Fidalgo and L. M. Ilharco, "The defect structure of sol-gel-derived silica/polytetrahydrofuran hybrid films by FTIR," *Journal of Non-Crystalline Solids*, vol. 283, no. 1-3, pp. 144–154, 2001.

[29] S. K. Young, W. L. Jarrett, and K. A. Mauritz, "Nafion®/ORMOSIL nanocomposites via polymer-in situ sol–gel reactions. 1. Probe of ORMOSIL phase nanostructures by ^{29}Si solid-state NMR spectroscopy," *Polymer*, vol. 43, no. 8, pp. 2311–2320, 2002.

[30] F. Rubio, J. Rubio, and J. L. Oteo, "A DSC study of the drying process of TEOS derived wet silica gels," *Thermochimica Acta*, vol. 307, no. 1, pp. 51–56, 1997.

[31] M. Bello, "Assessment of electrochemical methods for methanol crossover measurement through PEM of direct methanol fuel cell," *International Journal of Engineering & Technology*, vol. 11, pp. 92–111, 2011.

[32] M. Sluyters-Rehbach and J. H. Sluyters, *Comprehensive Treatise of Electrochemistry*, vol. 9, Plenum, New York, NY, USA, 1984.

[33] M. S. Rehbach and J. H. Sluyters, *Electroanalytical Chemistry*, vol. 4, Marcel Dekker, New York, NY, USA, 1970.

9

Hydrogen Reduction of Hematite Ore Fines to Magnetite Ore Fines at Low Temperatures

Wenguang Du,[1] Song Yang,[1] Feng Pan,[1] Ju Shangguan,[1] Jie Lu,[2] Shoujun Liu,[2] and Huiling Fan[1]

[1]Key Laboratory for Coal Science and Technology of Ministry of Education and Shanxi Province, Institute for Chemical Engineering of Coal, Taiyuan University of Technology, Taiyuan 030024, China
[2]College of Chemistry and Chemical Engineering, Taiyuan University of Technology, Taiyuan 030024, China

Correspondence should be addressed to Ju Shangguan; shanggj62@163.com

Academic Editor: Maria F. Carvalho

Surplus coke oven gases (COGs) and low grade hematite ores are abundant in Shanxi, China. Our group proposes a new process that could simultaneously enrich CH_4 from COG and produce separated magnetite from low grade hematite. In this work, low-temperature hydrogen reduction of hematite ore fines was performed in a fixed-bed reactor with a stirring apparatus, and a laboratory Davis magnetic tube was used for the magnetic separation of the resulting magnetite ore fines. The properties of the raw hematite ore, reduced products, and magnetic concentrate were analyzed and characterized by a chemical analysis method, X-ray diffraction, optical microscopy, and scanning electron microscopy. The experimental results indicated that, at temperatures lower than 400°C, the rate of reduction of the hematite ore fines was controlled by the interfacial reaction on the core surface. However, at temperatures higher than 450°C, the reaction was controlled by product layer diffusion. With increasing reduction temperature, the average utilization of hydrogen initially increased and tended to a constant value thereafter. The conversion of Fe_2O_3 in the hematite ore played an important role in the total iron recovery and grade of the concentrate. The grade of the concentrate decreased, whereas the total iron recovery increased with the increasing Fe_2O_3 conversion.

1. Introduction

Shanxi, a large coal and coke producing province in China, has a large amount of surplus coke oven gases (COGs) [1–6], which contain substantial amounts of H_2, CH_4, and CO. In 2014, the coke output of Shanxi was 87.22 million tons, which accounted for 18.29% of the national output. If 1t of coke can produce 430 Nm^3 of coke oven gas, of which approximately 200 Nm^3 is returned to a coke oven as a coking heat source, approximately 16~18 billion Nm^3 of COGs is produced just in Shanxi. Meanwhile, the apparent consumption of natural gas in China was 178.6 billion Nm^3 in 2014, which imports 63 billion Nm^3 of natural gas (approximately 33.9% external dependency). COG methanation for CH_4 enrichment can be regarded as a simple and highly efficient way of producing gas, which has attracted many specialists and scholars to research

and develop the process. However, the catalysts, which are required for both the reactivity and selectivity, are sensitive to sulfide and very expensive. Can we economically obtain enriched natural gas from surplus COGs?

Due to their natural abundance, low cost, and usefulness for a variety of applications, iron ores are considered some of the most promising and important resources for the future [7, 8]. The rapid development of the steel industry has resulted in the reduction of iron ore resources in recent years; these ores have been increasingly explored and utilized. At the same time, processed iron ores may be utilized as raw materials for the desulfurization sorbents used in gas purification; iron oxide is the main component of a renewable desulfurization sorbent for high-temperature coal gas desulfurization [9–12]. In addition, this new route uses iron ore as an oxygen carrier, which transfers oxygen from the combustion air to the fuel. Direct contact between the fuel and combustion air is prevented through the use of chemical looping combustion (CLC) [13–15].

However, using the iron ores is quite difficult due to their complex structure and nonuniform crystal size. A combination of low-temperature reduction roasting and low-intensity magnetic separation is considered a promising approach for increasing the usability of these ores [16, 17]. Low-temperature reduction roasting can convert the weakly magnetic iron minerals (Fe_2O_3, $FeCO_3$, and FeS) into a strongly magnetic phase (Fe_3O_4), which can be easily separated by a low-intensity magnetic field. In addition, magnetization, which can be readily performed at low operating costs, is used to reduce the hematite (Fe_2O_3) to magnetite (Fe_3O_4). H_2, CO, CH_4, and coal can be used as reducing agents [18, 19], and the corresponding reduction reactions should proceed as follows:

$$3Fe_2O_3 + H_2 \longrightarrow 2Fe_3O_4 + H_2O \quad (1)$$

$$3Fe_2O_3 + CO \longrightarrow 2Fe_3O_4 + CO_2 \quad (2)$$

$$6Fe_2O_3 + C \longrightarrow 4Fe_3O_4 + CO_2 \quad (3)$$

Compared to coal-based reduction, gas-based reduction can be performed at lower reduction temperatures [20] and typically results in higher quality concentrates, which have lower levels of carbon deposits. H_2, CO, and CH_4 are usually used as reducing agents in these gas-based reductions [21–23].

As such, our research group has proposed an innovative process, which combines low-temperature reduction magnetization of intractable hematite with the production of substitute natural gases (SNGs) from COGs. This combined process was used to remove H_2, CO, and H_2S, thereby enriching the methane gas contained in the COG and converting hematite to easily separated magnetite. The ultimate aim of this work was to use low grade hematite ore and the surplus COG comprehensively to obtain enriched SNG and high grade magnetite economically. Meanwhile, the whole process is described in Figure 1.

This paper, as a preliminary study of the innovative process, mainly describes the process of low-temperature hydrogen reduction of hematite ore and magnetic separation of the resulting magnetite ore, since hydrogen, which is the main component of COG, is an efficient reducing agent. The effect of this reduction on the properties of the magnetite ore is discussed.

2. Materials and Methods

2.1. Materials. The raw hematite ore used in this work was obtained from Guangling County, Shanxi province, China. Guangling County is located northeast of the Shanxi province. Guangling hematite ore is a Shanxi-type hematite ore.

2.2. Analysis and Characterization Method. The chemical compositions of the raw hematite ore, reduced ore samples, and concentrate ore samples were analyzed according to the

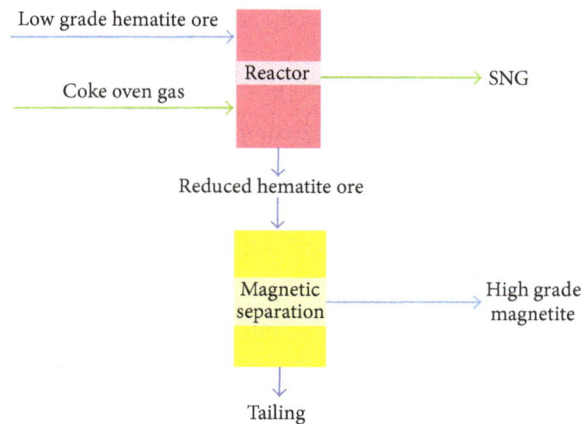

FIGURE 1: The schematic diagram of use for the low grade hematite ore and the surplus coke oven gas to produce SNGs and high grade magnetite.

National Standards of China number GB 6730.5-2008 and number GB 6730.8-2008.

The crystalline phases of the aforementioned samples were determined via X-ray diffraction (XRD) (using Cu-Kα radiation, scanning rate of 8°/min, and sweeping range of 5°–85°).

The structural characteristics of the raw hematite ore were examined using an optical microscope and a scanning electron microscope (SEM).

2.3. Experimental Procedure. A total of 100 kg of the raw hematite ore sample was first crushed using a jaw crusher and then sieved to a size of 2 mm in our laboratory. One kilogram of each of the 12 samples of the crushed hematite ore was then reduced to magnetite ore, under a total volumetric gas rate of 120 L/h, using a gas mixture of 50% H_2–50% N_2. The reduction process was performed in a fixed-bed reactor, and a stirring apparatus was used to enable full contact between the solids and the gas. The samples were reduced for 1, 1.5, 2, and 2.5 h, at 400°C. Similarly, the samples were reduced for 0.5, 1, 1.5, and 2 h at both 450°C and 500°C.

The total Fe content (TFe) and Fe^{2+} content (TFeO) of each sample were determined using a chemical analysis method. The titanium trichloride-potassium dichromate titration method was used to determine the TFe of each sample; the ferric chloride-potassium dichromate titration method was used to determine the TFeO. Fe_3O_4 contains one Fe^{2+} and two Fe^{3+} ions. Therefore, if all the hematite in the raw ore sample is reduced to Fe_3O_4, then the TFe and TFeO of the reduced sample are related through by TFeO = 3/7TFe. The conversion of Fe_2O_3 ($X_{Fe_2O_3}$) to Fe_3O_4 in each reduced sample can then be calculated by determining the TFe and TFeO of each reduced sample. $X_{Fe_2O_3}$ was calculated from

$$X_{Fe_2O_3}\% = \frac{7W_{TFeO}}{3W_{TFe}} \times 100\%. \quad (4)$$

The grain sizes of the newly generated magnetite were calculated from

$$D = \frac{Rl}{\beta \cos q},\tag{5}$$

where D is the distance between the atomic layers in the crystals, R is Scherer's constant, l is the wavelength of the X-ray radiation, q is the diffraction angle, and β is the full width at half maximum.

As for the low-intensity magnetic separation of the reduced samples, the optimal grinding fineness and magnetic field intensity should be chosen such that the optimal magnetic separation index is obtained. In this paper, reduced samples with different degrees of grinding fineness and magnetic field intensities were obtained by varying the grinding time and the working electrical current of the Davis tube magnetic separator, respectively. The experiments aimed at optimizing the grinding fineness and magnetic field intensity were performed for samples reduced at 500°C for 0.5 h or 2.0 h. The reduced samples were dry milled in a rod mill for various times and then separated by a laboratory Davis tube magnetic separator, with a working electrical current of 2 A, in order to determine the best grinding fineness. Furthermore, in order to determine the optimum magnetic field intensity, reduced samples with the best grinding fineness were separated by a laboratory Davis tube magnetic separator operating at various electrical currents.

Each of the 12 reduced samples was dry milled in a rod mill for 15 min and then separated by a laboratory Davis tube magnetic separator (model: XCGSφ50), operating at a working electrical current of 2 A (magnetic field intensity: 0.156 T). The grades and weights of the magnetic concentrates were determined, and the recovery of iron (R_{Fe}), which is the index of magnetic separation, was calculated during the low-intensity magnetic separation; R_{Fe}, the amount of iron recovered in the final concentrate, was calculated from

$$R_{Fe} = \frac{m_1 \times T_{Fe1}}{m_2 \times T_{Fe2}} \times 100\%,\tag{6}$$

where R_{Fe}, m_1, T_{Fe1}, m_2, and T_{Fe2} are the amount of iron recovered during the low-intensity magnetic separation, quality of the concentrate, iron content of the concentrate, quality of the reduced samples, and iron content of the reduced samples, respectively.

3. Results and Discussions

3.1. Properties of the Raw Ore Samples. The mineral composition, main chemical composition, and iron distribution of the raw hematite samples are listed in Tables 1, 2, and 3, respectively; the crystalline phase is shown in Figure 2.

The Guangling hematite ore sample consisted of 58.83% Fe_2O_3, 19.53% SiO_2, 6.75% Al_2O_3, 0.21% FeO, 2.46% CaO, 1.38% MgO, 0.11% Na_2O, and 3.65% K_2O. Harmful elements, such as phosphorus and sulfur, were present in only low quantities. This hematite ore contained a small amount of magnetite; hematite is the precious form of iron minerals. The gangue minerals were composed mainly of quartz and

TABLE 1: Mineral composition of the hematite ore.

Components	Content (wt.%)
Hematite	53
Limonite	4
Quartz	20
Mica	5
Kaolinite	8
Barite	Trace

TABLE 2: Chemical composition of the hematite ore.

Components	Content (wt.%)
TFe	41.14
Fe_2O_3	58.83
FeO	0.21
SiO_2	19.53
CaO	2.46
Al_2O_3	6.75
MgO	1.38
Na_2O	0.11
K_2O	3.65
S	0.18
P	0.66
LOI	3.95

*LOI: loss on ignition.

TABLE 3: Iron distribution of the hematite ore.

Components	Content (wt.%)	Fraction (wt.%)
Magnetic iron	0.057	0.14
Siderite	0.057	0.14
Iron sulfide	0.057	0.14
Iron silicate	0.50	1.21
Hematite and limonite	40.4	98.37
TFe	41.07	100

small amounts of mica and kaolinite. Furthermore, there was a 3.95% loss upon ignition for the Guangling hematite ore. As Figure 2 shows, the Guangling hematite ore was composed mainly of hematite and the nonmetallic mineral, quartz.

The optical micrographs in Figure 3 reveal that the Guangling hematite ore mainly had taxitic structure, disseminated structure, and similar oolitic structure; the hematite mixed with quartz or mica, hematite, quartz, and clay (mainly kaolinite) formed an oolitic-like structure, quartz formed the core of the similar oolitic structure, and hematite and clay (mainly kaolinite) formed a concentric circle in the similar oolitic structure.

The corresponding SEM images (Figure 4) confirmed that the hematite ore consisted mainly of scaly, acicular, cryptocrystalline, and metasomatic textures, with a sand consolidation structure (rarely, with an oolitic structure). The acicular and scaly hematite ores are bright white, whereas quartz shows a grey color. The disseminated structure of the hematite was very fine.

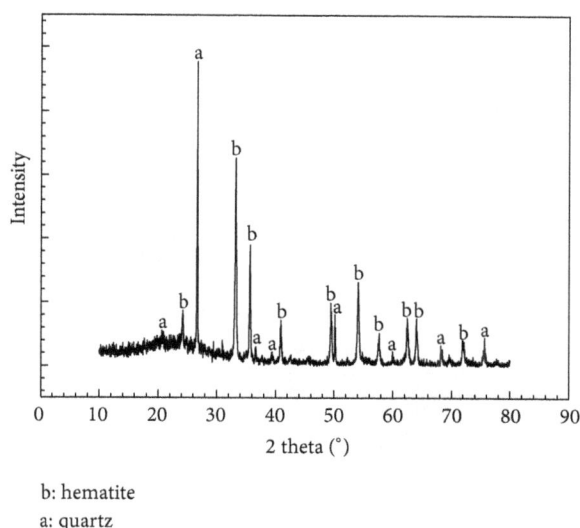

b: hematite
a: quartz

FIGURE 2: XRD pattern of the raw hematite ore.

Therefore, the hematite ore used in this study was a typical complex, fine-grained, marine-sediment-refractory hematite ore.

3.2. Reducing Magnetization of the Raw Hematite Ore

3.2.1. Thermodynamics of Reducing Fe_2O_3 with H_2. The Gibbs free energy function method was adopted for the thermodynamic calculation of the H_2-reduction of Fe_2O_3 [24, 25]; the results are shown in Figure 5.

Three reduction products were formed, namely, Fe_3O_4, FeO, and Fe. Moreover, Fe_2O_3 reacted with H_2 to form Fe_3O_4 at reduction temperatures of 400°C–600°C and H_2 concentrations of <65 vol.%.

3.2.2. Effect of the Reduction Temperature and Time on the Conversion of Fe_2O_3. Short, low-temperature reduction processes can reduce the cost of magnetizing the hematite ore. The dependence of the Fe_2O_3 conversion in the hematite ore on the reduction time (Figure 6) was investigated at the temperatures of 400°C, 450°C, and 500°C.

As Figure 6 shows, at 400°C, the amount of Fe_2O_3 converted increased linearly with the increasing reduction time. At reduction temperatures of 450°C or 500°C, the amount of Fe_2O_3 converted increased rapidly at the beginning of the reaction and slowly thereafter. The shrinking unreacted core model stipulates that there is a sharp boundary between the reacted (magnetite: Fe_3O_4) and the unreacted (hematite: Fe_2O_3) parts of the particle [26]. The hematite is reduced to magnetite (Fe_3O_4) via the diffusion of hydrogen through the product layer on the boundary (reacted-unreacted interface). As reduction proceeds, the boundary eventually recedes to the center, the hematite is exhausted, and the thickness of the product layer (magnetite layer) increases. The product layer diffusion and the interfacial reaction on the surface of the core constitute the rate-limiting steps of the process [25, 26].

The interfacial reaction on the surface of the core proceeded very slowly at low reduction temperatures. In fact, the rate of the interfacial reaction is significantly lower than that of the hydrogen diffusion process. Therefore, the interfacial reaction constituted the rate-limiting step of the process. Since the reduction temperature was constant and very low, the reaction proceeded extremely slowly; the conversion of Fe_2O_3 was proportional to the reduction time.

When the reduction temperature was high, the interfacial reaction on the core surface proceeded rapidly and at a significantly higher rate than that of the hydrogen diffusion process. Diffusion into the product layer was, therefore, the rate-limiting step of the process. During the beginning of reduction, the reaction proceeded rapidly since the product layer was very thin, and the diffusion resistance of hydrogen was correspondingly small. As the reduction proceeded, the thickness of the product layer increased the diffusion resistance of hydrogen, which in turn led to a sharp decrease in the reaction rate with the increasing reduction time.

The aforementioned results show that the transition of hematite (Fe_2O_3) to magnetite (Fe_3O_4) proceeded according to a shrinking unreacted core model. When the reduction temperature was low, the reaction rate was controlled by the interfacial reaction on the surface of the core; when the reduction temperature was high, the reaction rate was controlled by diffusion into the product layer.

On average, ~18.6%, 23.1%, and 23.4% of the hydrogen were utilized during the reduction process at 400°C, 450°C, and 500°C, respectively. This indicates that the average utilization of hydrogen increased initially and tended, subsequently, to a constant value with the increasing reduction temperature.

XRD was used to characterize 12 samples, which were reduced for 1 h, 1.5 h, 2 h, and 2.5 h, at the respective reduction temperatures of 400°C, 450°C, and 500°C, as shown in Figures 7, 8, and 9. As the figures show, the intensity of the peaks corresponding to hematite and magnetite decreases and increases, respectively, with the increasing reduction time. This indicates that the rate of reaction increased with

(a)

(b)

(c)

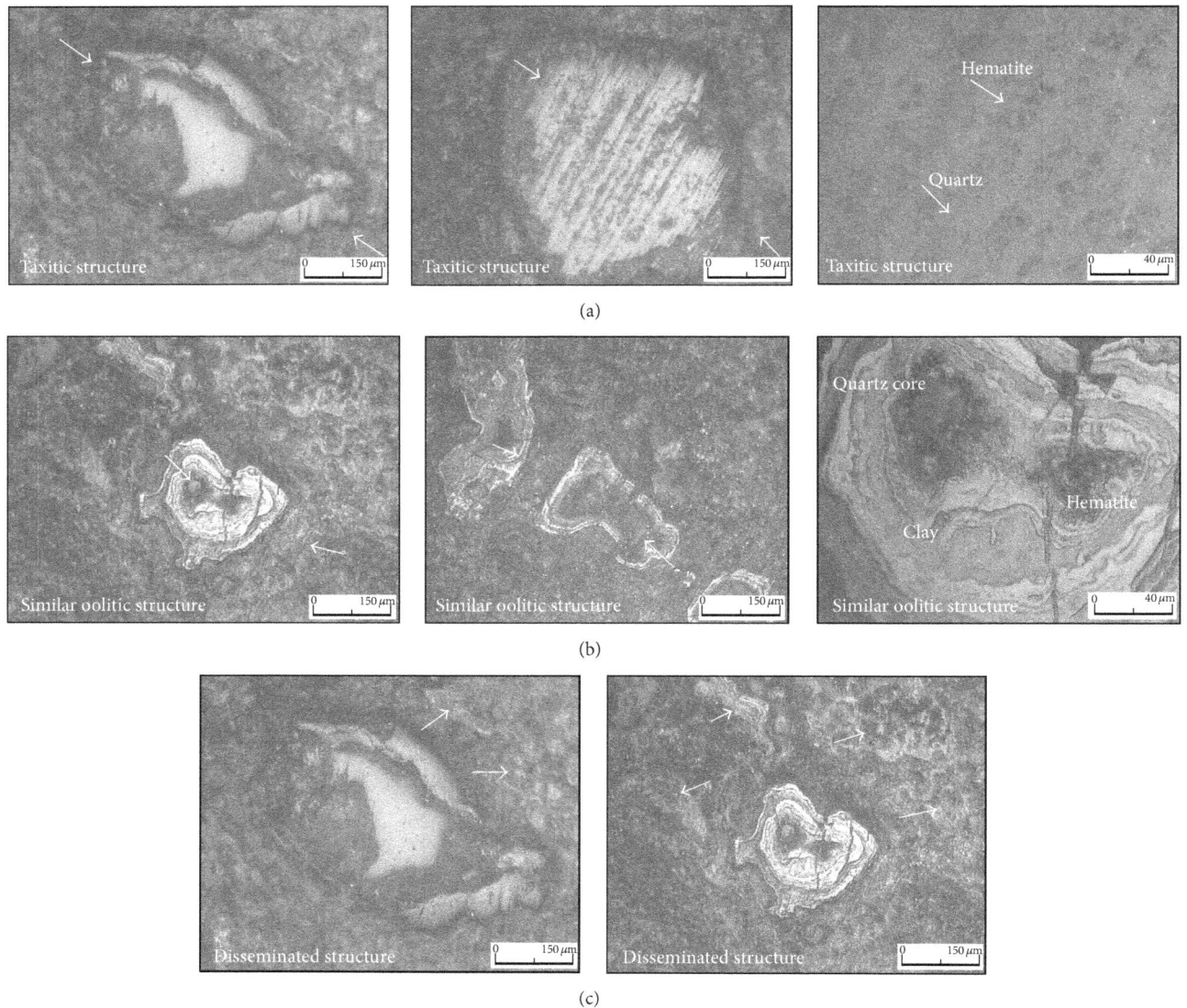

FIGURE 3: Optical microscope images of the raw hematite ore.

the increasing temperature. Moreover, nearly all the hematite was reduced to magnetite during the process.

3.3. *Mechanism of Reduction Magnetization Process.* The grain size of the newly generated magnetite is shown in Figure 9.

Figure 10 shows that the grain size of the newly generated magnetite increased initially and then decreased slightly and increased thereafter. Based on the structural characteristics of the hematite, this grain size trend also indicates that the transition of hematite (Fe_2O_3) to magnetite (Fe_3O_4) followed a shrinking unreacted core model [27–29]. This transition progressed in three steps [30].

Step 1. H_2 molecules diffused through the air and were absorbed on the surface of hematite; these H_2 molecules were then activated. These active H_2 molecules combined with

O^{2-} from the Fe_2O_3 lattice, thereby generating H_2O [31]. Moreover, the two electrons released, during the reaction, reduced Fe^{3+} to Fe^{2+}:

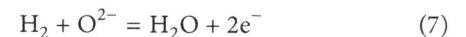

$$H_2 + O^{2-} = H_2O + 2e^- \tag{7}$$

$$2Fe^{3+} + 2e^- = 2Fe^{2+} \tag{8}$$

Step 2. The new generation of Fe^{2+} combined with Fe_2O_3, thereby producing Fe_3O_4 through lattice reconstruction [31, 32]:

$$4Fe_2O_3 + Fe^{2+} + 2e^- = 3Fe_3O_4 \left(Fe_2O_3 \bullet FeO\right) \tag{9}$$

Step 3. During the above process, the reduction reaction extended continuously to the inner layer, and the entire hematite particle was completely reduced to magnetite [32].

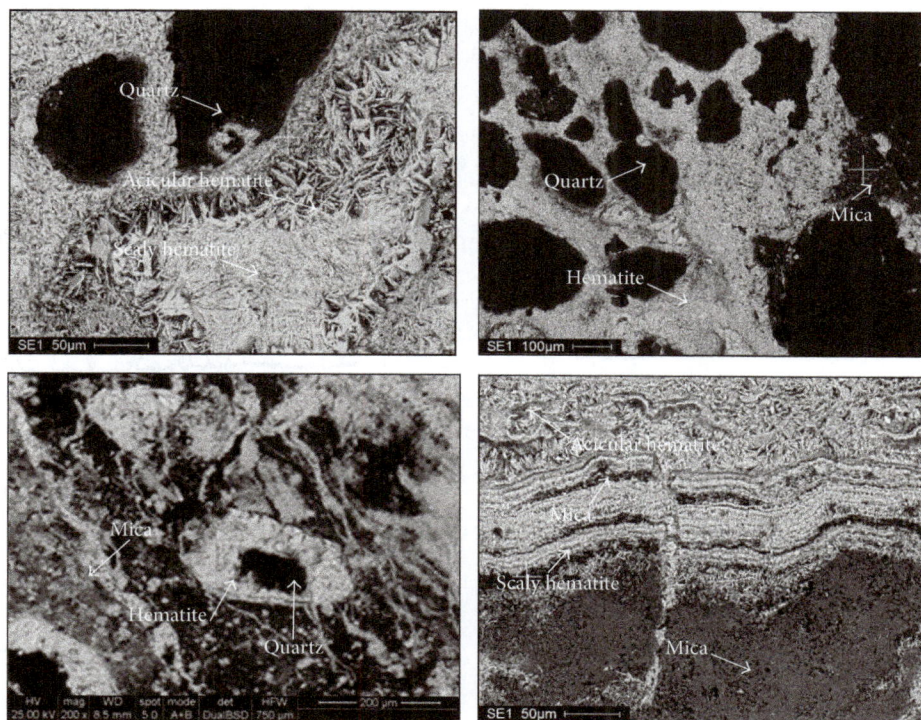

FIGURE 4: SEM images of the raw hematite ore.

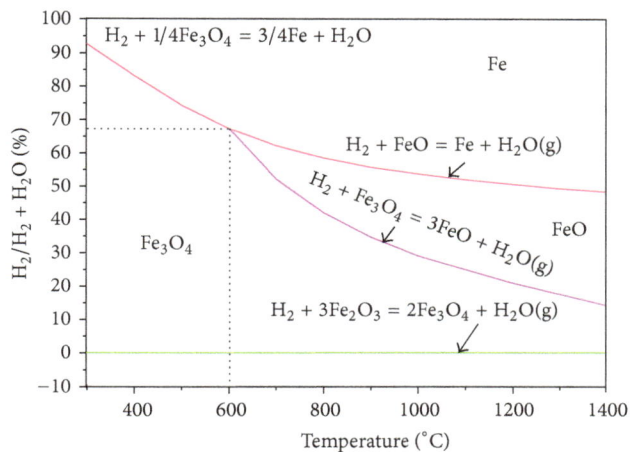

FIGURE 5: Gas-phase equilibrium composition of H_2-reduced iron oxide.

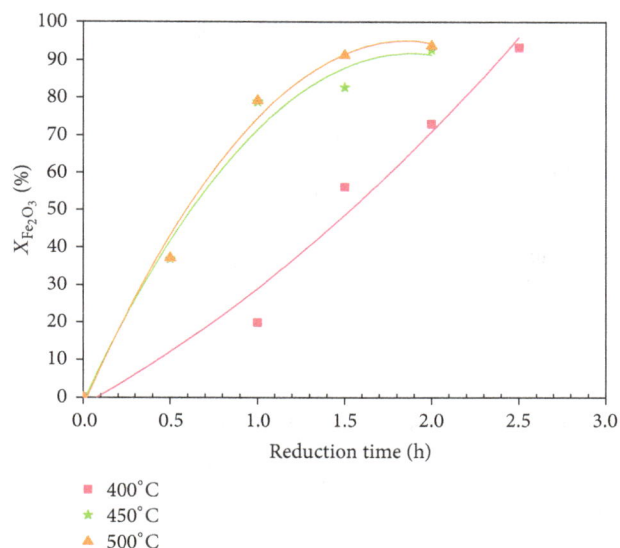

FIGURE 6: Effect of reaction time on the conversion of Fe_2O_3 at different temperatures.

Figure 11 shows the SEM images of samples reduced for 1 h and 2.5 h at 400°C as well as for 0.5 h and 2 h at 500°C.

The grain sizes of the new-generation magnetite increased with the increasing reaction time. Furthermore, the amount of Fe_2O_3 converted and the number of grains generated increased with the increasing reaction temperature and time.

3.4. Low-Intensity Magnetic Separation of the Reduced Samples

3.4.1. Optimization of the Grinding Fineness and Magnetic Field Intensity. To obtain a high grade concentrate with

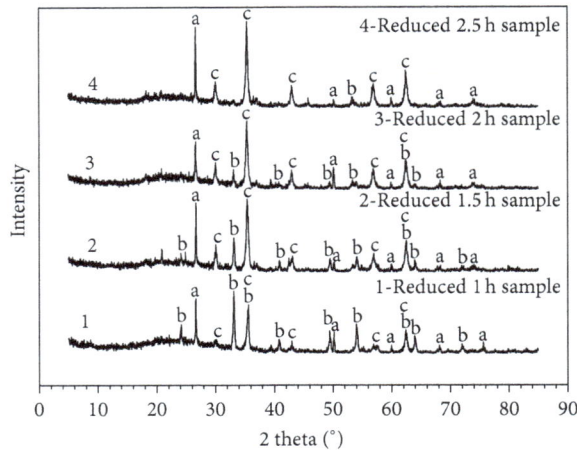

b: hematite
c: magnetite
a: quartz

FIGURE 7: XRD patterns of samples reduced at 400°C.

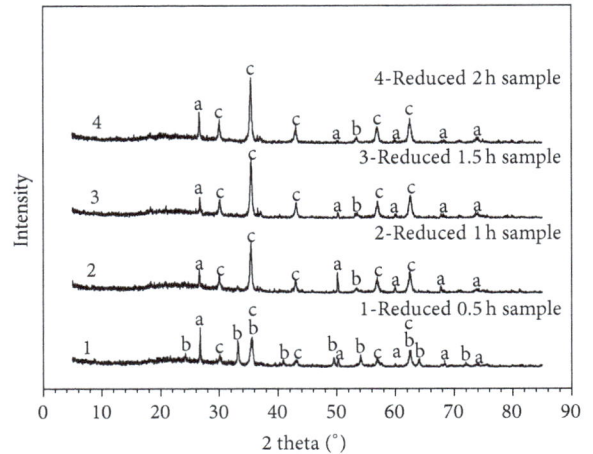

b: hematite
c: magnetite
a: quartz

FIGURE 9: XRD patterns of samples reduced at 500°C.

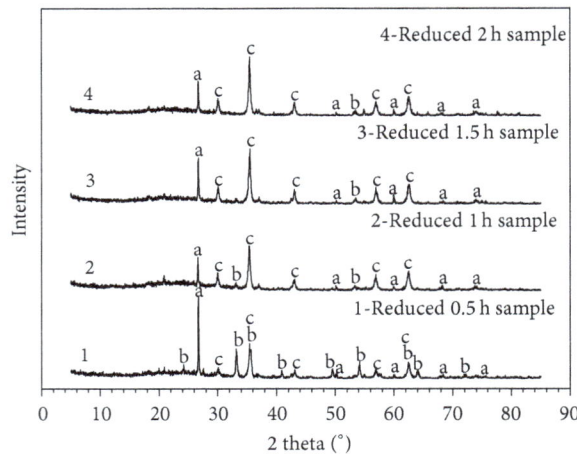

b: hematite
c: magnetite
a: quartz

FIGURE 8: XRD patterns of samples reduced at 450°C.

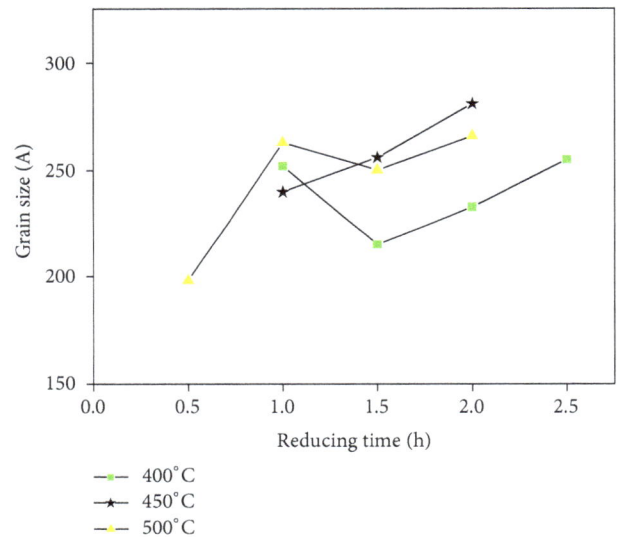

FIGURE 10: Grain size of the new-generation magnetite (Fe_3O_4).

excellent Fe recovery, experiments on the grinding fineness of the ore were performed prior to magnetic separation of the reduced samples. The results obtained for the samples reduced at 500°C for 0.5 h and 2.0 h are shown in Figures 12–15.

The figures reveal an optimum grinding time of 15 min and a corresponding grinding fineness and iron recovery of −200 mesh and 70.78%, respectively. In addition, the optimum magnetic field intensity (0.156 T) for magnetic separation occurred at an electrical current of 2 A.

The grain size of the hydrogen-reduced new-generation magnetite influenced the dissociation of the core and, therefore, played an important role in the subsequent magnetic separation. Comparing Figure 9 with Figures 14 and 15 reveals that the increases in the grain size for the former and latter

figures for the new-generation magnetite were favorable and unfavorable, respectively, to the dissociation of the ore.

3.4.2. Effect of Reduction on the Magnetic Separation. The relationship between the grade of the concentrate and the conversion of Fe_2O_3 (in the hematite ore) at a given reduction temperature was determined by analyzing the total iron content; the results are shown in Figure 16.

As Figure 16 shows, the grade of the concentrate decreased sharply with the increasing amount of converted Fe_2O_3. This indicates that the Fe_2O_3 conversion played an important role in determining the grade of the concentrate.

According to the shrinking unreacted core model, the reduction of hematite (Fe_2O_3) to magnetite (Fe_3O_4) proceeded from the outer surface to the center of the hematite

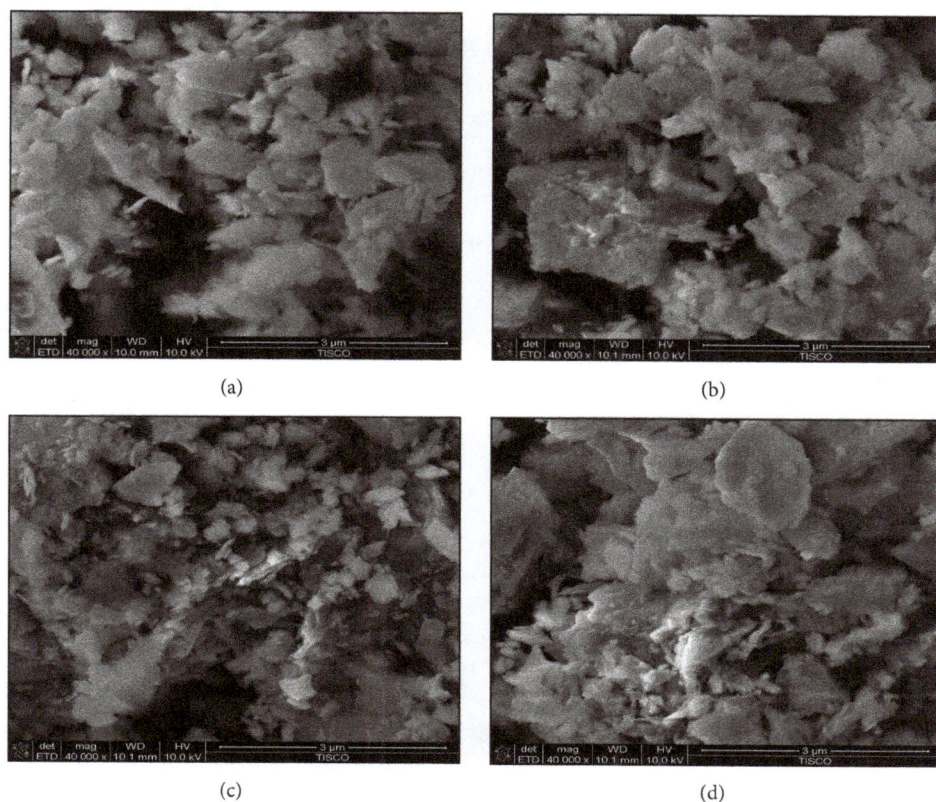

FIGURE 11: SEM images of samples reduced for (a) 1 h at 400°C, (b) 2.5 h at 400°C, (c) 0.5 h at 500°C, and (d) 2 h at 500°C.

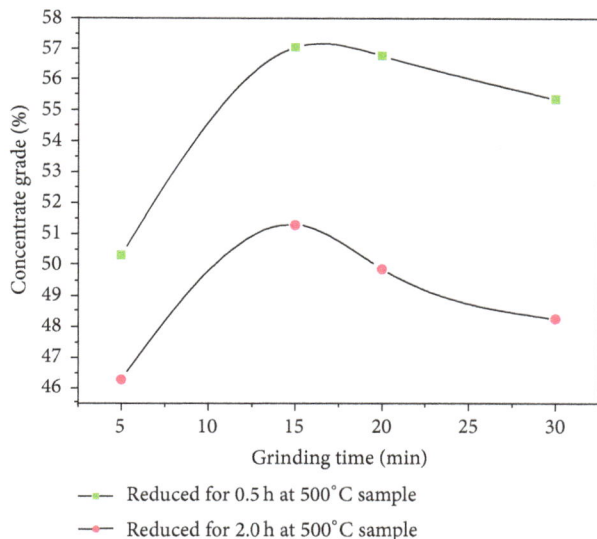

FIGURE 12: Effect of grinding time on the grade of the concentrate.

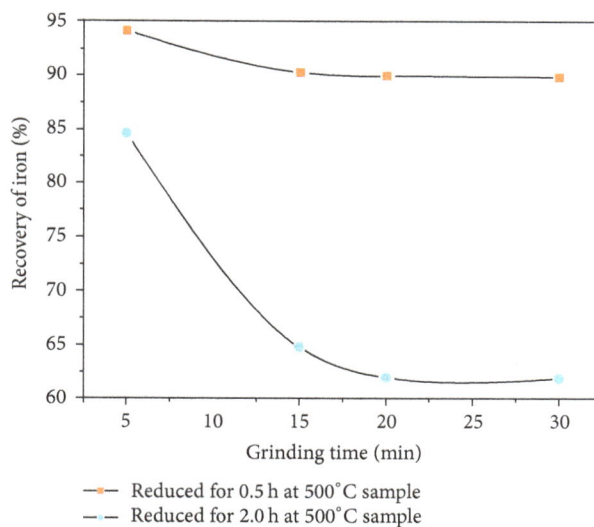

FIGURE 13: Effect of grinding time on the recovery of iron.

ore particle. The product layer (magnetite layer) thickened with the progression of reduction, and the unreacted gangue in the hematite ore particle was inevitably encapsulated by the new-generation magnetite (Fe_3O_4). This enclosure hindered the dissociation of the gangue and the newly generated magnetite (Fe_3O_4), and, as a result, the grade of the magnetic concentrate decreased with the increasing amount of converted Fe_2O_3.

The dependence of the iron recovery on the conversion of Fe_2O_3 at specific reduction temperatures was determined by analyzing the total iron content in both the hematite and concentrate ores; the results are shown in Figure 17.

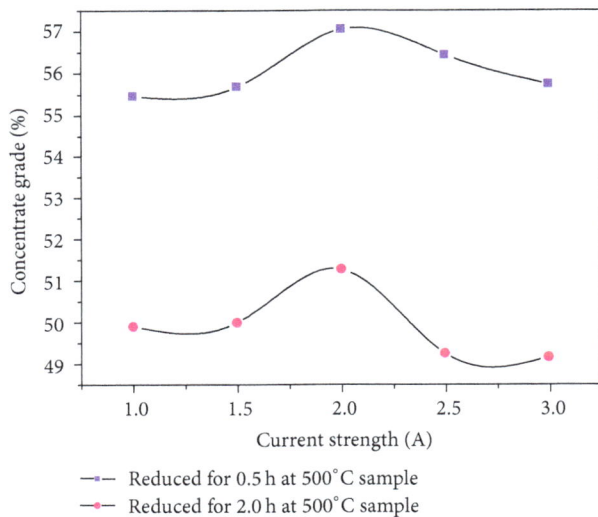

FIGURE 14: Effect of current strength on the concentrate grade.

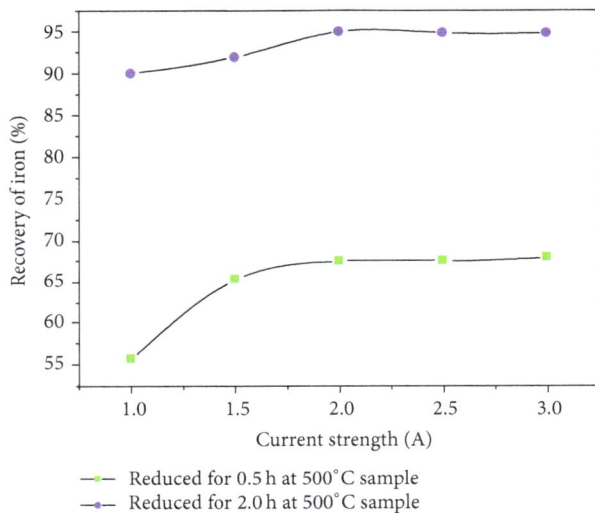

FIGURE 15: Effect of grinding time on the recovery of iron.

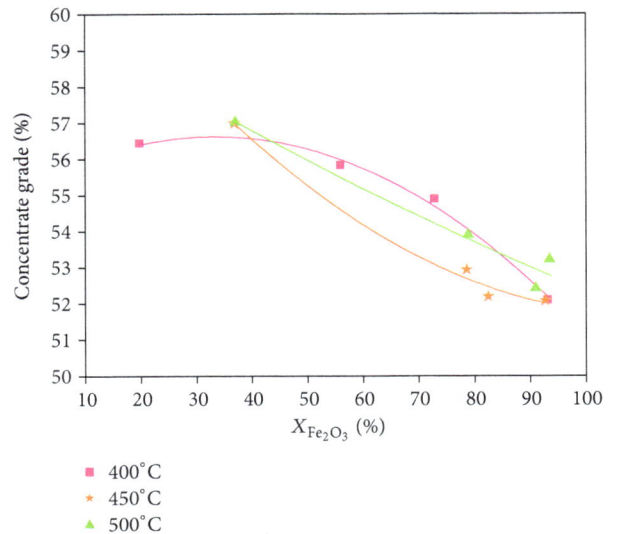

FIGURE 16: Dependence of the concentrate grade on the conversion of Fe_2O_3.

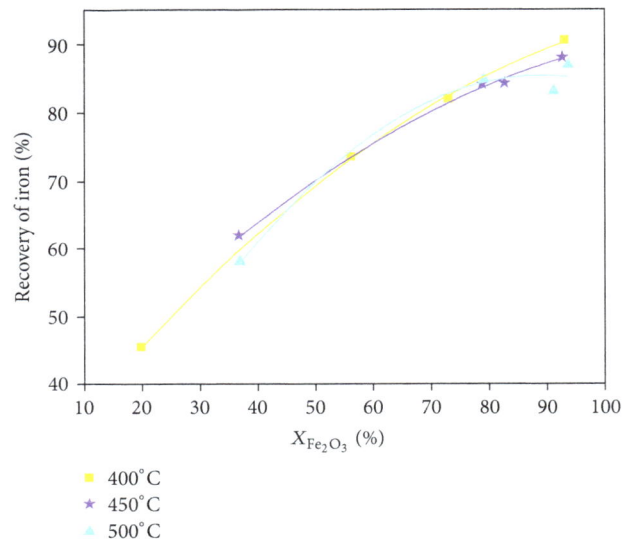

FIGURE 17: Dependence of iron recovery on the conversion of Fe_2O_3.

As Figure 17 shows, the total iron recovery increased significantly with the increasing amount of converted Fe_2O_3. This indicates that the Fe_2O_3 conversion played an important role in the recovery of iron.

The amount of hematite reduced to magnetite increased with the increasing amount of converted Fe_2O_3 and resulted, in turn, in increased total iron recovery. Therefore, the conversion of Fe_2O_3 in the hematite ore had a significant effect on the total iron recovery and concentrate grade. In fact, the concentrate grade decreased, whereas the total iron recovery increased, with the increasing Fe_2O_3 conversion.

The grade of the concentrate reduced at 450°C for 30 min and milled in a rod mill for 15 min could be improved (to grade: 56.99%, iron recovery rate: 61.93%) by performing a simple magnetic separation process, using a working electrical current of 2 A (magnetic field intensity: 0.156 T); this process could also be used to improve the iron recovery rate of the concentrate (to grade: 52.06%, iron recovery rate: 90.5%), which was reduced at 400°C for 150 min and rod-milled for 15 min.

4. Conclusions

We conclude the following:

(1) One kilogram of hematite (−2 mm) was reduced under a total volumetric gas rate of 120 L/h (gas mixture composition: 50% H_2–50% N_2). At reduction temperatures lower than 400°C, the rate of the reduction reaction of the hematite ore fines was controlled by the interfacial reaction on the core surface. However, at temperatures higher than 450°C,

the reaction rate was controlled by the product layer diffusion. With an increasing reduction temperature, the average utilization of hydrogen increased initially and tended to a constant value thereafter.

(2) The grade of the concentrate decreased and the total iron recovery increased with increasing Fe_2O_3 conversion. The grade of the concentrate, which was reduced at 450°C for 30 min and rod-milled for 15 min, was improved (to grade: 56.99%, iron recovery rate: 61.93%) through a simple magnetic separation process using a working electrical current of 2 A (magnetic field intensity: 0.156 T); this process was also used to improve the iron recovery rate (to grade: 52.06%, iron recovery rate: 90.5%) of the concentrate, which was reduced at 400°C for 150 min and rod-milled for 15 min.

(3) During the reduction magnetization, the grain size of the new-generation magnetite increased initially and then decreases slightly and increased thereafter. The former and latter increases in the grain size were favorable and unfavorable, respectively, to the dissociation of the hematite ore. In addition, the transition of hematite (Fe_2O_3) to magnetite (Fe_3O_4) followed a shrinking unreacted core model.

Competing Interests

The authors declare that they have no conflict of interests.

Acknowledgments

The authors thank the colleagues from Taiyuan University of Technology for their assistance during this work. This work was sponsored by the Taiyuan Green Coke Clean Energy Co., Ltd. (China).

References

[1] W.-H. Chen, M.-R. Lin, T.-S. Leu, and S.-W. Du, "An evaluation of hydrogen production from the perspective of using blast furnace gas and coke oven gas as feedstocks," *International Journal of Hydrogen Energy*, vol. 36, no. 18, pp. 11727–11737, 2011.

[2] R. Razzaq, C. Li, and S. Zhang, "Coke oven gas: availability, properties, purification, and utilization in China," *Fuel*, vol. 113, pp. 287–299, 2013.

[3] J. M. Bermúdez, A. Arenillas, R. Luque, and J. A. Menéndez, "An overview of novel technologies to valorise coke oven gas surplus," *Fuel Processing Technology*, vol. 110, pp. 150–159, 2013.

[4] W. G. Du, S. J. Liu, J. Shangguan, F. Gao, and J. Chen, "Method for producing natural gas by reduction magnetizing hematite with coke-oven gas," P.R.China Patent, CN102311821A.

[5] W.-H. Chen, M.-R. Lin, A. B. Yu, S.-W. Du, and T.-S. Leu, "Hydrogen production from steam reforming of coke oven gas and its utility for indirect reduction of iron oxides in blast furnace," *International Journal of Hydrogen Energy*, vol. 37, no. 16, pp. 11748–11758, 2012.

[6] J. Lu, S. Liu, J. Shangguan, W. Du, F. Pan, and S. Yang, "The effect of sodium sulphate on the hydrogen reduction process of nickel laterite ore," *Minerals Engineering*, vol. 49, pp. 154–164, 2013.

[7] P. Pourghahramani and E. Forssberg, "Effects of mechanical activation on the reduction behavior of hematite concentrate," *International Journal of Mineral Processing*, vol. 82, no. 2, pp. 96–105, 2007.

[8] M. Challenor, P. Gong, D. Lorenser et al., "Iron oxide-induced thermal effects on solid-state upconversion emissions in NaYF4:Yb,Er nanocrystals," *ACS Applied Materials and Interfaces*, vol. 5, no. 16, pp. 7875–7880, 2013.

[9] H. Fan, K. Xie, J. Shangguan, F. Shen, and C. Li, "Effect of calcium oxide additive on the performance of iron oxide sorbent for high-temperature coal gas desulfurization," *Journal of Natural Gas Chemistry*, vol. 16, no. 4, pp. 404–408, 2007.

[10] H.-L. Fan, J. Shangguan, L.-T. Liang, C.-H. Li, and J.-Y. Lin, "A comparative study of the effect of clay binders on iron oxide sorbent in the high-temperature removal of hydrogen sulfide," *Process Safety and Environmental Protection*, vol. 91, no. 3, pp. 235–243, 2013.

[11] N. Jordan, A. Ritter, A. C. Scheinost, S. Weiss, D. Schild, and R. Hübner, "Selenium(IV) uptake by maghemite (γ-Fe_2O_3)," *Environmental Science and Technology*, vol. 48, no. 3, pp. 1665–1674, 2014.

[12] E. Potapova, I. Carabante, M. Grahn, A. Holmgren, and J. Hedlund, "Studies of collector adsorption on iron oxides by in situ ATR-FTIR spectroscopy," *Industrial and Engineering Chemistry Research*, vol. 49, no. 4, pp. 1493–1502, 2010.

[13] Z. Yu, C. Li, Y. Fang, J. Huang, and Z. Wang, "Reduction rate enhancements for coal direct chemical looping combustion with an iron oxide oxygen carrier," *Energy & Fuels*, vol. 26, no. 4, pp. 2505–2511, 2012.

[14] Z. Yu, C. Li, X. Jing et al., "Effects of CO_2 atmosphere and K_2CO_3 addition on the reduction reactivity, oxygen transport capacity, and sintering of CuO and Fe_2O_3 oxygen carriers in coal direct chemical looping combustion," *Energy and Fuels*, vol. 27, no. 5, pp. 2703–2711, 2013.

[15] Z. Yu, C. Li, X. Jing et al., "Catalytic chemical looping combustion of carbon with an iron-based oxygen carrier modified by K_2CO_3: catalytic mechanism and multicycle tests," *Fuel Processing Technology*, vol. 135, pp. 119–124, 2015.

[16] C. Li, H. Sun, J. Bai, and L. Li, "Innovative methodology for comprehensive utilization of iron ore tailings: part 1. The recovery of iron from iron ore tailings using magnetic separation after magnetizing roasting," *Journal of Hazardous Materials*, vol. 174, pp. 71–77, 2010.

[17] C. Li, H. Sun, J. Bai, and L. Li, "Innovative methodology for comprehensive utilization of iron ore tailings: part 2: the residues after iron recovery from iron ore tailings to prepare cementitious material," *Journal of Hazardous Materials*, vol. 174, pp. 78–83, 2010.

[18] W. K. Jozwiak, E. Kaczmarek, T. P. Maniecki, W. Ignaczak, and W. Maniukiewicz, "Reduction behavior of iron oxides in hydrogen and carbon monoxide atmospheres," *Applied Catalysis A: General*, vol. 326, no. 1, pp. 17–27, 2007.

[19] H. L. Gilles and C. W. Clump, "Reduction of iron ore with hydrogen in a direct current plasma jet," *Industrial and Engineering Chemistry*, vol. 9, no. 2, pp. 194–207, 1970.

[20] A. Mcgeorge Jr., A. H. Hixson, and K. A. Krieger, "Low temperature gaseous reduction of iron ore in the presence of alkali," *I and E C Process Design and Development*, vol. 1, no. 3, pp. 217–225, 1962.

[21] J. Zieliński, I. Zglinicka, L. Znak, and Z. Kaszkur, "Reduction of Fe_2O_3 with hydrogen," *Applied Catalysis A: General*, vol. 381, no. 1-2, pp. 191–196, 2010.

[22] K.-S. Kang, C.-H. Kim, K.-K. Bae et al., "Reduction and oxidation properties of Fe_2O_3/ZrO_2 oxygen carrier for hydrogen production," *Chemical Engineering Research and Design*, vol. 92, no. 11, pp. 2584–2597, 2014.

[23] N. J. Welham, "Activation of the carbothermic reduction of manganese ore," *International Journal of Mineral Processing*, vol. 67, no. 1-4, pp. 187–198, 2002.

[24] A. Pineau, N. Kanari, and I. Gaballah, "Kinetics of reduction of iron oxides by H2. Part II. Low temperature reduction of magnetite," *Thermochimica Acta*, vol. 456, no. 2, pp. 75–88, 2007.

[25] A. Pineau, N. Kanari, and I. Gaballah, "Kinetics of reduction of iron oxides by H2. Part I: low temperature reduction of hematite," *Thermochimica Acta*, vol. 447, no. 1, pp. 89–100, 2006.

[26] J. P. Martins and F. Margarido, "The cracking shrinking model for solid-fluid reactions," *Materials Chemistry and Physics*, vol. 44, no. 2, pp. 156–169, 1996.

[27] X. Liu, F. Song, and Z. Wen, "A novel dimensionless form of unreacted shrinking core model for solid conversion during chemical looping combustion," *Fuel*, vol. 129, pp. 231–237, 2014.

[28] D. da Rocha, E. Paetzold, and N. Kanswohl, "The shrinking core model applied on anaerobic digestion," *Chemical Engineering and Processing: Process Intensification*, vol. 70, pp. 294–300, 2013.

[29] Y. Yu and C. Qi, "Magnetizing roasting mechanism and effective ore dressing process for oolitic hematite ore," *Journal Wuhan University of Technology, Materials Science Edition*, vol. 26, no. 2, pp. 176–181, 2011.

[30] W. V. Schulmeyer and H. M. Ortner, "Mechanisms of the hydrogen reduction of molybdenum oxides," *International Journal of Refractory Metals and Hard Materials*, vol. 20, no. 4, pp. 261–269, 2002.

[31] K. Higuchi and R. H. Heerema, "Influence of sintering conditions on the reduction behaviour of pure hematite compacts," *Minerals Engineering*, vol. 16, no. 5, pp. 463–477, 2003.

[32] H. Veeramani, D. Aruguete, N. Monsegue et al., "Low-temperature green synthesis of multivalent manganese oxide nanowires," *ACS Sustainable Chemistry & Engineering*, vol. 1, no. 9, pp. 1070–1074, 2013.

10

Synthesis of Gold Nanoparticles Stabilized in Dextran Solution by Gamma Co-60 Ray Irradiation and Preparation of Gold Nanoparticles/Dextran Powder

Phan Ha Nu Diem,[1,2] Doan Thi Thu Thao,[3] Dang Van Phu,[4] Nguyen Ngoc Duy,[4] Hoang Thi Dong Quy,[3] Tran Thai Hoa,[2] and Nguyen Quoc Hien[4]

[1]*Dong Nai University, 4 Le Quy Don, Bien Hoa City, Vietnam*
[2]*College of Sciences, Hue University, 77 Nguyen Hue, Hue City, Vietnam*
[3]*University of Science, Vietnam National University in Ho Chi Minh City, 227 Nguyen Van Cu, Ho Chi Minh City, Vietnam*
[4]*Research and Development Center for Radiation Technology, Vietnam Atomic Energy Institute, 202A Street 11, Linh Xuan Ward,*
 Thu Duc District, Ho Chi Minh City, Vietnam

Correspondence should be addressed to Nguyen Quoc Hien; hien7240238@yahoo.com

Academic Editor: Philippe Dugourd

Gold nanoparticles (AuNPs) in spherical shape with diameter of 6–35 nm stabilized by dextran were synthesized by γ-irradiation method. The AuNPs were characterized by UV-Vis spectroscopy and transmission electron microscopy. The influence of pH, Au^{3+} concentration, and dextran concentration on the size of AuNPs was investigated. Results indicated that the smallest AuNPs size (6 nm) and the largest AuNPs size (35 nm) were obtained for pH of 1 mM Au^{3+}/1% dextran solution of 5.5 and 7.5, respectively. The smaller Au^{3+} concentration favored smaller size and conversely the smaller dextran concentration favored bigger size of AuNPs. AuNPs powders were prepared by spay drying, coagulation, and centrifugation and their sizes were also evaluated. The purity of prepared AuNPs powders was also examined by energy dispersive X-ray (EDX) analysis. Thus, the as-prepared AuNPs stabilized by biocompatible dextran in solution and/or in powder form can be potentially applied in biomedicine and pharmaceutics.

1. Introduction

Synthetic polymers typically polyvinyl alcohol [1], polyvinyl pyrrolidone [2, 3], polyethylene glycol [4], sodium poly-acrylate [5, 6], and surfactants [7, 8] have been commonly used as stabilizers in the bottom-up approach of synthesis of gold nanoparticles (AuNPs). In addition, biopolymers such as chitosan [9–11], alginate [12, 13], hyaluronan [14, 15], gum arabic [16, 17], protein [18, 19], gelatin [20], dextran [21], and glucomannan [22] have been also used as stabilizers of colloidal AuNPs solution. On the other hand, materials based on functionalized AuNPs are regarded as one of the most important nanocomposites especially due to their biological and catalytic activities [23, 24]. For example, assembly of chromanol group on AuNPs could efficiently enhance the activity of the vitamin E-derived antioxidant

[25]; aptamer-conjugated AuNPs showed the advantages in detection of cancer cells by colorimetric assay [26], AuNPs supported on MgO for CO oxidation [27].

Various methods for the synthesis of AuNPs through bottom-up approach, that is, reduction of Au^{3+} ions in solution, have been reported [11] and the most common method is chemical reduction of gold salt precursor using chemical reducing agents. Compared with other methods, gamma Co-60 ray irradiation has been considered as an effective method with several advantages as described by Hien et al. [15]. In addition, gamma Co-60 ray irradiation has been also considered as a green method to noble metal nanoparticle synthesis [28]. Dextran polysaccharide composed of repeated monomeric glucose units with a predominance of 1,6-α-D-glucopyranosyl linkages and annual world production is of about 500 metric tons [29]. Dextran is readily soluble in water and electrolytes with excellent stability and it has wide range of use in food, medical related areas, and biological functions, especially for blood flow improvement by reduction of blood viscosity and inhibition of erythrocyte aggregation [30, 31]. Therefore, dextran stabilized AuNPs are promising to use for intravenous administration as X-ray contrast agent [16, 32], cancer therapy [33, 34] including photothermal cancer therapy [35, 36], and for other purposes of application as well [23].

In the present study, AuNPs were synthesized by gamma Co-60 ray irradiation method using dextran as stabilizer and hydroxyl radical ($\cdot$OH) scavenger. The influence of pH, Au^{3+} and dextran concentrations in aqueous solution on the size of AuNPs was studied. AuNPs powders from AuNPs/dextran solution were prepared by spray drying, coagulation, and centrifugation methods. The purity and size of the AuNPs powders were also examined.

2. Experiments

2.1. Materials and Chemicals. Hydrogen tetrachloroaurate trihydrate ($HAuCl_4 \cdot 3H_2O$) was obtained from Merck, Germany. Dextran (MW 60,000–90,000) was purchased from Himedia, India. Other chemicals were of analytical grade and used as received. Distilled water was used in all experiments. Glassware was treated with regia solution (1V HNO_3 : 3V HCl), washed with distilled water, and dried.

2.2. Preparation of Au^{3+}/Dextran Solution and γ-Irradiation. Two stock solutions particularly 10 mM Au^{3+} and 4% dextran were prepared by dissolving $HAuCl_4 \cdot 3H_2O$ and dextran into water. Au^{3+}/dextran solutions were prepared by pouring Au^{3+} solution into dextran solution with desired different concentration while stirring for about 10 min. pH of Au^{3+}/dextran solution was adjusted to 5.5, 6.5, and 7.5 by 0.5% NH_4OH solution. Then, the prepared Au^{3+}/dextran solutions of 100 ml were put into glass bottles with plastic cap. Irradiation of Au^{3+}/dextran solutions was carried out on the gamma Co-60 irradiator SVST Co-60/B (Hungary) with dose rate of 1.1 kGy/h at VINAGAMMA Center, Ho Chi Minh City, at ambient condition with required doses (~6 kGy for reduction of 1 mM Au^{3+}) as reported in our previous paper [15].

2.3. Characterization of AuNPs/Dextran. The absorption spectra of AuNPs/dextran solutions obtained after γ-irradiation were taken on an UV-Vis spectrophotometer model UV-2401PC (Shimadzu, Japan). The size and size distribution of the AuNPs were characterized from TEM images on transmission electron microscope (TEM) model JEM1010 (JEOL, Japan) and statistically calculated from about 300 particles. The AuNPs/dextran solution from 1 mM Au^{3+}/1% dextran sample was used to prepare AuNPs powder by spray drying on spray dryer model ADL311 (Yamato, Japan) and by centrifugation on ultra centrifuged machine model Ultra 5.0 (Hanil Science Industrial, Korea) at 30,000 rpm for 20 min. After centrifugation, the AuNPs precipitate was isolated and dried in forced air oven (DNF 410, Yamato, Japan) at 60°C.

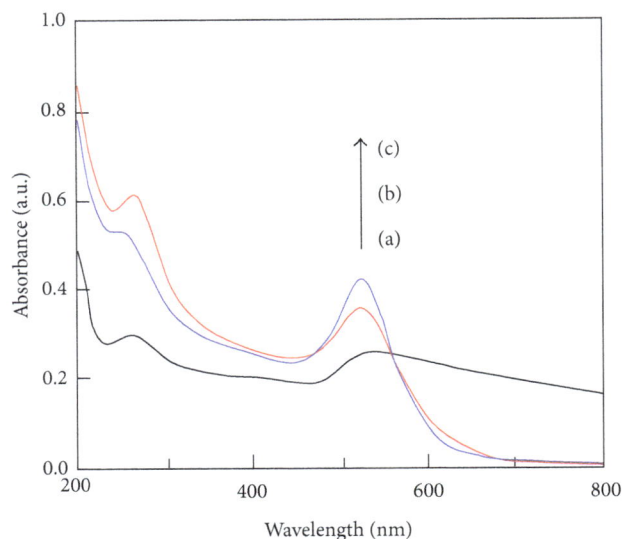

FIGURE 1: UV-Vis spectra of 1 mM AuNPs/1% dextran solutions at different pH: 5.5 (a), 6.5 (b), and 7.5 (c).

The AuNPs powder was also prepared by coagulation with solvent mixture of ethanol/acetone with equal volume ratio. For coagulation of AuNPs/dextran, 8 volumes of solvent mixture were used to mix with 1 volume of AuNPs/dextran solution. The resulting AuNPs/dextran coagulate was filtered and washed several times with solvent and dried in forced air oven at 60°C. The purity of AuNPs/dextran powders was assessed by energy dispersive X-ray spectroscopy (EDX) on a JEOL 6610 LA. The UV-Vis spectra and TEM images of AuNPs powders redissolved in water were also taken.

3. Results and Discussions

After irradiation, the color of Au^{3+}/dextran solution turned from yellow to ruby-red which is surface plasmon resonance characterized for AuNPs [11, 15]. The mechanism of reduction of Au^{3+} to $Au°$ by γ-irradiation method through the reducing species which resulted from water radiolysis was described by Henglein [1].

3.1. Effect of pH. Results of UV-Vis spectra in Figure 1 indicated that the maximum absorption wavelengths (λ_{max}) of AuNPs/dextran solutions were of 543.5, 526.0, and 520.5 nm for pH 5.5, 6.5, and 7.5, respectively.

Results of TEM images and size distribution histograms in Figure 2 indicated that the size of AuNPs decreased with the increase of pH value and particularly the particle sizes were of 35.5, 24.1, and 6.2 nm for pH 5.5, 6.5, and 7.5, respectively. It is clear that slightly basic medium (pH 7.5) was a suitable condition to synthesize AuNPs stabilized by dextran with smaller size (6.2 nm) and narrower particles size distribution (Figures 2(C) and 2(c)) compared to lower pH (6.5) (Figures 2(B) and 2(b)) or acidic medium pH 5.5 (Figures 2(A) and 2(a)) by γ-irradiation method. The obtained results suggested that adjustment of pH of Au^{3+}/dextran

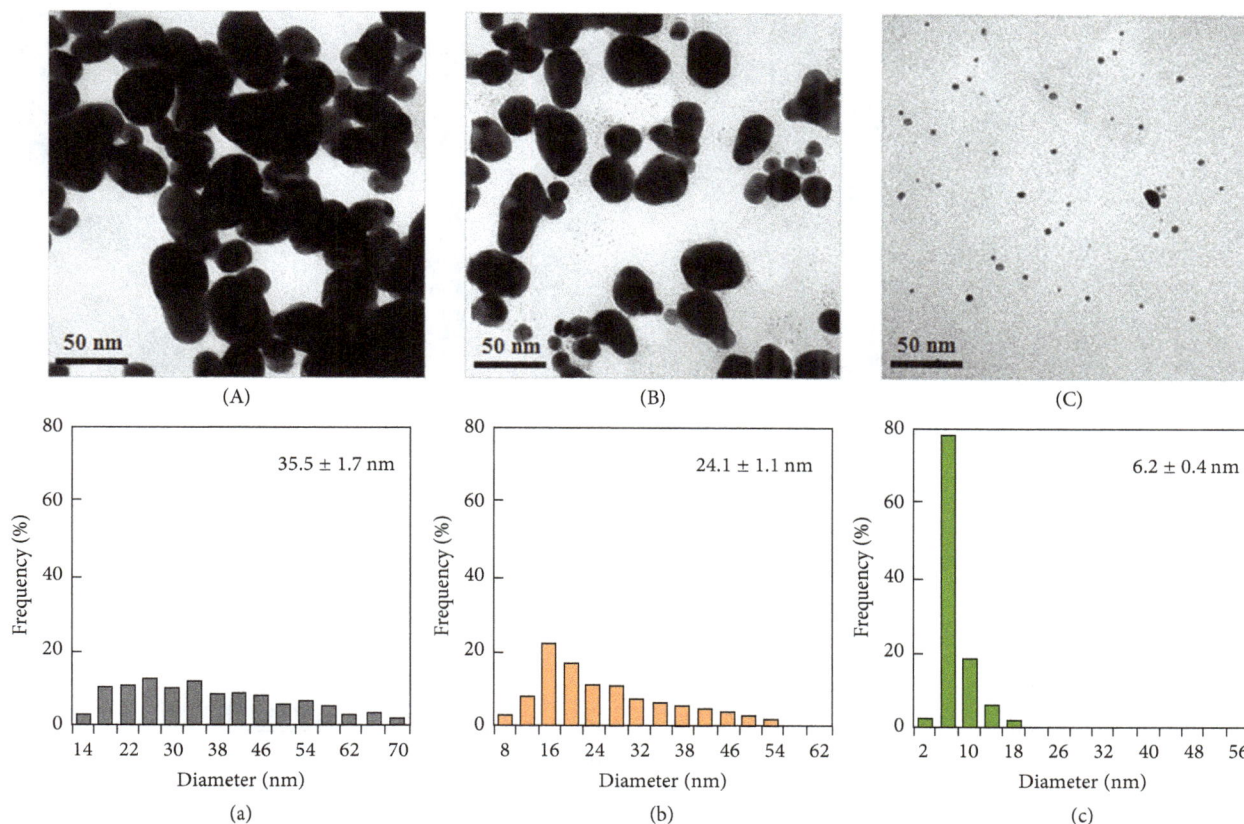

FIGURE 2: TEM images and size distribution histograms of AuNPs/dextran at different pH: 5.5 (A, a), 6.5 (B, b), and 7.5 (C, c).

solution before irradiation significantly affected the size of AuNPs. The same results of the pH effect were also reported by Phu et al. on radiolytic synthesis of AgNPs stabilized by chitosan [37]. Particularly, the sizes of AgNPs were of 15.0 and 7.3 nm for pH 3 and 6, respectively.

3.2. Effect of Au^{3+} Concentration.

In this experiment, AuNPs were synthesized from Au^{3+}/detxran solutions containing 1% (w/v) dextran with various Au^{3+} concentrations particularly 0.5, 1.0, and 2.0 mM. The pH of Au^{3+}/dextran solutions was adjusted to 7.5. The UV-Vis spectra in Figure 3 showed the λ_{max} at 519.5, 520.5, and 524.0 nm for Au^{3+} concentration of 0.5, 1.0, and 2.0 mM, respectively.

The TEM images and the size distribution histograms of the three AuNPs/dextran solutions with various Au^{3+} concentrations were shown in Figure 4. The sizes of AuNPs were of 5.8, 6.3, and 11.9 nm for Au^{3+} concentrations of 0.5, 1.0, and 2.0 mM, respectively. Thus, as Au^{3+} concentration increased the sizes of the AuNPs were bigger and λ_{max} shifted to longer wavelengths. The increase of AuNPs size with the increasing of Au^{3+} concentration was also reported in previous articles for AuNPs/PVP [3], AuNPs/alginate [13], and AuNPs/hyaluronan [15]. The reason for this phenomenon may be due to the development of clusters and the agglomeration among AuNPs when the ratio of stabilizer and Au^{3+} concentration is not high enough [15].

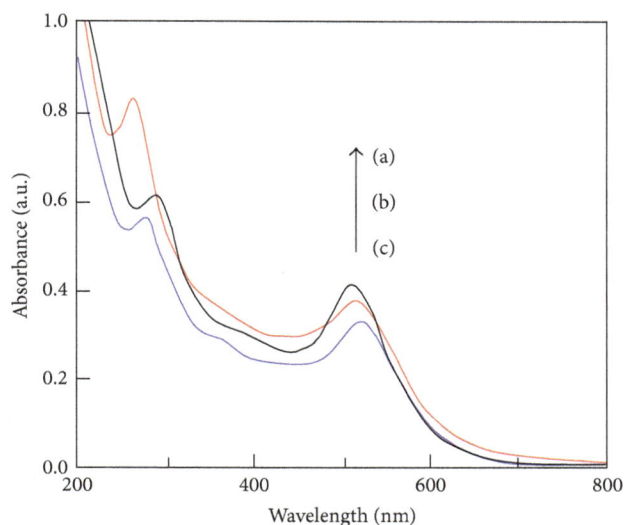

FIGURE 3: UV-Vis spectra of AuNPs/dextran solutions with various Au^{3+} concentrations: 0.5 (a), 1.0 (b), and 2.0 mM (c).

3.3. Effect of Dextran Concentration.

In this experiment, the Au^{3+} concentration was fixed at 1 mM while the dextran concentrations were of 0.5, 1.0, and 2.0% (w/v) and the pH of Au^{3+}/dextran solutions was also adjusted to 7.5. Results of UV-Vis spectra in Figure 5 showed λ_{max} at 522.5, 520.5, and

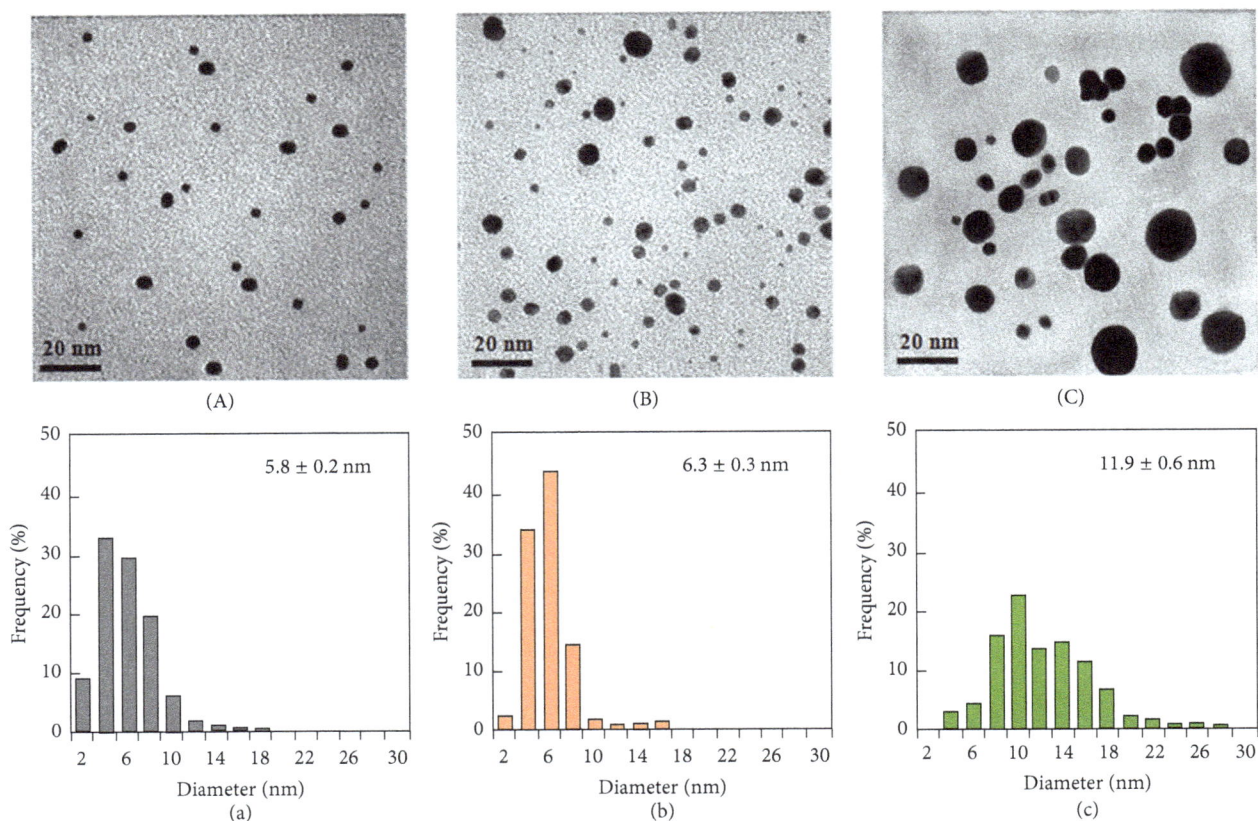

FIGURE 4: TEM images and size distribution histograms of AuNPs/dextran at various Au^{3+} concentrations: 0.5 (A, a), 1.0 (B, b), and 2.0 mM (C, c).

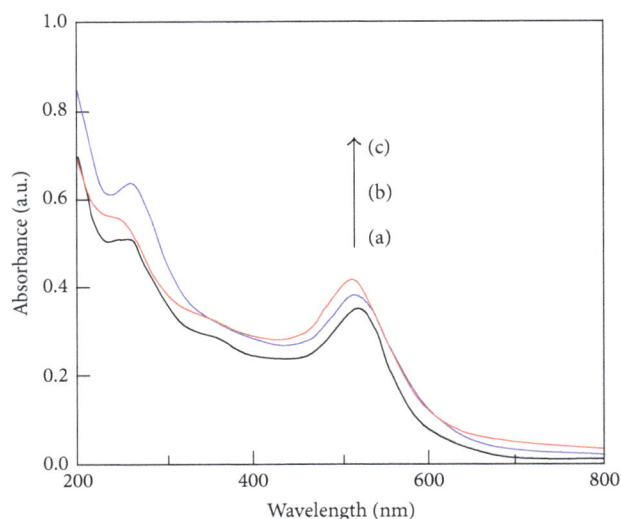

FIGURE 5: UV-Vis spectra of AuNPs/dextran solutions from different dextran concentrations: 0.5 (a), 1.0 (b), and 2.0% (c).

519.0 nm for the dextran concentrations of 0.5, 1.0, and 2.0%, respectively.

The sizes of AuNPs decreased with the increase of dextran concentration particularly the sizes of 8.0, 6.2, and 5.2 nm

for dextran concentrations of 0.5, 1.0, and 2.0%, respectively (Figure 6). Thus, the decrease in the size of AuNPs was not much significant as the increase of dextran concentration from 0.5 to 2.0%.

Dextran has good ability to stabilize AuNPs due to the chain of dextran which consists of –OH and –COR groups which stabilize AuNPs thought steric and electrostatic effects [38]. Simultaneously, dextran acted as free radical scavenger via hydrogen abstraction by $^{\bullet}$H and $^{\bullet}$OH released during radiolysis of water [11, 13, 15]. Therefore, $^{\bullet}$OH scavenger such as alcohols is not necessary to add into Au^{3+}/dextran solution as in the case of Au^{3+}/PVP to prepare colloidal AuNPs solution by γ-irradiation method [3]. As described by Hien et al. [15], mass production of colloidal AuNPs solution can be suitably carried out by gamma Co-60 ray irradiation method. The production rate (Q) of a gamma Co-60 irradiator can be calculated by (1) [39] as follows:

$$Q\,(\text{kg/h}) = \frac{(3{,}600 \times P \times f)}{D}, \tag{1}$$

where D is the absorbed dose (kGy), f is the radiation energy utilization efficiency, and P is source power (kW) (100 kCi is equivalent to 1.48 kW).

Assuming that one gamma Co-60 irradiator with activity of 300 kCi (eqv. 4.44 kW) is used for irradiation of Au^{3+}/dextran solution with Au^{3+} concentration of 1 mM, at

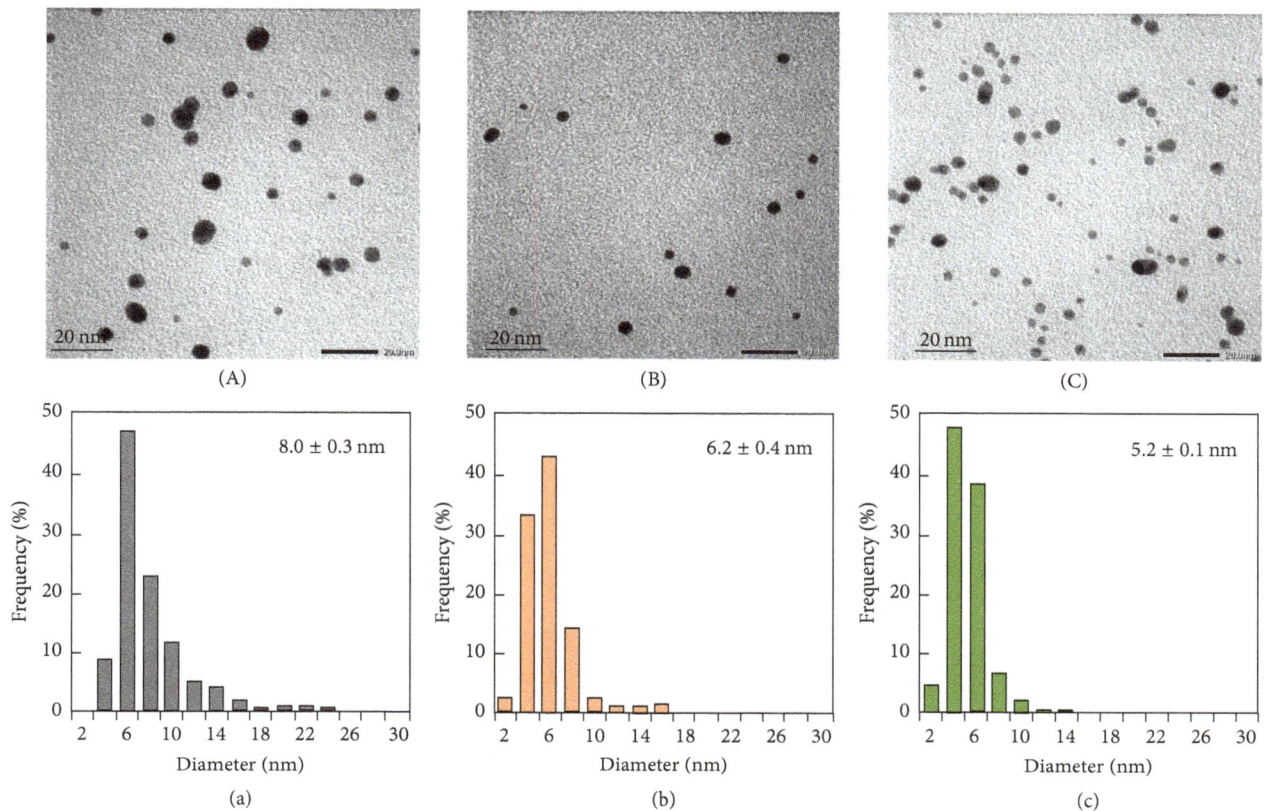

FIGURE 6: TEM images and size distribution histogram of AuNPs/dextran at different dextran concentrations: 0.5 (A, a), 1.0 (B, b), and 2.0% (C, c).

FIGURE 7: Photograph of AuNPs/dextran solution (a), spray drying (b), coagulation (c), and centrifugation (d) of AuNPs powders.

dose of 6 kGy and with $f = 0.35$, then the production rate of AuNPs/dextran solution calculated according to (1) will be of 932 kg/h (~932 litters/h). Besides mass production, the obtained colloidal AuNPs solution is of high purity without residues of excessive chemical reductant and by-products from chemical reductant after reduction process that were typically analyzed for using trisodium citrate as reducing agent [40].

3.4. *Characteristics of AuNPs Powders.* Photograph of AuNPs/dextran solution and powders was presented in Figure 7. The AuNPs powders are brownish-pink to dark-brownish in color depending on the powdering method.

Figure 8 presented the UV-Vis spectra of original AuNPs/dextran solution and AuNPs solutions from AuNPs powders. The values of λ_{max} and the sizes of AuNPs were shown in Table 1.

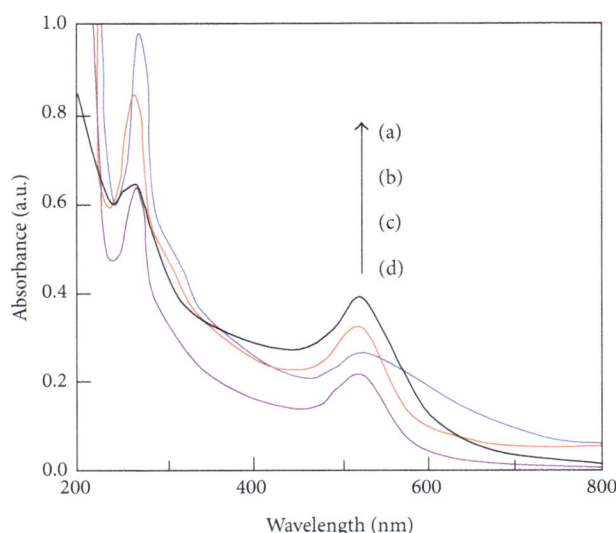

FIGURE 8: UV-Vis spectra of original AuNPs/dextran solution (a), and AuNPs/dextran solutions from AuNPs/dextran powders: spray drying (b), centrifugation (c), and coagulation (d).

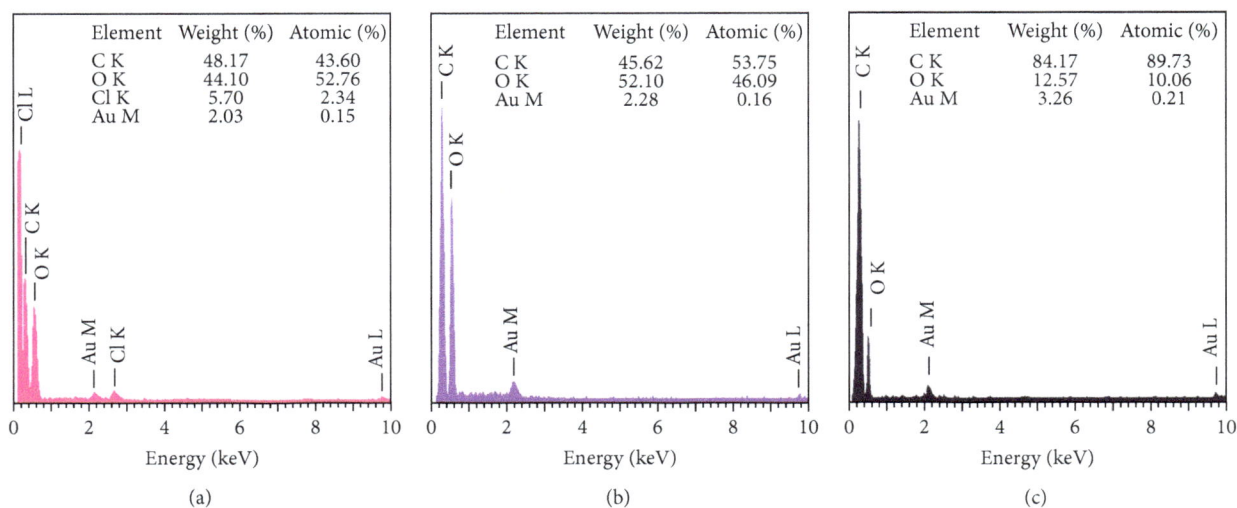

FIGURE 9: EDX spectra of AuNPs/dextran powders: spray drying (a), coagulation (b), and centrifugation (c).

TABLE 1: The values of λ_{max} and the size of AuNPs (d) from AuNPs/dextran solution and powders prepared by different methods.

AuNPs sample	λ_{max}, nm	d, nm
Solution	520.5	6.2 ± 0.4
Spray drying	522.0	7.5 ± 0.5
Coagulation	522.5	8.0 ± 0.3
Centrifugation	523.5	11.2 ± 0.5

It is interesting to note that the value of λ_{max} and the size of AuNPs powders made from spray drying and coagulation were almost not much changed in comparison with those of original AuNPs/dextran solution, but these parameters increased for the AuNPs made from centrifugation

(Table 1). The reason may be due to aggregation of AuNPs which occurred during centrifugation with ultrahigh speed (30,000 rpm). To the best of our knowledge, no research of the effect of AuNPs powder preparation methods on the characteristics of AuNPs, particularly λ_{max} and the size, has been reported yet except the study of the purity of AuNPs synthesized by using trisodium citrate as reducing agent and purified by centrifugation method which was reported [40]. Nevertheless, the final form of AuNPs in this study was not powder but referred to as "treated AuNPs suspension."

Results of the EDX spectra in Figure 9 indicated that the AuNPs powders prepared by coagulation and centrifugation did not contain chlorine (Figures 9(b) and 9(c)); however the AuNPs powder prepared by spray drying was contaminated by 5.7% chlorine (Figure 9(a)). Thus, it can be deduced that AuNPs powders prepared by coagulation and centrifugation were of high purity. In other words, the AuNPs/dextran powders obtained by coagulation and centrifugation methods were effectively purified. However, chlorine content in AuNPs powder should be analyzed with more precise methods.

The stability of colloidal AuNPs solution depends on various factors such as pH, dielectric constant, and concentration of ligand around the particles [19]. Furthermore, the stability of colloidal AuNPs solution stored at low temperature is better than at high ones [40]. In general, polysaccharides such as chitosan, hyaluronan, alginate, and dextran with oxygen rich structures in hydroxyl and ether groups lead to tightly binding polysaccharides to nanoparticles via steric and electrostatic interaction [38]. However, for a long time of storage, the AuNPs in colloidal AuNPs solution will be gradually aggregated to form bigger sizes due to sedimentation phenomenon of AuNPs in colloidal system. In addition, storage and transportation of colloidal AuNPs solution are not always convenient. In fact, it was observed that several samples of the synthesized AuNPs/dextran solutions stored at normal condition were randomly attacked by fungi. Therefore, in those aspects AuNPs powder is much more suitable. Besides, AuNPs/dextran powder holds great potential for biomedical and pharmaceutical applications; especially those AuNPs/dextran powders can be sterilized suitably by radiation method (gamma Co-60 ray, electron beam) [41] for use in intravenous administration for X-ray contrast imaging, photothermal cancer therapy, and in other purposes of application as well.

4. Conclusions

Dextran stabilized AuNPs with diameter in the range of 6–35 nm were synthesized by gamma Co-60 ray irradiation of Au^{3+} in dextran solution. The gamma Co-60 ray irradiation method for production of AuNPs can be favorably carried out on large scale. The AuNPs sizes can be controlled by varying pH, Au^{3+}, and dextran concentrations. AuNPs powders with the almost unchanged size were also prepared by spray drying, coagulation, and centrifugation methods. Due to the biocompatibility of dextran, the purity of AuNPs/dextran colloidal solution as well as the AuNPs/dextran powders and unique attributes of AuNPs, AuNPs/dextran can be potentially applied in biomedicine and pharmaceutics especially in X-ray contrast imaging and photothermal cancer therapy.

Conflicts of Interest

The authors declare that they have no conflicts of interest.

Acknowledgments

This research is funded by Vietnam National Foundation for Science and Technology Development (NAFOSTED) under Grant no. "104.06-2014.87."

References

[1] A. Henglein, "Radiolytic preparation of ultrafine colloidal gold particles in aqueous solution: Optical spectrum, controlled growth, and some chemical reactions," *Langmuir*, vol. 15, no. 20, pp. 6738–6744, 1999.

[2] R. Seoudi, A. A. Fouda, and D. A. Elmenshawy, "Synthesis, characterization and vibrational spectroscopic studies of different particle size of gold nanoparticle capped with polyvinylpyrrolidone," *Physica B: Condensed Matter*, vol. 405, no. 3, pp. 906–911, 2010.

[3] T. Li, H. G. Park, and S.-H. Choi, "γ-Irradiation-induced preparation of Ag and Au nanoparticles and their characterizations," *Materials Chemistry and Physics*, vol. 105, no. 2-3, pp. 325–330, 2007.

[4] G. Zhang, Z. Yang, W. Lu et al., "Influence of anchoring ligands and particle size on the colloidal stability and in vivo biodistribution of polyethylene glycol-coated gold nanoparticles in tumor-xenografted mice," *Biomaterials*, vol. 30, no. 10, pp. 1928–1936, 2009.

[5] J. Belloni, M. Mostafavi, H. Remita, J.-L. Marignier, and M.-O. Delcourt, "Radiation-induced synthesis of mono- and multimetallic clusters and nanocolloids," *New Journal of Chemistry*, vol. 22, no. 11, pp. 1239–1255, 1998.

[6] I. Hussain, M. Brust, A. J. Papworth, and A. I. Cooper, "Preparation of acrylate-stabilized gold and silver hydrosols and gold-polymer composite films," *Langmuir*, vol. 19, no. 11, pp. 4831–4835, 2003.

[7] T. K. Sau, A. Pal, N. R. Jana, Z. L. Wang, and T. Pal, "Size controlled synthesis of gold nanoparticles using photochemically prepared seed particles," *Journal of Nanoparticle Research*, vol. 3, no. 4, pp. 257–261, 2001.

[8] J. Duy, L. B. Connell, W. Eck, S. D. Collins, and R. L. Smith, "Preparation of surfactant-stabilized gold nanoparticle-peptide nucleic acid conjugates," *Journal of Nanoparticle Research*, vol. 12, no. 7, pp. 2363–2369, 2010.

[9] D. R. Bhumkar, H. M. Joshi, M. Sastry, and V. B. Pokharkar, "Chitosan reduced gold nanoparticles as novel carriers for transmucosal delivery of insulin," *Pharmaceutical Research*, vol. 24, no. 8, pp. 1415–1426, 2007.

[10] C. Sun, R. Qu, H. Chen et al., "Degradation behavior of chitosan chains in the 'green' synthesis of gold nanoparticles," *Carbohydrate Research*, vol. 343, no. 15, pp. 2595–2599, 2008.

[11] N. N. Duy, D. X. Du, D. Van Phu, L. A. Quoc, B. D. Du, and N. Q. Hien, "Synthesis of gold nanoparticles with seed enlargement size by γ-irradiation and investigation of antioxidant activity," *Colloids and Surfaces A: Physicochemical and Engineering Aspects*, vol. 436, pp. 633–638, 2013.

[12] G. Sonavane, K. Tomoda, and K. Makino, "Biodistribution of colloidal gold nanoparticles after intravenous administration: effect of particle size," *Colloids and Surfaces B: Biointerfaces*, vol. 66, no. 2, pp. 274–280, 2008.

[13] N. Tue Anh, D. Van Phu, N. Ngoc Duy, B. Duy Du, and N. Quoc Hien, "Synthesis of alginate stabilized gold nanoparticles by γ-irradiation with controllable size using different Au3+ concentration and seed particles enlargement," *Radiation Physics and Chemistry*, vol. 79, no. 4, pp. 405–408, 2010.

[14] H. Lee, K. Lee, I. K. Kim, and T. G. Park, "Synthesis, characterization, and in vivo diagnostic applications of hyaluronic acid immobilized gold nanoprobes," *Biomaterials*, vol. 29, no. 35, pp. 4709–4718, 2008.

[15] N. Q. Hien, D. Van Phu, N. N. Duy, and L. A. Quoc, "Radiation synthesis and characterization of hyaluronan capped gold nanoparticles," *Carbohydrate Polymers*, vol. 89, no. 2, pp. 537–541, 2012.

[16] V. Kattumuri, K. Katti, and S. Bhaskaran, "Gum arabic as a phytochemical construct for the stabilization of gold nanoparticles: in vivo pharmacokinetics and X-ray-contrast-imaging studies," *Small*, vol. 3, no. 2, pp. 333–341, 2007.

[17] C.-C. Wu and D.-H. Chen, "Facile green synthesis of gold nanoparticles with gum arabic as a stabilizing agent and reducing agent," *Gold Bulletin*, vol. 43, no. 4, pp. 234–240, 2010.

[18] P. Ravindra, "Protein-mediated synthesis of gold nanoparticles," *Materials Science and Engineering: B Advanced Functional Solid-State Materials*, vol. 163, no. 2, pp. 93–98, 2009.

[19] A. Akhavan, H. R. Kalhor, M. Z. Kassaee, N. Sheikh, and M. Hassanlou, "Radiation synthesis and characterization of protein stabilized gold nanoparticles," *Chemical Engineering Journal*, vol. 159, no. 1-3, pp. 230–235, 2010.

[20] M. P. Neupane, S. J. Lee, I. S. Park et al., "Synthesis of gelatin-capped gold nanoparticles with variable gelatin concentration," *Journal of Nanoparticle Research*, vol. 13, no. 2, pp. 491–498, 2011.

[21] S. Nath, C. Kaittanis, A. Tinkham, and J. M. Perez, "Dextran-coated gold nanoparticles for the assessment of antimicrobial susceptibility," *Analytical Chemistry*, vol. 80, no. 4, pp. 1033–1038, 2008.

[22] Z. Gao, R. Su, R. Huang, W. Qi, and Z. He, "Glucomannan-mediated facile synthesis of gold nanoparticles for catalytic reduction of 4-nitrophenol," *Nanoscale Research Letters*, vol. 9, no. 1, pp. 1–8, 2014.

[23] M.-C. Daniel and D. Astruc, "Gold nanoparticles: assembly, supramolecular chemistry, quantum-size-related properties, and applications toward biology, catalysis, and nanotechnology," *Chemical Reviews*, vol. 104, no. 1, pp. 293–346, 2004.

[24] S. F. Chen, J. P. Li, K. Qian et al., "Large scale photochemical synthesis of M@TiO$_2$ nanocomposites (M = Ag, Pd, Au, Pt) and their optical properties, CO oxidation performance, and antibacterial effect," *Nano Research*, vol. 3, no. 4, pp. 244–255, 2010.

[25] Z. Nie, K. J. Liu, C.-J. Zhong et al., "Enhanced radical scavenging activity by antioxidant-functionalized gold nanoparticles: A novel inspiration for development of new artificial antioxidants," *Free Radical Biology & Medicine*, vol. 43, no. 9, pp. 1243–1254, 2007.

[26] C. D. Medley, J. E. Smith, Z. Tang, Y. Wu, S. Bamrungsap, and W. Tan, "Gold nanoparticle-based colorimetric assay for the direct detection of cancerous cells," *Analytical Chemistry*, vol. 80, no. 4, pp. 1067–1072, 2008.

[27] S. A. C. Carabineiro, N. Bogdanchikova, A. Pestryakov, P. B. Tavares, L. S. G. Fernandes, and J. L. Figueiredo, "Gold nanoparticles supported on magnesium oxide for CO oxidation," *Nanoscale Research Letters*, vol. 6, article no. 435, pp. 1–6, 2011.

[28] M. Chandra and P. K. Das, "Green routes to noble metal nanoparticle synthesis," *International Journal of Green Nanotechnology: Physics and Chemistry*, vol. 1, no. 1, pp. P10–P25, 2009.

[29] A. N. de Belder, "Dextran, Industrial gums: polysaccharides and their derivatives," in *A. N. de Belder*, R. L. Whistler and J. L. BeMiller, Eds., pp. 399–425, Academic Press, 1993.

[30] A. N. de Belder, "Medical applications of dextran and its derivative, polysaccharides in medicinal applications," in *A. N. de Belder*, S. Dumitriu, Ed., pp. 505–523, Marcel Dekker, New York, NY, USA, 1996.

[31] K. P. Bankura, D. Maity, M. M. R. Mollick et al., "Synthesis, characterization and antimicrobial activity of dextran stabilized silver nanoparticles in aqueous medium," *Carbohydrate Polymers*, vol. 89, no. 4, pp. 1159–1165, 2012.

[32] J. F. Hainfeld, D. N. Slatkin, T. M. Focella, and H. M. Smilowitz, "Gold nanoparticles: a new X-ray contrast agent," *British Journal of Radiology*, vol. 79, no. 939, pp. 248–253, 2006.

[33] O. S. Muddineti, B. Ghosh, and S. Biswas, "Current trends in using polymer coated gold nanoparticles for cancer therapy," *International Journal of Pharmaceutics*, vol. 484, no. 1-2, pp. 252–267, 2015.

[34] R. Geetha, T. Ashokkumar, S. Tamilselvan, K. Govindaraju, M. Sadiq, and G. Singaravelu, "Green synthesis of gold nanoparticles and their anticancer activity," *Cancer Nanotechnology*, vol. 4, no. 4-5, pp. 91–98, 2013.

[35] Y. Ren, H. Qi, Q. Chen, and L. Ruan, "Thermal dosage investigation for optimal temperature distribution in gold nanoparticle enhanced photothermal therapy," *International Journal of Heat and Mass Transfer*, vol. 106, pp. 212–221, 2017.

[36] C. Yao, L. Zhang, J. Wang et al., "Gold nanoparticle mediated phototherapy for cancer," *Journal of Nanomaterials*, vol. 2016, Article ID 5497136, 29 pages, 2016.

[37] D. V. Phu, V. T. K. Lang, N. T. K. Lan et al., "Synthesis and antimicrobial effects of colloidal silver nanoparticles in chitosan by γ-irradiation," *Journal of Experimental Nanoscience*, vol. 5, no. 2, pp. 169–179, 2010.

[38] L. Huang, M. Zhai, J. Peng, L. Xu, J. Li, and G. Wei, "Synthesis, size control and fluorescence studies of gold nanoparticles in carboxymethylated chitosan aqueous solutions," *Journal of Colloid and Interface Science*, vol. 316, no. 2, pp. 398–404, 2007.

[39] K. Makuuchi, *An introduction to radiation vulcanization of natural rubber latex*, T.R.I. Global Co. Ltd, Bangkok, Thailand, 2003.

[40] S. K. Balasubramanian, L. Yang, L.-Y. L. Yung, C.-N. Ong, W.-Y. Ong, and L. E. Yu, "Characterization, purification, and stability of gold nanoparticles," *Biomaterials*, vol. 31, no. 34, pp. 9023–9030, 2010.

[41] ISO 11137 – 2:2013, Sterilization of health care products – Radiation – Part 2: Establishing the sterilization dose.

11

Robust Nonlinear Regression in Enzyme Kinetic Parameters Estimation

Maja Marasović,[1] Tea Marasović,[2] and Mladen Miloš[1]

[1]*Faculty of Chemistry and Technology, University of Split, Ruđera Boškovića 35, 21000 Split, Croatia*
[2]*Faculty of Electrical Engineering, Mechanical Engineering and Naval Architecture, University of Split,*
 Ruđera Boškovića 32, 21000 Split, Croatia

Correspondence should be addressed to Tea Marasović; tmarasov@fesb.hr

Academic Editor: Murat Senturk

Accurate estimation of essential enzyme kinetic parameters, such as K_m and V_{max}, is very important in modern biology. To this date, linearization of kinetic equations is still widely established practice for determining these parameters in chemical and enzyme catalysis. Although simplicity of linear optimization is alluring, these methods have certain pitfalls due to which they more often then not result in misleading estimation of enzyme parameters. In order to obtain more accurate predictions of parameter values, the use of nonlinear least-squares fitting techniques is recommended. However, when there are outliers present in the data, these techniques become unreliable. This paper proposes the use of a robust nonlinear regression estimator based on modified Tukey's biweight function that can provide more resilient results in the presence of outliers and/or influential observations. Real and synthetic kinetic data have been used to test our approach. Monte Carlo simulations are performed to illustrate the efficacy and the robustness of the biweight estimator in comparison with the standard linearization methods and the ordinary least-squares nonlinear regression. We then apply this method to experimental data for the tyrosinase enzyme (EC 1.14.18.1) extracted from *Solanum tuberosum*, *Agaricus bisporus*, and *Pleurotus ostreatus*. The results on both artificial and experimental data clearly show that the proposed robust estimator can be successfully employed to determine accurate values of K_m and V_{max}.

1. Introduction

Enzymes are molecules that act as biological catalysts and are responsible for maintaining virtually all life processes. Most enzymes are proteins, although a few are catalytic RNA molecules. Like all catalysts, enzymes increase the rate of chemical reactions without themselves undergoing any permanent chemical change in a process. They achieve their effect by temporarily binding to the substrate and, in doing so, lowering the activation energy needed to convert it to a product. The study of the rate at which an enzyme works is called enzyme kinetics and it is often regarded as one of the most fascinating research areas in biochemistry [1].

Mathematically, the relationship between substrate concentration and reaction rate under isothermal conditions for many of enzyme-catalyzed reactions can be modeled by the Michaelis-Menten equation [2]:

$$v = \frac{V_{max}s}{s + K_m}, \tag{1}$$

where v denotes a reaction rate, s is a substrate concentration, V_{max} is the maximum initial velocity, which is theoretically attained when the enzyme has been "saturated" by an infinite concentration of a substrate, and K_m is the Michaelis constant, representing a measure of affinity of the enzyme-substrate interaction. By definition, K_m is equal to the concentration of the substrate at half maximum initial velocity. The Michaelis constant, K_m, is an intrinsic parameter of enzyme-catalyzed reactions and it is significant for its biological function [3].

Three most common methods, available in the literature, for determining the parameters of Michaelis-Menten equation based on a series of measurements of velocity v as a function of substrate concentration, are Lineweaver-Burk plot, also known as the double reciprocal plot, Eadie-Hofstee plot, and Hanes-Woolf plot. All three of these methods are linearized models that transform the original Michaelis-Menten equation into a form which can be graphed as a straight line.

Lineweaver-Burk [4] (LB) plot, still the most popular and favored plot amongst the researchers, is defined by an equation:

$$\frac{1}{v} = \frac{1}{V_{\max}} + \frac{K_m}{V_{\max}}\frac{1}{s}. \tag{2}$$

The y-intercept in this plot is $1/V_{\max}$, the x-intercept in second quadrant represents $-1/K_m$, and the slope of the line is $K_m/V_{\max}$.

Eadie-Hofstee [5] (EH) plot is a semireciprocal plot of v versus v/s. The linear equation has the following form:

$$v = V_{\max} - K_m\frac{v}{s}, \tag{3}$$

where the y-intercept is $V_{\max}$ and the slope is K_m.

In Hanes-Woolf [6] (HW) plot, s/v is plotted against s. The linear equation is given by

$$\frac{s}{v} = \frac{K_m}{V_{\max}} + \frac{1}{V_{\max}}s, \tag{4}$$

where the y-intercept is $K_m/V_{\max}$ and the slope is $1/V_{\max}$.

In all of the above-described linear transformations, linear regression is used to estimate the slope and intercept of the straight line and afterwards K_m and $V_{\max}$ are computed from the straight line parameters. Although these methods are very useful for data visualization and are still widely employed in enzyme kinetic studies, each of them possesses certain deficiencies, which make them prone to errors. For instance, Lineweaver-Burk plot has the disadvantage of compressing the data points at high substrate concentrations into a small region and emphasizing the points at lower substrate concentrations, which are often the least accurate [7]. The V-intercept in Lineweaver-Burk plot is equivalent to inverse of $V_{\max}$ due to which any small error in measurement gets magnified. Similarly, the Eadie-Hofstee plot has the disadvantage that v appears on both axes; thus, any experimental error will also be present in both axes. In addition, experimental errors or uncertainties are propagated unevenly and become larger over the abscissa thereby giving more weight to smaller values of v/s. Hanes-Woolf plot is the most accurate of the three; however, its major drawback is that again neither ordinate nor abscissa represents independent values: both are dependent on substrate concentration.

In order to reduce the errors due to the linearization of parameters, Wilkinson [8] proposed the use of least-squares nonlinear regression for more accurate estimation of enzyme kinetic parameters. Nonlinear regression allows direct determination of parameter values from untransformed data points. The process starts with initial estimates and then iteratively converges on parameter estimates that provide the best fit of the underlying model to the actual data points [9, 10]. The algorithms used include the Levenberg-Marquardt method, the Gauss-Newton method, the steepest-descent method, and simplex minimization. Numerous software packages, such as Excel, MATLAB, and GraphPrism, nowadays include readily available routines and scripts to perform nonlinear least-squares fitting [11, 12].

Least-squares nonlinear regression has been criticized for its performance in dealing with experimental data. This is mainly due to the fact that implicit assumptions related with nonlinear regression are in general not met in the context of deviations that appear as a result of biological errors (e.g., variations in the enzyme preparations due to oxidation or contaminations) and/or experimental errors (e.g., variations in measured volume of substrates and enzymes, imprecisions of the instrumentation). With the presence of outliers or influential observations in the data, the ordinary least-squares method can result in misleading values for the parameters of the nonlinear regression and estimates may no longer be reliable [13].

In this paper, we propose the use of robust nonlinear regression estimator based on modified Tukey's biweight function for determining the parameters of Michaelis-Menten equation using experimental measurements in enzyme kinetics. The main idea is to fit a model to the data that gives resilient results in the presence of influential observations and/or outliers. To the best of our knowledge, this is the first study that examines the use of this technique for application in Michaelis-Menten enzyme analysis. We employ Monte Carlo simulations to validate the efficacy of the proposed procedure in comparison with the ordinary least-squares method and Eadie-Hofstee, Hanes-Woolf and Lineweaver-Burk plots. In addition, we illustrate the viability of our method by estimating the kinetic parameters of tyrosinase, an important enzyme widely distributed in microorganisms, animals, and plants, responsible for melanin production in mammal and enzymatic browning in plants, extracted from potato and two edible mushrooms.

The remainder of the paper is organized as follows. Section 2 provides a brief overview of the robust estimation model. Section 3 describes the experimental setup used in this research and the diagnostics that will be used to evaluate the effectiveness of the proposed procedure in determination of enzyme kinetic parameters. Results are discussed in Section 4. Finally, Section 5 summarizes the paper with a few concluding remarks.

2. Robust Nonlinear Regression

Nonlinear regression, same as linear regression, relies heavily on the assumption that the scatter of data around the ideal curve follows, at least approximately, a Gaussian or normal distribution. This assumption leads to the well-known regression goal: to minimize the sum of the squares of the vertical distances (a.k.a residuals) between the points and the curve. In practice, however, this assumption does not always hold true. The analytical data often contains outliers that can play havoc with standard regression methods based on the normality assumption, causing them to produce more or less strongly biased results, depending on the magnitude of deviation and/or sensitivity of the procedure. It is not unusual to find an average of 10% of outlying observations in data set of some processes [14].

Outliers are most commonly thought to be extreme values which are a result of measurement or experimental errors. Barnett and Lewis [15] provide a more cautious definition of the term outlier, describing it as the observation (or subset of observations) that appears to be inconsistent with the remainder of the dataset. This definition also includes the observations that do not follow the majority of the data, such as values that have been measured correctly but are, for one reason or another, far away from other data values, while the formulation "appears to be inconsistent" reflecting the subjective judgement of the observer whether or not an observation is declared to be outlying.

The ordinary least-squares (OLS) estimate $\widehat{\beta}_{LS}$ of the parameter vector $\beta = [\beta_1, \beta_2, \ldots, \beta_n]$ is obtained as the solution of the problem:

$$\widehat{\beta}_{LS} = \arg\min_{\beta} \frac{1}{n} \sum_{i=1}^{n} r_i(x, \beta)^2$$

$$= \arg\min_{\beta} \frac{1}{n} \sum_{i=1}^{n} (y_i - f(x_i, \beta))^2, \tag{5}$$

where n denotes the number of observations, $x = [\mathbf{x_1}, \mathbf{x_2}, \ldots, \mathbf{x_n}]^T$ is a $n \times p$ matrix, whose rows are p-dimensional vectors of predictor variables (or regressors), $y = [y_1, y_2, \ldots, y_n]^T$ is a $n \times 1$ vector of responses, and $f(x_i, \beta)$ is model function. Since all data points are attributed the same weights, OLS implicitly favors the observations with very large residuals and, consequently, the estimated parameters end up distorted if outliers are present.

In order to achieve robustness in coping with the problem of outliers, Huber [16] introduced a class of so-called M-estimators, for which the sum of function ρ of the residuals is minimized. The resulting vector of parameters $\widehat{\beta}_M$ estimated by an M-estimator is then

$$\widehat{\beta}_M = \arg\min_{\beta} \sum_{i=1}^{n} \rho\left(\frac{r_i}{\sigma}\right). \tag{6}$$

The residuals are standardized by a measure of dispersion σ to guarantee scale equivariance (i.e., independence with respect to the measurement units of the dependent variable). Function $\rho(\cdot)$ must be even, nondecreasing for positive values, and less increasing than the square.

The minimization in (6) can always be done directly. However, often it is simpler to differentiate ρ function with respect to β and solve for the root of the derivative. When this differentiation is possible, the M-estimator is said to be of ψ-type. Otherwise, the M-estimator is said to be of ρ-type.

Let $\psi = \rho'$ be the derivative of ρ. Assuming σ is known and defining weights $w_i = \psi(r_i/\sigma)/r_i$, the estimates $\widehat{\beta}_M$ can be obtained by solving the system of equations:

$$\sum_{i=1}^{n} w_i^2 r_i^2 = 0. \tag{7}$$

The weights are dependent upon the residuals, the residuals are dependent upon the estimated coefficients, and the estimated coefficients are dependent upon the weights. Hence, to solve for M-estimators, an iteratively reweighted least-squares (IRLS) algorithm is employed. Starting from some

initial estimates $\widehat{\beta}^{(0)}$, at each iteration t until it converges, this algorithm computes the residuals $r_i^{(t-1)}$ and the associated weights $w_i^{(t-1)} = w[r_i^{(t-1)}]$ from the previous iteration and yields new weighted least-squares estimates.

2.1. Objective Function. Several ρ functions can be used. Here we opted for Tukey's biweight [17] or bisquare function defined as

$$\rho(z_i) = \begin{cases} \frac{c^2}{6}\left(1 - \left[1 - \left(\frac{z_i}{c}\right)^2\right]^3\right) & \text{if } |z_i| \leq c \\ \frac{c^2}{6} & \text{if } |z_i| > c \end{cases}, \tag{8}$$

where c is a tuning constant and $z_i = r_i/\sigma$.

The corresponding $\psi(z_i)$ function is

$$\psi(z_i) = \begin{cases} z_i\left[1 - \left(\frac{z_i}{c}\right)^2\right]^2 & \text{if } |z_i| \leq c \\ 0 & \text{if } |z_i| > c \end{cases}. \tag{9}$$

Tukey's biweight estimator has a smoothly redescending ψ function that prevents extreme outliers to affect the calculation of the biweight estimates by assigning them a zero weighting. As can be seen in Figure 1, the weights for the biweight estimator decline as soon as z departs from 0 and are 0 for $|z| > c$. Smaller values of c produce more resistance to outliers, but at the expense of lower efficiency when the errors are normally distributed. The tuning constant is generally picked to give reasonably high efficiency in normal case; in particular $c = 4.685$ produces a 95% efficiency when the errors are normal, while guaranteeing resistance to contamination of up to 10% of outliers.

In an application, an estimate of the standard deviation of the errors is needed in order to use these results. Usually a robust measure of spread is used in preference to the standard deviation of the residuals. A common approach is to take $\widehat{\sigma} = \text{MAD}/0.6745$, where MAD is the median absolute deviation. Despite having the best possible breakout point of 50%, the MAD is not without its weaknesses. It exhibits superior statistical efficacy for the contaminated data (i.e., the data that contains extreme scores); however, when the data approaches a normal distribution, the MAD is only 37% efficient. Furthermore, it is ill-suited for asymmetrical distributions, since it attaches equal importance to positive and negative deviations from location estimate.

Hence, the scale parameter σ is computed using Rousseeuw-Croux estimator Q_n [18]:

$$Q_n = d\left\{|x_i - x_j|; \; i < j\right\}_{(k)}, \tag{10}$$

where d is a calibration factor and $k = \binom{h}{2} \approx \binom{n}{2}/4$, where $h = n/2 + 1$ is roughly half the number of observations. The estimator Q_n has the optimal 50% breakdown point; it is equally suitable for both symmetrical and asymmetrical distributions and considerably more efficient (about 82%) than the MAD under a Gaussian distribution.

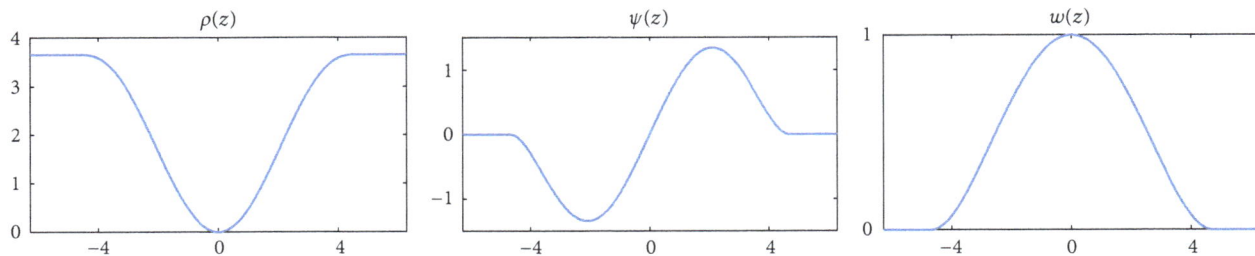

FIGURE 1: Tukey's biweight estimator objective, ψ, and weight functions for $c = 4.685$.

3. Experimental Setup

To illustrate the efficacy of the proposed approach we use both artificial data, generated from the Monte Carlo simulations, and the experimental data for the tyrosinase enzyme (EC 1.14.18.1) extracted from three different sources.

3.1. Monte Carlo Simulations. Simulation studies are useful for gaining insight into the examined algorithms strengths and weaknesses, such as robustness, against number of variable factors. There are three outlier scenarios and a total of 18 different situations considered in this research. The data sets are generated from the model:

$$y_i = \frac{(\theta_1 x_i)}{(\theta_2 + x_i)} + \epsilon_i, \quad i = 1, 2, \ldots, n, \tag{11}$$

where regression coefficients are fixed $\theta = [80 \quad 240]$ for each $i = 1, 2, \ldots, n$. The explanatory variables x_i are set to $100 \cdot i$ and a zero mean unit variance random number with Gaussian density ϵ_i is added as measurement error.

The factors considered in this simulation are (1) level of outlier contamination: 10% or 20%, (2) sample size: small ($n = 10$), medium ($n = 20$), or large ($n = 50$), and (3) distances of outliers from clean observations: 10 standard deviations, 50 standard deviations, or 100 standard deviations. There are 1200 replications for each scenario and all simulations are carried out in MATLAB. The 3 scenarios and the 18 situations considered in this research are summarized in Table 1.

These simulated data are then used to estimate the values of K_m and V_{max} using different fitting techniques. The mean estimated values of K_m and V_{max} for a particular scenario and fitting technique are subsequently calculated by averaging K_m and V_{max} values obtained in each of the 1200 trials. The estimator efficacy is assessed in terms of its bias, precision, and accuracy. Bias is defined as an absolute difference between mean estimated parameter values and known parameter values:

$$\text{Bias} = \left| \frac{\sum_{j=1}^{N} \theta_{ij}}{N} - \widehat{\theta}_i \right|, \tag{12}$$

where N is the total number of replications in the simulated scenario. The term precision refers to the absence of random errors or variability. It is measured by the coefficient of

TABLE 1: The 18 situations considered in the simulations.

Scenario	Situation	Sample size	Outliers	σ
1	1	10	10%	10
	2		20%	10
	3		10%	50
	4		20%	50
	5		10%	100
	6		20%	100
2	7	20	10%	10
	8		20%	10
	9		10%	50
	10		20%	50
	11		10%	100
	12		20%	100
3	13	50	10%	10
	14		20%	10
	15		10%	50
	16		20%	50
	17		10%	100
	18		20%	100

variation (C_v), that is, the standard deviation expressed as a percentage of the mean:

$$C_v = \frac{\sigma}{\theta_{ij}} \cdot 100. \tag{13}$$

The prediction accuracy is defined as the overall distance between estimated values and true values. The accuracy is measured by a normalized mean squared error (NMSE), that is, the mean of the squared differences between the estimated and the known parameter values normalized by a mean of estimated data:

$$\text{NMSE} = \frac{1}{N} \sum_{j=1}^{N} \left(\frac{\theta_{ij} - \widehat{\theta}_i}{\overline{\theta_{ij}}} \right)^2, \tag{14}$$

where again N is the total number of replications in the simulated scenario.

3.2. Enzyme Data Sets. Tyrosinase (EC 1.14.18.1) is a ubiquitous enzyme responsible for melanization in animals and plants [19, 20]. In the presence of molecular oxygen,

TABLE 2: Input experimental kinetic data sets.

Concentration [μmol]	Reaction rate [μmol/min]		
	Ab	Po	Potato
25	0.0243	1.1×10^{-4}	0.0004
50	0.0292	2.6×10^{-4}	0.0011
100	0.0546	2.9×10^{-4}	0.0018
250	0.1388	3.7×10^{-4}	0.0023
500	0.1726	6.9×10^{-4}	0.0030
1000	0.2374	13.1×10^{-4}	0.0101
2500	0.3023	26.9×10^{-4}	0.0124
5000	0.3395	49.6×10^{-4}	0.0375
7500	0.3515	50.3×10^{-4}	0.0485
10000	0.3652	60.7×10^{-4}	0.0569

this enzyme catalyzes hydroxylation of monophenols to *o*-diphenols (cresolase activity) and their subsequent oxidation to *o*-quinones (catecholase activity). The latter products are unstable in aqueous solution, further polymerizing to undesirable brown, red, and black pigments. Tyrosinase has attracted a lot of attention with respect to its biotechnological applications [21], due to its attractive catalytic ability, as the catechol products are useful as drugs or drug synthons.

For the purposes of present study, tyrosinase was extracted from potato *(Solanum tuberosum)* and two species of common edible mushrooms: *Agaricus bisporus* (Ab) and *Pleurotus ostreatus* (Po). All the source materials were purchased from the local green market in Split, Croatia. Enzyme extraction was prepared with 100 mL of cold 50 mM phosphate buffer (pH 6.0) per 50 g of a source material. The homogenates were centrifuged at 5000 rpm for 30 min and supernatant was collected. The sediments were mixed with cold phosphate buffer and were allowed to sit in cold condition with occasional shaking. Then the sediments containing buffer were centrifuged once again to collect supernatant. These supernatants were subsequently used as sources of enzyme.

The tyrosinase activity was determined spectrophotometrically at room temperature ($t = 25°C$) and $\lambda = 475$ nm, measuring the conversion of L-DOPA to red coloured oxidation product dopachrome [22]. The reaction mixture—obtained after adding a 50 μL of enzyme extract to a cuvette containing 1.2 mL of 50 mM phosphate buffer (pH 6.0) and 0.8 mL of 10 mM L-DOPA—was immediately shaken and the increase in absorbance was measured for 3 minutes. The change in the absorbance was proportional to the enzyme concentration. The initial rate was calculated from the linear part of the recorded progress curve. One unit of enzyme was defined as the amount which catalyzed the transformation of 1 μmol of L-DOPA to dopachrome per minute under the above conditions. The dopachrome extinction coefficient at 475 nm was 3600 M^{-1} cm^{-1}.

To determine the values of K_m and V_{max} for tyrosinase, experimental kinetic data, summarized in Table 2, was gathered by measuring enzyme activity in a cuvette where 50 μL of enzyme solution was added to 2 mL of 50 mM phosphate

buffer (pH 6.0) containing various concentrations of L-DOPA (0–10 mM). In this case, the estimator performance is evaluated by computing mean absolute error (MAE), that is, the mean of the absolute differences between the observed reaction rate, v_i and the expected reaction rate, calculated using estimates of K_m and V_{max}, at a concentration, s_i:

$$\text{MAE} = \frac{1}{n} \sum_{i=1}^{n} \left(v_i - \frac{V_{max} \cdot s_i}{s_i + K_m} \right), \quad (15)$$

where n is the number of experimental data points. Mean absolute error is regularly employed quality measure that provides an objective assessment of how well the various estimated values of K_m and V_{max} fit the untransformed experimental data.

4. Results and Discussion

4.1. Parameter Estimation Using Simulated Data. Figures 2–5 provide the summary of our simulation outcomes for different sample sizes, different contamination levels, and different outlier distances. By examining the simulation results, it is evident that modified robust Tukey's biweight estimator outperforms all other four alternative fitting techniques with respect to bias, coefficient of variation, and normalized mean square error, yielding both accurate and precise estimates of K_m and V_{max} at all test conditions. For example, looking at the set of values obtained for a small sample size ($n = 10$) with a minimal level of contamination present in the data and minimal outlier scatter (Situation 1, Table 1), we observe that as per the RNR estimator the average estimated values of K_m and V_{max} are 240.08 ± 14.39 and 80 ± 1.5. When EH, HW, and LB plots were used, the data produced 224.38±38.01, 241.02± 44.08, and 237.47 ± 52.05, respectively, as average estimates of K_m and 78.18±4.46, 79.8±4.85 and 79.42±6.15, respectively, as V_{max}. When OLS estimator was applied, the corresponding average values of K_m and V_{max} were 238.71 ± 40.04 and 79.86 ± 4.39, respectively. If the reported standard deviations are scaled by dividing them with an appropriate mean, the resulting coefficients of variation of K_m and V_{max} are 16.9 and 5.7% (EH), 18.3% and 6.1% (HW), 21.9% and 7.7% (LB), 16.8 and 5.5% (OLS), and 6% and 1.9% (RNR), respectively. Thus, it is revealed that, though all three Hanes-Woolf, Lineweaver-Burke, and ordinary least-squares methods have a low bias (Figures 2 and 4) and produce the results that are in a close proximity of the values obtained by robust nonlinear regression method, their estimates are much more imprecise and as such are of a limited utility. Figure 6 shows the plots of fitted reaction curves for the randomly selected replications of situations with small sample size (Situations 1–6).

For a medium sample size ($n = 20$) with the same levels of contamination and outlier scatter (Situation 7, Table 1), the analysis for the RNR approach yielded almost identical results, that is, average K_m and V_{max} estimates as 239.68 ± 9.3 and 80 ± 0.68, respectively. The EH, HW, and LB plots estimated the average K_m as 228.83 ± 26.39, 238.01 ± 38.12, and 238.52±49.21, respectively, and V_{max} as 79.09±2.12, 79.6± 2.73, and 79.6±4.07, respectively. The average kinetic parameters, that is, K_m and V_{max}, obtained by the OLS method were

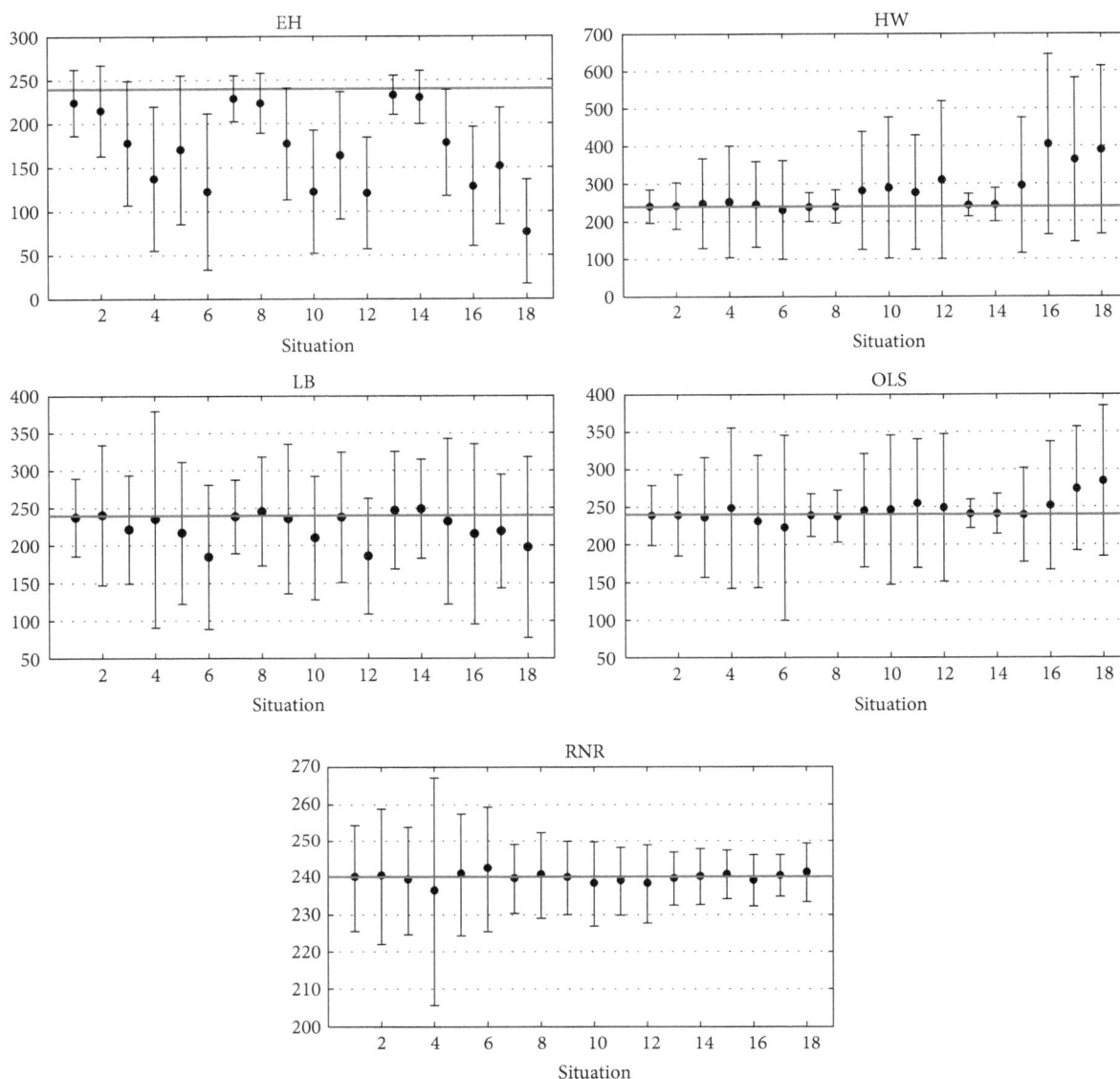

FIGURE 2: Mean estimated values (black dots) and standard deviations of Michaelis constant, K_m, for different simulated scenarios. Red lines denote true parameter value.

238.75±28.43 and 79.88±2.05, respectively. Again, if the standard deviations are scaled by an appropriate mean, the resulting coefficients of variation of K_m and V_{max} are 11.5% and 2.7% (EH), 16% and 3.4% (HW), 20.6% and 5.1% (LB), 11.9% and 2.6% (OLS), and 3.9% and 0.8% (RNR), respectively.

Similarly, for a large sample size (n = 50) with the same levels of contamination and outlier scatter (Situation 13, Table 1), the values of K_m and V_{max} are estimated as 232.67 ± 22.38 and 79.71 ± 0.98 (EH), 243.2 ± 29.88 and 79.99 ± 1.11 (HW), 246.97±78.33 and 80.15±3.43 (LB), 240.68±19.19 and 80.08 ± 0.86 (OLS), and 239.62 ± 7.19 and 80.01 ± 0.31 (RNR), respectively. In this case, the resulting coefficients of variation of K_m and V_{max} are 9.6% and 1.2% (EH), 12.3% and 1.4% (HW), 31.7% and 4.3% (LB), 8% and 1.1% (OLS), and 3% and 0.4% (RNR), respectively. It should be noted that, all the while in all of the aforementioned cases, the Eadie-Hofstee method

has coefficients of variation that are highly comparable to those based on ordinary least-squares method; the EH estimated K_m and V_{max} values are much further away from the true values than the estimates obtained by Hanes-Woolf, ordinary least-squares, and robust nonlinear regression methods.

With the increase of the contamination level and the outlier scatter, the average estimates of K_m and V_{max} values as per linear plots and the OLS method begin to deviate significantly. However, the modified robust Tukey's biweight estimator is able to keep the errors in check and produce the results that are highly comparable and much closer to the true parameter values.

Numerically, by looking at the estimated values, it is hard to tell which of the selected estimators has the overall best performance; nevertheless, with the help of the normalized mean square error method we can see the values of parameters for

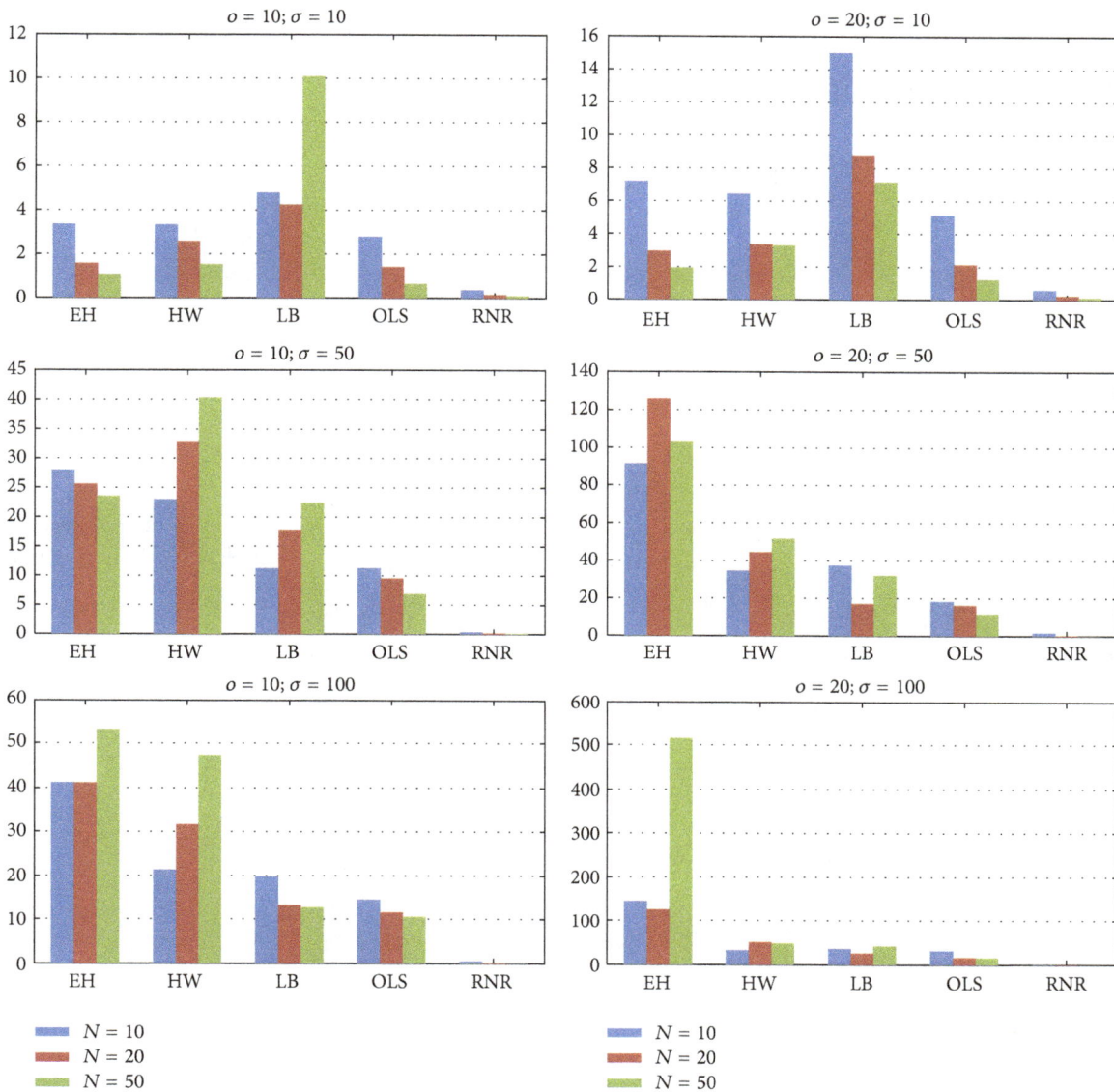

FIGURE 3: Normalized mean square error of K_m for different scenarios.

which the error is minimum. Thus, from Figures 3 and 5, we may say that the best (most accurate) values of K_m and $V_{\max}$ are obtained in situation 17, for which the minimum error values are 0.0541 and 8.5×10^{-4}, respectively (as obtained by RNR). This proves that the estimated values are more or less similar to the true values. The worst error values of 1.6982 and 0.1517 for RNR method are obtained in situation 4 (Figures 3 and 5). In all other situations, the normalized mean square errors for RNR method are less than 1 and 0.1, respectively (Figures 3 and 5). This shows the credibility and the robustness of the proposed modified Tukey's biweight estimator relative to other methods when outliers or influential observations are present in the data. If we compare the robust nonlinear regression method with ordinary least-squares method, we find that the RNR method normalized mean square errors

are on average more than 10 times lower than the normalized mean square errors produced by the OLS method.

4.2. Parameter Estimation Using Experimental Data.

The viability of the proposed robust estimator was also tested by using the experimental kinetic data for tyrosinase enzyme. The corresponding K_m and $V_{\max}$ values, produced by different estimation models, are given in Table 3. Upon closer inspection and analysis of these values, it can be observed that, in case of Ab mushroom and potato tyrosinase, the kinetic values, that is, K_m and $V_{\max}$, yielded by the RNR method (599 μmol and 0.38 μmol/min for Ab mushroom and 10740 μmol and 0.118 μmol/min for potato, resp.) are much closer to the values yielded by HW plot (555 μmol and 0.381 μmol/min for Ab mushroom and

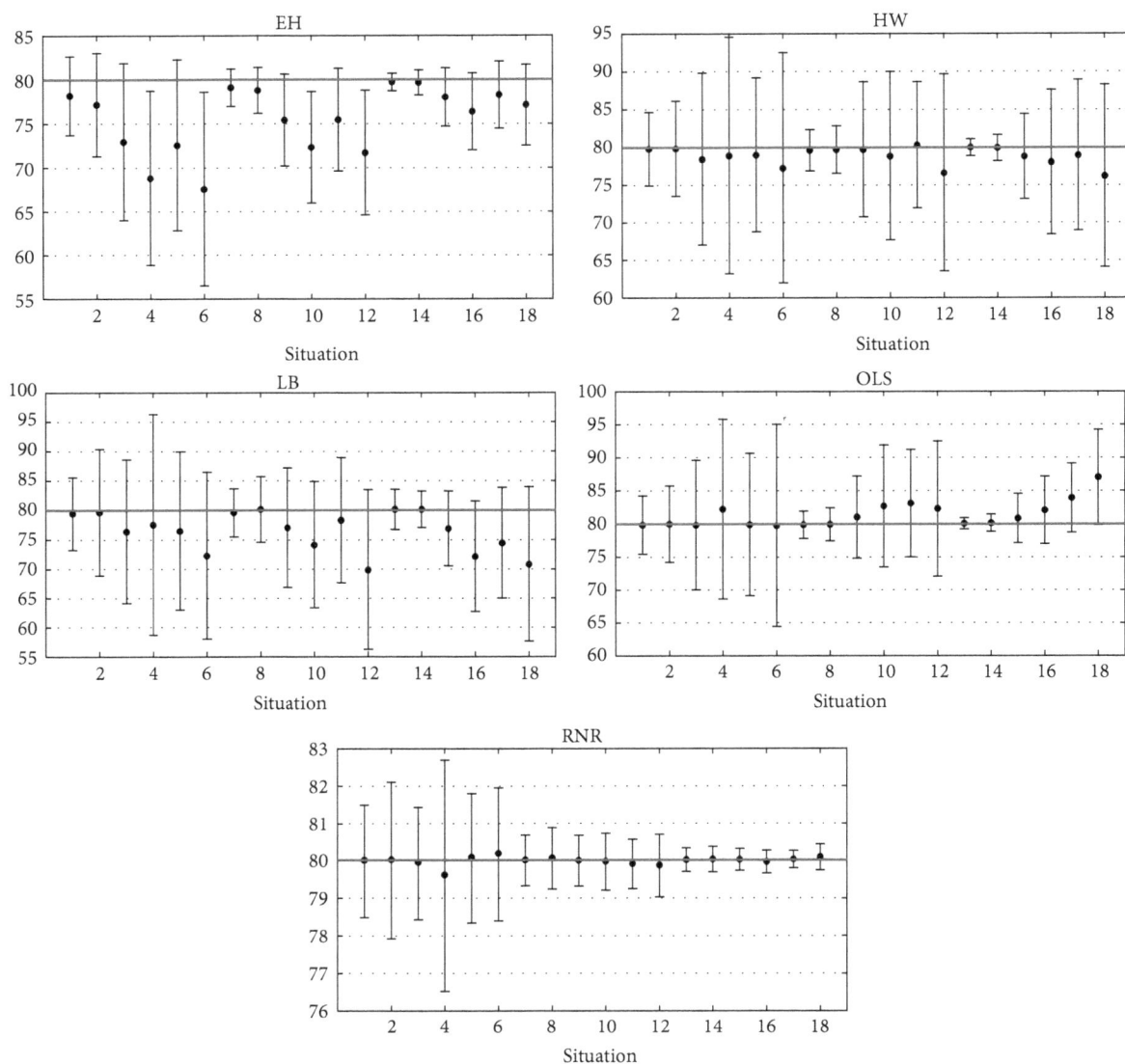

FIGURE 4: Mean estimated values (black dots) and standard deviations of maximum initial velocity, V_{max} for different simulated scenarios. Red lines denote true parameter value.

TABLE 3: Kinetic parameters, K_m and V_{max}, values, and mean absolute errors for tyrosinase extracted from different sources, estimated using different methods.

Source	Parameter	Method				
		LB	HW	EH	OLS	RNR
Ab	K_m	263.09	555.18	414.09	544.79	599.41
	V_{max}	0.2484	0.3812	0.3470	0.3767	0.3798
	MAE	0.0511	0.0071	0.0137	0.0064	0.0048
Po	K_m	359.91	3592.1	992.27	5700.3	6630.5
	V_{max}	0.0017	0.0079	0.0042	0.0095	0.0099
	MAE	0.0013	2.6×10^{-4}	7.5×10^{-4}	1.7×10^{-4}	1.5×10^{-4}
Potato	K_m	1216.0	10659	2093.4	25720	10740
	V_{max}	0.0208	0.1104	0.0394	0.2086	0.1180
	MAE	0.0139	0.0022	0.0087	0.0018	0.0014

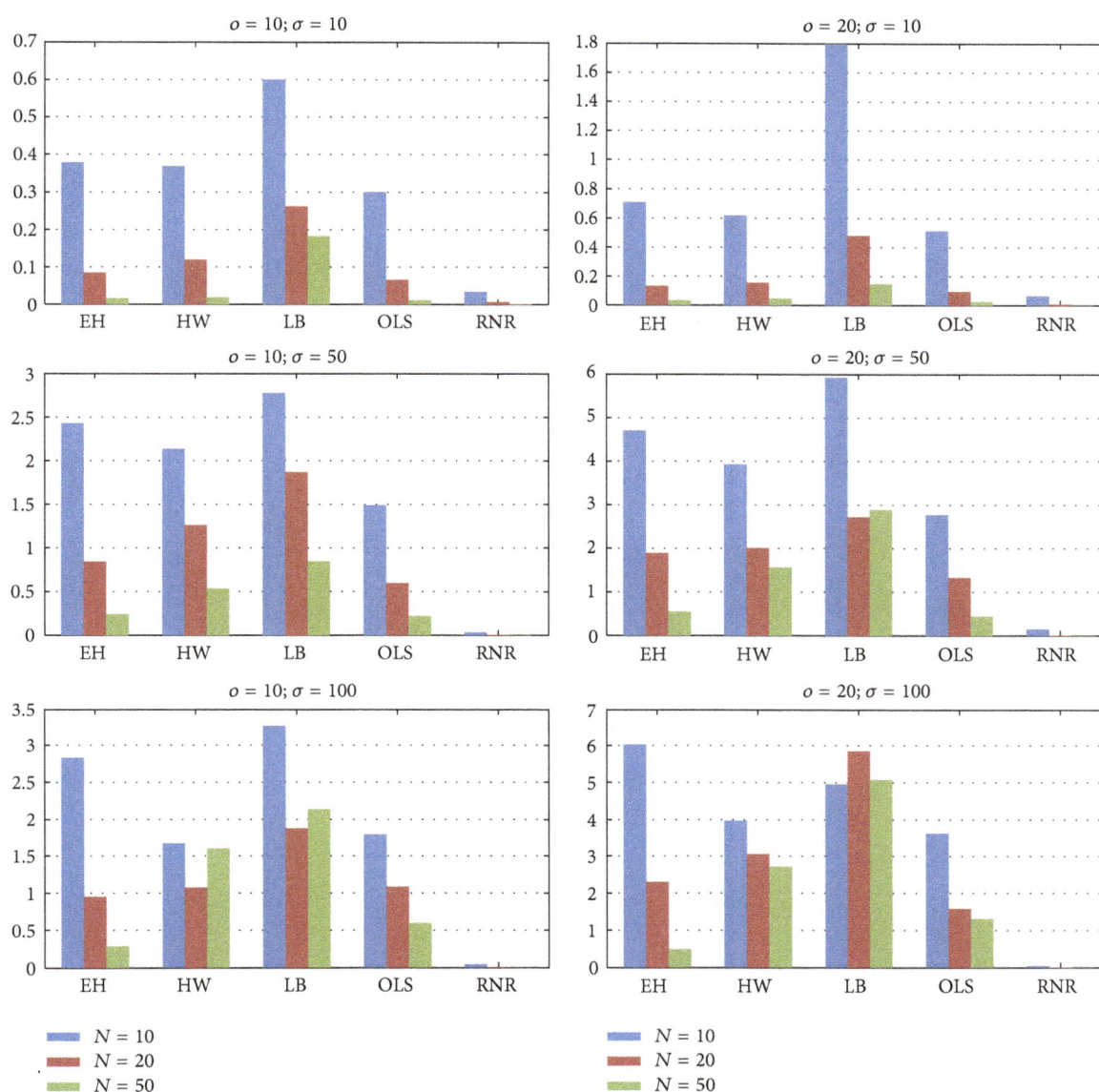

FIGURE 5: Normalized mean square error of $V_{\max}$ for different scenarios.

10659 μmol and 0.11 μmol/min for potato, resp.) than by OLS method (545 μmol and 0.376 μmol/min for Ab mushroom and 25720 μmol and 0.209 μmol/min for potato, resp.). Furthermore, it is interesting to note that the parameter values yielded by LB plot (K_m 263 μmol and $V_{\max}$ 0.248 μmol/min for Ab mushroom, K_m 360 μmol and $V_{\max}$ 0.002 μmol/min for Po mushroom, and K_m 1216 μmol and $V_{\max}$ 0.021 μmol/min for potato) for all three tyrosinase source are very far from the values yielded by other four estimation methods. Figures 7(a), 7(b), and 8(a) show the curves fitted to the experimental data using the modified Tukey's biweight estimator in comparison with the standard linearization methods and the ordinary least-squares non-linear regression. The mean absolute errors between the predicted reaction rates and the actual data are plotted in the right graph in Figure 8. Particularly, the mean errors for

RNR method are 0.0048, 1.5×10^{-4}, and 0.0014, respectively, which shows a good fit of the achieved model.

5. Conclusion

When an enzymatic reaction follows Michaelis-Menten kinetics, the equation for the initial velocity of reaction as a function of the substrate concentration is characterized by two parameters, the Michaelis constant, K_m, and the maximum velocity of reaction, $V_{\max}$. Up to this day, these parameters are routinely estimated using one of these different linearization models: Lineweaver-Burke plot ($1/v$ versus $1/s$), Eadie-Hofstee plot (v versus v/s), and Hanes-Wolfe plot (s/v versus v). Although the linear plots obtained by these methods are very illustrative and useful in analyzing the behavior of enzymes, the common problem they all share

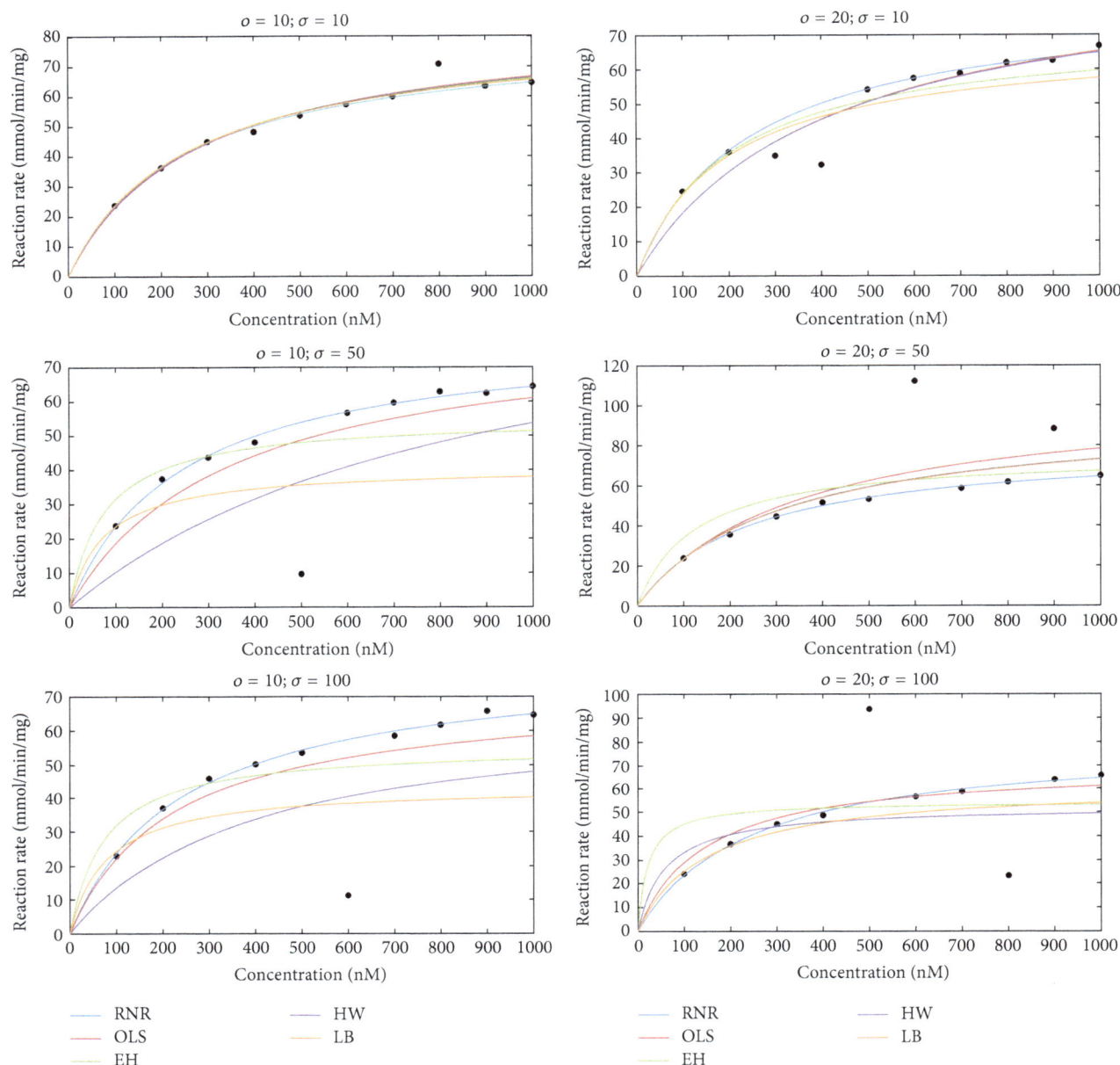

FIGURE 6: Curves fitted using different fitting techniques for random replications of situations 1–6.

is the fact that transformed data usually do not satisfy the assumptions of linear regression, namely, that the scatter of data around the straight line follows Gaussian distribution, and that the standard deviation is equal for every value of independent variable.

More accurate approximation of Michaelis-Menten parameters can be achieved through use of nonlinear least-squares fitting techniques. However, these techniques require good initial guess and offer no guarantee of convergence to the global minimum. On top of that, they are very sensitive to the presence of outliers and influential observations in the data, in which case they are likely to produce biased, inaccurate, and imprecise parameter estimates.

In this paper, a robust estimator of nonlinear regression parameters based on a modification of Tukey's biweight function is introduced. Robust regression techniques have received considerable attention in mathematical statistics literature, but they are yet to receive similar amount of attention by practitioners performing data analysis. Robust nonlinear regression aims to fit a model to the data so that the results are more resilient to the extreme values and are relatively consistent when the errors come from the high-tailed distribution. The experimental comparisons, using both real and synthetic kinetic data, show that the proposed robust nonlinear estimator based on modified Tukey's biweight function outperforms the standard linearization models and

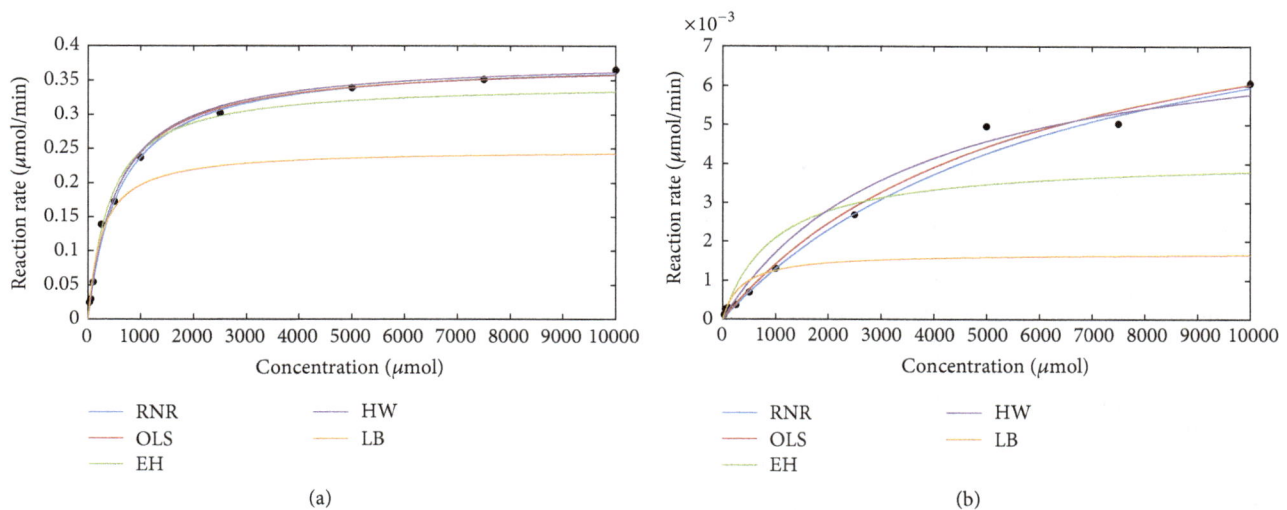

FIGURE 7: Curves fitted using different fitting techniques for tyrosinase extracted from *Agaricus bisporus* (a) and *Pleurotus ostreatus* (b) mushrooms.

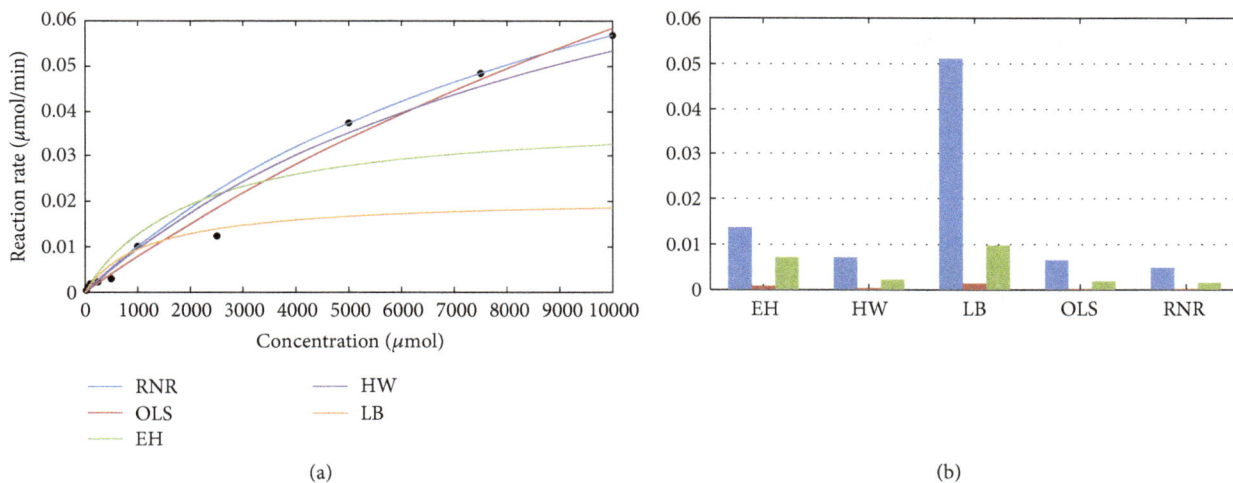

FIGURE 8: Curves fitted using different fitting techniques for tyrosinase extracted from potato (a). Mean absolute errors for different regression models used (b).

ordinary least-squares method and yields superior results with respect to bias, accuracy, and consistency, when there are outliers or influential observations present in the data.

Competing Interests

The authors declare that there is no conflict of interests regarding the publication of this paper.

Acknowledgments

The work reported in this paper was supported by Croatian Science Foundation Research Project "Investigation of Bioactive Compounds from Dalmatian Plants: Their Antioxidant, Enzyme Inhibition and Health Properties" (IP 2014096897).

References

[1] H. J. Fromm and M. Hargrove, *Essentials of Biochemistry*, Springer, Berlin, Germany, 2012.

[2] K. A. Johnson and R. S. Goody, "The original Michaelis constant: translation of the 1913 Michaelis-Menten paper," *Biochemistry*, vol. 50, no. 39, pp. 8264–8269, 2011.

[3] A. Cornish-Bowden, "The origins of enzyme kinetics," *FEBS Letters*, vol. 587, no. 17, pp. 2725–2730, 2013.

[4] H. Lineweaver and D. Burk, "The determination of enzyme dissociation constants," *Journal of the American Chemical Society*, vol. 56, no. 3, pp. 658–666, 1934.

[5] G. S. Eadie, "The inhibition of cholinesterase by physostigmine and prostigmine," *The Journal of Biological Chemistry*, vol. 146, pp. 85–93, 1942.

[6] C. S. Hanes, "Studies on plant amylases," *Biochemical Journal*, vol. 26, no. 5, pp. 1406–1421, 1932.

[7]   A. Ferst, *Structure and Mechanism in Protein Science: A Guide to Enzyme Catalysis and Protein Folding*, Freeman, 2000.

[8]   G. N. Wilkinson, "Statistical estimations in enzyme kinetics," *The Biochemical journal*, vol. 80, pp. 324–332, 1961.

[9]   H. Motulsky and A. Christopoulus, *Fitting Models to Biological Data Using Linear and Nonlinear Regression*, GraphPad Software Inc, 2003.

[10] C. Cobelli and E. Carson, *Introduction to Modeling in Physiology and Medicine*, Academic Press, 2008.

[11] S. R. Nelatury, C. F. Nelatury, and M. C. Vagula, "Parameter estimation in different enzyme reactions," *Advances in Enzyme Research*, vol. 2, no. 1, pp. 14–26, 2014.

[12] G. Kemmer and S. Keller, "Nonlinear least-squares data fitting in Excel spreadsheets," *Nature Protocols*, vol. 5, no. 2, pp. 267–281, 2010.

[13] C. Lim, P. K. Sen, and S. D. Peddada, "Robust nonlinear regression in applications," *Journal of the Indian Society of Agricultural Statistics*, vol. 67, no. 2, pp. 215–234, 291, 2013.

[14] R. A. Maronna, R. D. Martin, and V. J. Yohai, *Robust Statistics: Theory and Methods*, Wiley Series in Probability and Statistics, John Wiley & Sons, New York, NY, USA, 2006.

[15] V. Barnett and T. Lewis, *Outliers in statistical data*, Wiley Series in Probability and Mathematical Statistics: Applied Probability and Statistics, John Wiley & Sons, Ltd., Chichester, England, Third edition, 1994.

[16] P. J. Huber, "Robust estimation of a location parameter," *Annals of Mathematical Statistics*, vol. 35, no. 1, pp. 73–101, 1964.

[17] F. Mosteller and J. W. Tukey, *Data Analysis and Regression*, Addison-Weasley, 1977.

[18] P. J. Rousseeuw and C. Croux, "Alternatives to the median absolute deviation," *Journal of the American Statistical Association*, vol. 88, no. 424, pp. 1273–1283, 1993.

[19] M. R. Loizzo, R. Tundis, and F. Menichini, "Natural and synthetic tyrosinase inhibitors as antibrowning agents: an update," *Comprehensive Reviews in Food Science and Food Safety*, vol. 11, no. 4, pp. 378–398, 2012.

[20] S. Y. Lee, N. Baek, and T.-G. Nam, "Natural, semisynthetic and synthetic tyrosinase inhibitors," *Journal of Enzyme Inhibition and Medicinal Chemistry*, vol. 31, no. 1, pp. 1–13, 2016.

[21] K. U. Zaidi, A. S. Ali, S. A. Ali, and I. Naaz, "Microbial tyrosinases: promising enzymes for pharmaceutical, food bioprocessing, and environmental industry," *Biochemistry Research International*, vol. 2014, Article ID 854687, 16 pages, 2014.

[22] Z. Yang and F. Wu, "Catalytic properties of tyrosinase from potato and edible fungi," *Biotechnology*, vol. 5, no. 3, pp. 344–348, 2006.

12

Enthalpic Contribution of Ni(II) in the Interaction between Carbonaceous Material and Aqueous Solution

Liliana Giraldo[1] and Juan Carlos Moreno-Piraján[2]

[1]*Departamento de Química, Facultad de Ciencias, Universidad Nacional de Colombia, Sede Bogotá, Colombia*
[2]*Departamento de Química, Facultad de Ciencias, Universidad de Los Andes, Grupo de Investigación en Sólidos Porosos y Calorimetría, Bogotá, Colombia*

Correspondence should be addressed to Juan Carlos Moreno-Piraján; jumoreno@uniandes.edu.co

Academic Editor: Christophe Coquelet

Solid adsorbents were prepared from corn cob that was modified with a solution of HNO_3 6 M at different contact times. The solids are characterized by physical N_2 adsorption at 77 K to know their surface area by applying the BET model and surface chemistry is determined using the Bohem method. Once we have prepared the adsorbents we determine the immersion enthalpy, ΔH_{im}, of the solids in Ni(II) aqueous solutions of different concentrations between 20 and 800 mg·L^{-1}, with values for ΔH_{im} between 10.0 and 35.3 J·g^{-1}. From the results obtained for the immersion enthalpy in function of the ion Ni(II) concentration we calculate the contribution to the immersion enthalpy that corresponds to the ion when it is treated with the system adsorbent-solution as a mixture in which the solid, the solvent, and the adsorbate are involved. The solution thermodynamics allows for establishing the enthalpic changes that bring the ion in function of the concentration and the intensity of the interaction of solid-metal ion that is favored by the presence of acid groups in the solid.

1. Introduction

The process of interaction between a solid and a liquid can be considered, in a simplified manner as a mixture of the two components of the system that produce changes in the thermodynamic properties with regard to the components before the interaction and the thermodynamic properties as enthalpy, entropy, and energy of Gibbs can be determined for mixing process [1].

The purpose of the thermodynamic properties determination of an interaction between various components is to know the intensity of the property and how it is affected in function of the mixture. So you can find woks that are related to the determination of excess enthalpy for organic compounds [2], determinations of partial molar enthalpy in the components of different alloys [3], excess molar enthalpies for binary systems of solvent-solute [4], among others. The methodology for the calculation of the partial

thermodynamic properties, that is, the contribution of each component to the property thermodynamics total, can be extended to the solid-liquid interaction taking into account the particular characteristics of the solid and the interaction type that is being carried out. In this research the adsorption process of ion Ni(II) takes place on a carbonaceous material.

Myers and Monson [5] make an interesting discussion on the implementation of the solutions (mixtures) thermodynamics, to the adsorption process of a fluid component on a porous solid, and establish that by relatively simple way the formulation of mixtures thermodynamics can be extend to properties like the enthalpy differential for the adsorbent and the adsorbate.

1.1. Differential Enthalpy of the Mixture Components. The adsorbent-solution is a multicomponent system and for conditions of temperature T and pressure P specifically, the variation of a property thermodynamics considered X (as the enthalpy for this case) can be expressed as

$$dX = \left(\frac{\partial X}{\partial T}\right)_{P,n_i} dT + \left(\frac{\partial X}{\partial P}\right)_{T,n_i} dP + \left(\frac{\partial X}{\partial n_1}\right)_{T,P,n_j} dn_1$$
$$+ \left(\frac{\partial X}{\partial n_2}\right)_{T,P,n_j} dn_2 + \left(\frac{\partial X}{\partial n_3}\right)_{T,P,n_j} dn_3. \tag{1}$$

For the enthalpy system a similar expression is considered

$$dH = \left(\frac{\partial H}{\partial T}\right)_{P,n_i} dT + \left(\frac{\partial H}{\partial P}\right)_{T,n_i} dP$$
$$+ \left(\frac{\partial H}{\partial n_1}\right)_{T,P,n_j} dn_1 + \left(\frac{\partial H}{\partial n_2}\right)_{T,P,n_j} dn_2 \tag{2}$$
$$+ \left(\frac{\partial H}{\partial n_3}\right)_{T,P,n_j} dn_3,$$

where

$$\left(\frac{\partial H}{\partial n_i}\right)_{T,P,n_j} = \overline{H_i} \tag{3}$$

corresponds to the partial molar enthalpy of component i to T, P, and n_j constant. The contribution of each one of the system components to the total enthalpy is established in this manner. Thus it is possible to determine experimentally a change in the enthalpy that manifests for the process that corresponds to the adsorbent-solution interaction and define the change in the enthalpy differential for each component of the mixture $\Delta H_{\mathrm{DIF}_i}$, as the difference between the partial molar enthalpy of component i, $\overline{H_i}$, and the molar enthalpy of component i pure, $H_i^{\bullet}$:

$$\Delta H_{\mathrm{DIF}_i} = \overline{H_i} - H_i^{\bullet}. \tag{4}$$

It is noted that we cannot determine absolute values of enthalpy, but we can determine the difference between the enthalpic content of component in the system and the pure component. The change in the differential enthalpy, $\Delta H_{\mathrm{DIF}_i}$, is also expressed as

$$\Delta H_{\mathrm{DIF}_i} = \left(\frac{\partial \Delta H_{\exp}}{\partial n_i}\right)_{T,P,n_j}. \tag{5}$$

Each system component makes a contribution to the mixture enthalpy and therefore the change in the enthalpy differential can be determined, because this property is a suitable tool to describe mixture properties and the changes that occur when changing, for example the components quantities [6].

When we calculate the enthalpy differential for the components of a mixture, as a solution or an alloy [7], variations in the thermodynamic property are expressed by mol in the mixture; however in the case of a mixture as that described in this work which determines the change in the experimental enthalpy for the mixture of carbonaceous solids, water, and

ion Ni(II), the contribution cannot be calculated by mol but per gram of each system component.

In this work we determined the immersion enthalpy, ΔH_{im}, of the carbonaceous adsorbent materials in aqueous Ni(II) solutions of different concentrations and calculate the relationship of mass of Ni(II) with respect to the system mass, $X_{\mathrm{Ni(II)}}$, and the enthalpy change by μg of ion Ni(II), $\Delta H_{\exp}$, from which we calculate the change in the differential enthalpy of Ni(II) in the mixture, $\Delta H_{\mathrm{DIF}_{\mathrm{Ni(II)}}}$. Established are correlations between, $\Delta H_{\mathrm{DIF}_{\mathrm{Ni(II)}}}$, the total acid groups content, and the carboxylic groups content on the solids surface with the purpose of observing the difference in behavior of dilute and concentrated solutions of Ni(II).

2. Experimental and Methods

2.1. Preparation and Characterization of the Carbonaceous Solids. The carbonaceous solid adsorbents are prepared from corn cob that is dried in an oven at $100°C$ and 5 hours later is crushed to a particle size of 4-5 mm and it chars in a horizontal oven for 1 hour in a nitrogen atmosphere to $450°C$, to increase the material porosity.

The carbonized solid is subjected to oxidation with a HNO_3 solution 6 M at $60°C$ and at different contact times of 3, 6, and 9 hours with the purpose that the surface increases the oxygenated groups content. Once the time established elapses the solids are washed with distilled water until a pH around 6.5. The solids are called CC by corn cob for solid without treatment acid and CCox followed by the contact time with the HNO_3 solution.

The solids were characterized by means of the N_2 adsorption at 77 K for the surface area and for determining the oxygenated groups content the method proposed by Boehm [8] is followed, to put 100 mg of solid in contact with 50 mL of NaOH, Na_2CO_3, and $NaHCO_3$ solutions with concentration 0.1 M and 50 mL of HCl solution 0.1 M; mixtures are kept at a temperature of $25°C$, with constant agitation, for five days. Finally 10 mL aliquots of solutions are titled in contact with the solid.

2.2. Determination of the Immersion Enthalpy. It uses a heat conduction microcalorimeter to determine the immersion enthalpy of the carbonaceous solids in aqueous Ni(II) solutions, which are prepared from $Ni(NO_3)_2 \cdot 6H_2O$ of the tradmark Merck with concentrations between 20 and $800\ \mathrm{mgL^{-1}}$. 10 g of the solution at $25°C$ and a solid sample of 100 mg are added to the calorimetric cell (weight with accuracy of 0.1 mg) that is placed inside the cell in a glass ampoule and starts the potential registration output from the thermal sensor for a period of approximately 15 minutes taking potential readings every 20 seconds; it is necessary to perform the breaking of the glass ampoule, record the thermal effect generated, and continue with the potential readings for approximately 15 more minutes; finally the system is calibrated electrically [9].

TABLE 1: Physiochemical characterization of the carbonized oxidized solids.

Solid	Surface area (m^2g^{-1})	Carboxylic groups $(mmolg^{-1})$	Phenolic groups $(mmolg^{-1})$	Lactonic groups $(mmolg^{-1})$	Total acidity $(mmolg^{-1})$	Total basicity $(mmolg^{-1})$
CC	130	0.13	0.23	0.10	0.46	0.28
CCox3	122	0.26	0.16	0.15	0.57	0.12
CCox6	115	0.38	0.17	0.20	0.75	0.10
CCox9	110	0.46	0.42	0.38	1.26	0.07

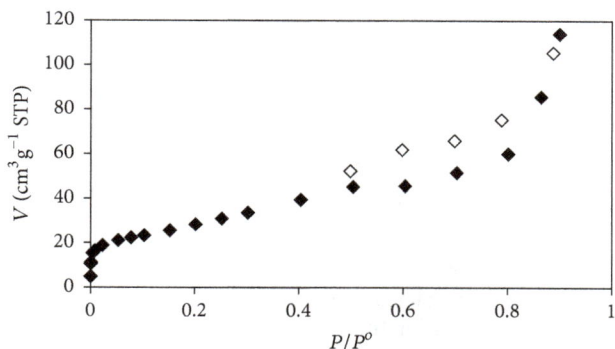

FIGURE 1: N_2 adsorption isotherm for solid carbonaceous CC.

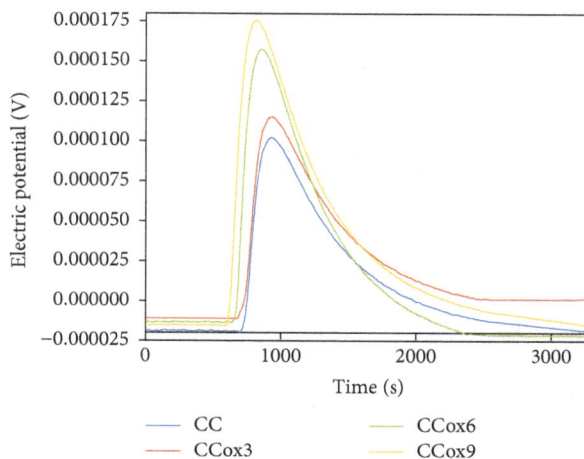

FIGURE 2: Immersion calorimetric curves of the carbonaceous solids in an aqueous ion Ni(II) solution of $100\,mgL^{-1}$.

3. Results and Discussion

Once the carbonaceous solids that from the corn cob were prepared, these are characterized by N_2 adsorption with the aim of knowing the surface area values and pores total volume using the BET model. Figure 1 shows the N_2 adsorption isotherm for solid CC which is carbonized in an inert atmosphere.

For the carbonized CC a type IV isotherm is observed with hysteresis cycle that presents a plateau at relatively high pressures; in the adsorption branch a slope is observed high near the saturation and the desorption occurs at intermediate pressures. This characteristics type corresponds to a hysteresis cycle type H3 that is associated with capillaries in the slit form open with parallel walls and capillaries with wide bodies and short and narrow neck [10]. Solids that are subjected to oxidation with the HNO3 solution presented similar isotherms and the apparent surface area for these was determined by the BET model.

Table 1 presents the results obtained for the solids characterization in regard to their surface area values and the oxygenated groups content that are generated by the oxidation process that occurs in the carbonized solids.

According to the results the carbonized solids show surface area of similar values around $110\,m^2g^{-1}$ that is slightly modified by the oxidation treatment because oxygen groups are generated in the surface that limit the incipient porosity produced by the carbonization process. It is observed that the greatest contact time between the solid and the nitric acid solution presents the solid with a higher content in groups of acid character, with a value of $1.26\,mmolg^{-1}$ and the smallest value for the total basicity. The results in regard to the surface groups content are comparable to a work in which activated carbons are prepared by chemical activation with KOH [11].

The immersion enthalpy of the carbonaceous solids with mentioned characteristics in aqueous Ni(II) solutions is determined in a wide concentration range in order to observe the behavior with regard to the enthalpy change for immersion in dilute and concentrated solutions. Figure 2 presents the calorimetric curves that are obtained for the solids to an ion Ni(II) concentration of $100\,mgL^{-1}$.

The peak in the potential curve in function of time is proportional to the heat quantity that is generated in the solid-liquid contact. It is observed that the solid that presents the highest peak when immersed in the ion solution is the one that was submitted for oxidation for 9 hours and that has the highest acid groups content that have the ability to interact with the Ni(II), showing that the effect in the surface modification increases the immersion enthalpy of the solid in the solution [12]. The effect that occurs in the immersion increases the potential of the thermal sensor indicating that the process is of exothermic character and involving the interactions that occur between the three mixture components to be studied.

From curves like the ones shown in Figure 2 calculate the immersion enthalpy of solids in solutions that are tested in conditions of low and high dilution with the purpose of comparing the enthalpy change of the immersion process. Figure 3 presents the relation between the immersion enthalpy and the mass of ion Ni(II) present in the mixture along with water and the carbonaceous solid.

To appreciate the Ni(II) contribution to the total immersion enthalpy the solid adsorbent quantities and water must

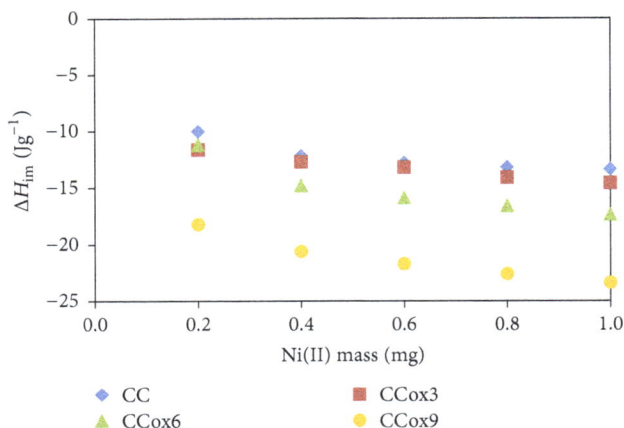

FIGURE 3: Immersion enthalpy of the carbonized solids in aqueous Ni(II) solutions in function of the ion mass.

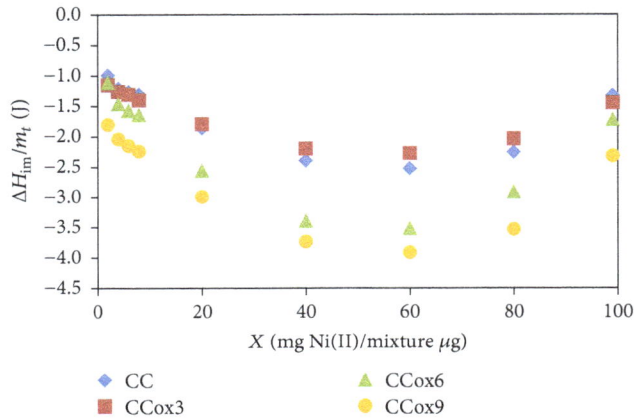

FIGURE 4: Immersion enthalpy per mixture mass as a function of the Ni(II) mass fraction.

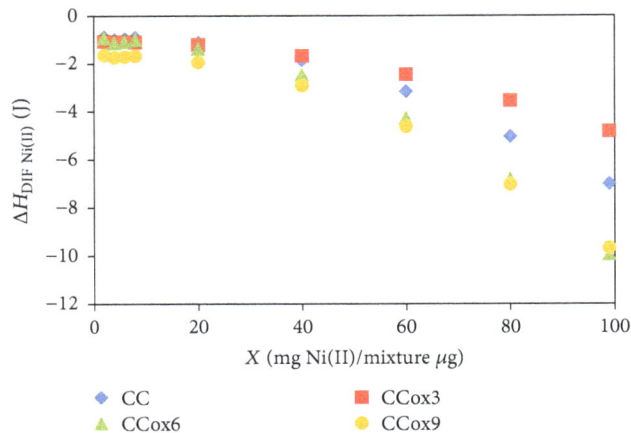

FIGURE 5: Ion Ni(II) differential enthalpy change in function of their mass fraction in the mixture.

Figure 4 shows curves for the four solids with peaks toward the mass fraction, $X_{Ni(II)}$, around 60 and indicates different behaviors for the low and the high ion concentrations that indicate different interactions in the ion adsorption process on the solid that generates a enthalpy total change for the mixture.

From Figure 4 calculate the change in the Ni(II) differential enthalpy, $\Delta H_{DIF_{Ni(II)}}$, which corresponds to the partial derivative of the immersion enthalpy with regard to the mass fraction, $(\partial \Delta H_{im}/\partial X_{Ni(II)})$, and represents the difference between the ion Ni(II) enthalpy when it is in the mix and the pure ion enthalpy. In this way it is stated that it is not possible to calculate the ion Ni(II) absolute enthalpy in the mixture but the difference between the two states. Figure 5 presents the results obtained for $\Delta H_{DIF_{Ni(II)}}$, in function of $X_{Ni(II)}$.

Figure 5 is an interesting result because it shows that when increasing the ion Ni(II) quantity in the mixture its contribution to the total enthalpy becomes greater. For the dilute concentrations in the three solids CC, CCox3, and CCox6, the values are similar and note that overlap for the carbonized CCox9 shows a greater change of enthalpy. For mass fractions of the ion greater than 20 (mg Ni(II)/μg mixed) it is noted that the Ni(II) differential enthalpy increases with a greater contribution to the CC carbonized without treatment with respect to the one that was oxidized during three hours, while for solids with greater oxidation time this enthalpic contribution is similar. In this way by the calculation of the change in the ion differential enthalpy, $\Delta H_{DIF_{Ni(II)}}$, it is useful to see the difference in the interactions and the influence of the superficial oxygen groups in the solid adsorbent.

It is known that the interaction between metal ions that are in aqueous solution and a solid adsorbent has different influences, including the oxygen groups content of the solid surface [15]. For this reason once it is estimated the $\Delta H_{DIF_{Ni(II)}}$ relates to the content of carboxylic groups on the solid surface and with the total acidity that this presents. Figure 6 shows the result of such relationships in a diluted Ni(II) solution with a mass ratio X of 2, to an intermediate concentration solution of ion, with an X 40, and a ion concentrated solution.

be kept constant that acts as a solvent of smaller amount of ions that have a positive electrical charge [13]. Figure 3 shows the negative values of enthalpy change and exothermic character, which are greater for the two solids that were in contact with the HNO_3 solution for 6 and 9 hours. The highest values of enthalpy indicate greater solid-liquid interaction, in that the solid to be oxidized during 3 hours shows a similar behavior to the carbonized one that had not been subjected to oxidation.

The ion Ni(II) contribution to the immersion enthalpy, ΔH_{im}, is calculated assuming that the system is a mixture of three components for which a mass relationship is established instead of a molar relationship (which would allow knowing the partial molar properties) since it does not know the moles of solid but if it is a mass, this defines the variable mass fraction, $X_{Ni(II)}$, as the Ni(II) mass in the solution on the total mixture mass. In the same way we can calculate the solid and the solvent contribution to the enthalpy total change; however the ion contribution is of greater interest since that is the adsorbate [14]. Figure 4 shows the relation of ΔH_{im}/mixture mass in function of the Ni(II) mass fraction for the range of concentrations which are studied.

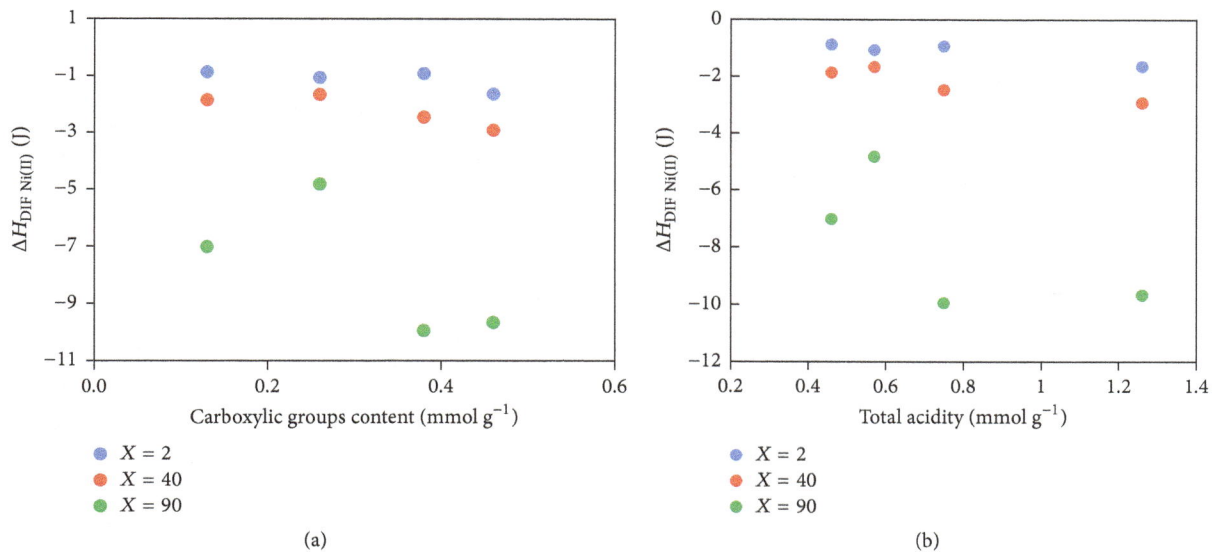

FIGURE 6: Ion Ni(II) differential enthalpy change, $\Delta H_{DIF_{Ni(II)}}$, in function of the oxygen content groups on the carbonized surface. (a) Carboxylic groups content, (b) total acidity.

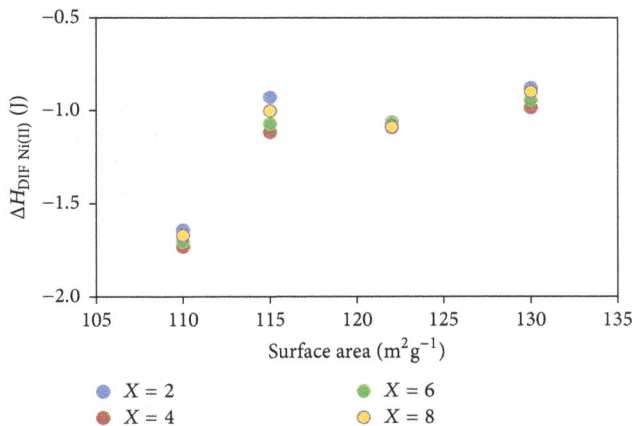

FIGURE 7: Ion Ni(II) differential enthalpy change in function of the carbonized surface area.

Relations show a similar behavior for the solutions with mass fractions of 2 and 40 for the change in the differential enthalpy ion are between -0.87 and -1.73 J but when the ion concentration increases the change in the enthalpy becomes larger and is marked by two trends: for the carbonized less oxidized solids the difference of enthalpy decreases and for solids with greater oxidation values the enthalpy differential is nearby; the above indicates that the solid interactions with ion in the mixture are intensified with the presence of oxygen groups, as shown by different works that determine the metal ions adsorption capacity in aqueous solution on solid adsorbents [15–17]. It is noted that $\Delta H_{DIF_{Ni(II)}}$ is sensitive to the ion quantity present in the mixture and to the solid surface chemistry.

As the other characteristic of the solids that generates an energetic contribution in the mixture the surface area in Figure 7 shows the relation between the $\Delta H_{DIF_{Ni(II)}}$ and the carbonized surface area.

The results show that the higher values for $\Delta H_{DIF_{Ni(II)}}$ are to the carbonized CCox9 that presented the lowest value of surface area ($110 \, m^2 g^{-1}$) and the highest oxygen groups content confirming the fact that the oxygen groups in the carbonized solids interact with the ion to a greater extent than the single attraction of the surface. For the carbonized CCox3 differential enthalpy values are obtained similar between -1.06 and -1.09 J for different quantities of the ion in the mix, which can indicate effects compensation on the solid-ion interaction.

As in the mixture that is considered the solid and the solvent quantities have remained constant, by application of the Gibbs-Duhem equation:

$$\left(X_{\text{sólido-solvente}}\right)\left(\frac{\partial \Delta H_{DIF_{\text{sólido-solvente}}}}{\partial X_{Ni(II)}}\right)$$
$$+ \left(X_{Ni(II)}\right)\left(\frac{\partial \Delta H_{DIF_{Ni(II)}}}{\partial X_{Ni(II)}}\right) = 0. \tag{6}$$

This expression indicates that the variations of the partial quantities, in this case the change in the enthalpy differential, with regard to the system composition are not independent of a temperature and constant pressure [13]. Using (6) we can calculate the differential enthalpy change for solid and solvent in the mixture and this in function of ion Ni(II) mass fraction presents a behavior contrary to the ion Ni(II) differential enthalpy change, $\Delta H_{DIF_{Ni(II)}}$, as what happens with the partial molar properties in the development of the mixture thermodynamics. Figure 8 presents the results obtained for these relations to the CC carbonized without acid treatment CC.

Presenting the two trends in the enthalpy differential change, most of the values show exothermic character, and note that the change in the ion differential enthalpy increases when that of the set solid-solvent decreases. Also shown

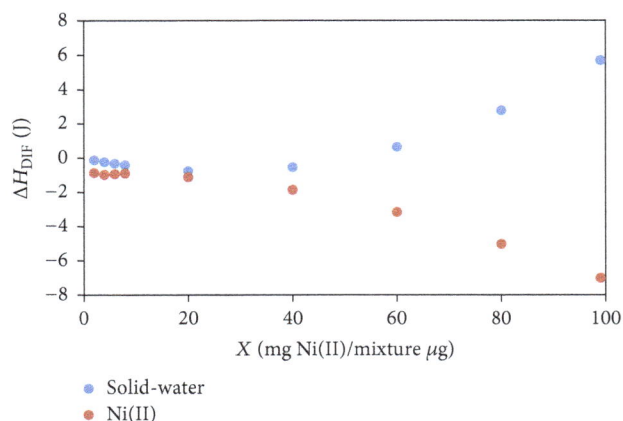

FIGURE 8: Change in the differential enthalpy (solid-solvent) and Ni(II) in function of the mass fraction of Ni(II) to the carbonized CC.

for low mass fractions changes in enthalpy are small and from a mass fraction of 40 the behaviors are larger and opposite.

4. Conclusions

Prepare four carbonized solids from corn cob with different oxygen groups contents in the surface. The surface area values of the solids are between 110 and 130 m^2g^{-1} and the total acidity is between 0.46 and 1.26 $mmolg^{-1}$.

From the determination of the immersion enthalpy of the carbonized solids in aqueous solutions of different Ni(II) concentration and assuming the relations of the mixtures thermodynamics, calculate the contribution of ion to the enthalpy total.

The change in the Ni(II) differential enthalpy, $\Delta H_{DIF_{Ni(II)}}$, in the carbonized is between -0.87 and -7.03 J which indicates an increase in the enthalpy change with the ion Ni(II) quantity.

$\Delta H_{DIF_{Ni(II)}}$ is related with the carboxylic groups content and with the total acidity on the surface of the solid. It is noted for a mass fraction of Ni(II) in the mixture of 90 that in the carbonized less oxidized solids the enthalpy difference decreases and for solids with greater oxidation values the enthalpy differential is nearby, which indicates that the interactions of solid with ion in the mixture are intensified with the presence of oxygen groups.

The enthalpic contribution is calculated for the whole solid-solvent in the CC carbonized without treatment acid, with values between -0.12 and 2.78 J with an opposite behavior to that of the ion Ni(II) enthalpic contribution in the mixture, complying with the Gibbs-Duhem equation.

Conflicts of Interest

The authors declare that they have no conflicts of interest.

Acknowledgments

The authors acknowledge the framework agreement between the University of the Andes and the National University of Colombia and the act of agreement established by the Departments of Chemistry of both universities. The authors also acknowledge Project 3580 of Bank of Republic of Colombia for the partial fund to carry out this research.

References

[1] M. Y. Klotz and R. M. Rosenberg, "Chemical thermodynamics," in *Basic Concepts and Methods*, John Wiley & Sons Inc., Hoboken, NJ, USA, 7th edition, 2008.

[2] I. Alonso, J. A. González, I. Garcia, and J. C. Cobos, "Thermodynamics of (ketone + amine) mixtures. Part XI. Excess molar enthalpies at T = 298.15 K for the (1-propanol + N,N,N-triethylamine+2-butanone) system," *Journal of Chemical Thermodynamics*, vol. 69, pp. 6–11, 2014.

[3] H. Arslan, "Determinations of enthalpy and partial molar enthalpy in the alloys Bi-Cd-Ga-In-Zn, Bi-Cd-Ga-Zn and Au-Cu-Sn," *Materials Chemistry and Physics*, vol. 153, pp. 384–389, 2015.

[4] R. Zhang, J. Chen, and J. Mi, "Excess molar enthalpies for binary mixtures of different amines with water," *Journal of Chemical Thermodynamics*, vol. 89, pp. 16–21, 2015.

[5] A. L. Myers and P. A. Monson, "Physical adsorption of gases: The case for absolute adsorption as the basis for thermodynamic analysis," *Adsorption*, vol. 20, no. 4, pp. 591–622, 2014.

[6] R. S. Neyband and H. Zarei, "A combined experimental and computational investigation of excess molar enthalpies of (nitrobenzene + alkanol) mixtures," *Journal of Chemical Thermodynamics*, vol. 80, pp. 119–123, 2015.

[7] A. Dębski and W. Gąsior, "Calorimetric studies and thermodynamic properties of liquid Ag-Ca alloys," *Journal of Chemical Thermodynamics*, vol. 77, pp. 159–166, 2014.

[8] J. Rouquerol, F. Rouquerol, P. Llewellyn, G. Maurin, and K. S. W. Sing, *Adsorption by Powders and Porous Solids: Principles, Methodology and Applications*, Oxford, London, UK, 2nd edition, 2013.

[9] J. Kaźmierczak, P. Nowicki, and R. Pietrzak, "Sorption properties of activated carbons obtained from corn cobs by chemical and physical activation," *Adsorption*, vol. 19, no. 2-4, pp. 273–281, 2013.

[10] J. C. P. Melo, E. C. Silva Filho, S. A. A. Santana, and C. Airoldi, "Synthesized cellulose/succinic anhydride as an ion exchanger. Calorimetry of divalent cations in aqueous suspension," *Thermochimica Acta*, vol. 524, no. 1-2, pp. 29–34, 2011.

[11] P. Atkins and J. Paula, *Physical Chemistry*, Oxford University Press, London, UK, 8th edition, 2008.

[12] A. R. Cestari, E. F. S. Vieira, R. C. Silva, and M. A. S. Andrade, "Direct determinations of energetic parameters at chitosan/Cr(VI) interfaces by means of immersion heat-conduction microcalorimetry," *Journal of Colloid and Interface Science*, vol. 352, no. 2, pp. 491–497, 2010.

[13] C. Moreno-Castilla, M. A. Álvarez-Merino, L. M. Pastrana-Martínez, and M. V. López-Ramón, "Adsorption mechanisms of metal cations from water on an oxidized carbon surface," *Journal of Colloid and Interface Science*, vol. 345, no. 2, pp. 461–466, 2010.

[14] V. C. Srivastava, I. D. Mall, and I. M. Mishra, "Competitive adsorption of cadmium(II) and nickel(II) metal ions from aqueous solution onto rice husk ash," *Chemical Engineering and Processing: Process Intensification*, vol. 48, no. 1, pp. 370–379, 2009.

[15] R. Fonseca-Correa, L. Giraldo, and J. C. Moreno-Piraján, "Tri-valent chromium removal from aqueous solution with phys-ically and chemically modified corncob waste," *Journal of Analytical and Applied Pyrolysis*, vol. 101, pp. 132–141, 2013.

[16] H. P. Boehm, "Some aspects of the surface chemistry of carbon blacks and other carbons," *Carbon*, vol. 32, no. 5, pp. 759–769, 1994.

[17] F. Stoeckli and T. A. Centeno, "On the characterization of micro-porous carbons by immersion calorimetry alone," *Carbon*, vol. 35, no. 8, pp. 1097–1100, 1997.

13

Polymer Electrolytes Based on Borane/Poly(ethylene glycol) Methyl Ether for Lithium Batteries

Ali Murat Soydan[1] and Recep Akdeniz[2]

[1]*Institute of Energy Technologies, Gebze Technical University, Cayirova, Gebze, 41400 Kocaeli, Turkey*
[2]*Department of Materials Science and Engineering, Gebze Technical University, Cayirova, Gebze, 41400 Kocaeli, Turkey*

Correspondence should be addressed to Ali Murat Soydan; asoydan@gtu.edu.tr

Academic Editor: Ewa Schab-Balcerzak

This work presents a different approach to preparing polymer electrolytes having borate ester groups for lithium ion batteries. The polymers were synthesized by reaction between poly(ethylene glycol) methyl ether (PEGME) and BH_3-THF complex. Molecular weight of PEGMEs was changed with different chain lengths. Then the polymer electrolytes comprising boron were prepared by doping of the matrices with CF_3SO_3Li at various molar ratios with respect to EO to Li and they are abbreviated as PEGMEX-B-Y. The identification of the PEGME-borate esters was carried out by FTIR and ^1H NMR spectroscopy. Thermal properties of these electrolytes were investigated via thermogravimetric analysis (TGA) and differential scanning calorimetry (DSC). The ionic conductivity of these novel polymer electrolytes was studied by dielectric-impedance spectroscopy. Lithium ion conductivity of these electrolytes was changed by the length of PEGME as well as the doping ratios. They exhibit approximate conductivities of 10^{-4} S·cm^{-1} at 30°C and 10^{-3} S·cm^{-1} at 100°C.

1. Introduction

Lithium ion batteries have been widely used as energy storage devices due to their high volumetric (W h L^{-1}) and gravimetric energy (W h kg^{-1}) density and high cycle life [1, 2], but their safety problems should be solved, especially for the high energy density applications such as portable phones and electric vehicles [1–3]. These drawbacks of lithium ion batteries are mainly caused by the use of organic liquid electrolytes due to high volatility, leakage, or flammability [4]. Solid polymer electrolyte seems to be an alternative to solve the safety problem, but the low ionic conductivity at room temperature is not enough to cover the required performance. The other alternative is aqueous electrolyte lithium ion batteries which promise higher safety due to nonflammability or toxicity of water [5, 6]. Narrow electrochemical stability window of water (1.23 V) as a consequence of low energy density and instability of electrode materials prevents its applications. Although water in salt electrolyte system seems to be promising due to increased working voltage, this technology is still premature and the salts that are used are too expensive for any application [7, 8]. Gel polymer electrolytes (GPEs) has higher ionic conductivity than solid polymer electrolyte, better stability than liquid electrolyte, and higher operating voltage than aqueous electrolytes yielding a different method to solve the problems of those energy storage systems [9–11].

Generally, GPE systems have been produced by immersing the polymer in a polar solvent where it dissolves the salt and enhances ionic conductivity. GPEs at lithium anode/electrolyte interface inhibit dendrite formation on lithium anode. Previously, GPEs based on poly(vinylidene fluoride) (PVdF) [12], poly(vinylidene fluoride cohexafluoropropylene) P(VdF-co-HFP) [5, 13], poly(methyl methacrylate) (PMMA) [14, 15], and poly(acrylonitrile) (PAN) [16] have been reported by several groups.

Another method of synthesis of GPEs was the direct production of PEG branched macromolecules that bear flexible side chains for ion transfer. For example, polymer boron containing branched chain systems were synthesized from boron trioxide, poly(ethylene glycol) (PEG), and methoxy end-capped PEG [17]. They have inserted a Lewis acid in

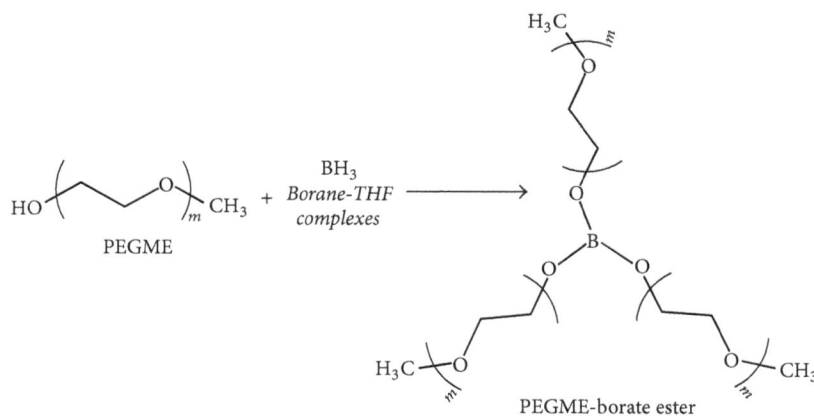

FIGURE 1: Synthesis scheme of PEGME-borate ester polymers.

the branched structure of a poly(ethylene oxide) (PEO) and then the transfer of species between the electrolyte and the electrode is improved as a result of ion-pairing in the electrolyte [18, 19]. Lewis acid boron and Lewis-base oxygen interact with both cations and anions in the electrolytes and contribute to the ion-pairing in the electrolytes [20–23]. Borane or borate derivatives can also improve the thermal stability of the electrolytes [24]. Lithium batteries or ultracapacitors include these kinds of stable branched electrolytes either directly or imbibing into porous nanofibers that have been produced via electrospinning.

In the present work, a more practical and innovative method was proposed for the synthesis of high ionic conductivity and stable boron containing PEG branched electrolytes for faster charge-discharging applications. Branched PEG-borate esters were produced with different lengths. To have high ionic conductivity, the branched materials were doped with CF_3SO_3Li at several stoichiometric ratios. After drying the samples, they were characterized by Fourier transformed infrared (FTIR), differential scanning calorimetry (DSC), thermogravimetric analysis (TGA), and AC impedance spectroscopy.

2. Experimental

2.1. Materials.
BH_3/THF complex and PEGME (average Mn = 350, 550, and 750 g/mol) were purchased from Aldrich. Dimethylsulfoxide (DMSO, analytical reagent) was received from Merck. CF_3SO_3Li 97% was bought from Alfa Aesar and stored in a dry atmosphere.

2.2. Synthesis of the Graft Copolymer Electrolytes.
The PEG-borate esters were produced via the method as presented in Figure 1, by the reaction of poly(ethylene glycol) methyl ether, $CH_3O(CH_2CH_2O)_nH$, of various molecular weights X = 350, 550, and 750 g/mol, and borane tetrahydrofuran complex solution, BH_3/THF. Firstly, PEGME was added to DMSO, and then BH_3/THF was dropwise admixed into the solution. The resulting solution was stirred under N_2 atmosphere and refluxed at 100°C for 24 h. After that DMSO was removed by rotary evaporator at 80°C, under vacuum. Finally, transparent

pale yellowish liquid polymer (for X = 350 g/mol) and gels (for X = 550 g/mol and X = 750 g/mol) were obtained. All the materials were kept in a Ar-filled glove box (moisture content below 1 ppm).

All the electrolytes were prepared by doping of branched polymers with lithium trifluoromethanesulfonate (CF_3SO_3Li, 97%) at several stoichiometric ratios. For example, the prepared PEGME-borate esters and lithium trifluoromethanesulfonate (CF_3SO_3Li, 97%) were transferred into a flask and stirred for about 4 h. The concentration of CF_3SO_3Li was adjusted so that the molar ratio of lithium atoms to ether oxygen atoms in the PEGME-borate ester is [EO] : [Li] = 25, 50, 75, and 100. The final homogeneous viscous solutions were casted on Teflon plates.

2.3. Characterizations.
A Bruker Alpha-P in ATR system was used to record the IR spectra within the range of 4000–400 cm^{-1} with a resolution 4 cm^{-1}. Prior to Fourier transform infrared (FTIR) spectra measurements, materials were dried under vacuum and stored in a glove box.

Thermogravimetric analysis (TGA) polymer electrolytes were examined with a Perkin Elmer Pyris 1. The samples (~5 mg) were heated from RT to 750°C under inert atmosphere at a heating rate of 10°C min^{-1}. Differential scanning calorimetry (DSC) (Perkin Elmer Pyris 1 instrument) was used to study glass transition temperature and melting temperatures. The temperature of the samples was reduced from RT to −50°C, kept isothermally for 10 min, and then heated up to 150°C. Then the samples were cooled to −50°C and reheated to 180°C. All thermograms were recorded at a rate of 10°C under inert gas flow. All DSC experiments were done in duplicate and the thermograms shown refer to the final heating.

Novocontrol dielectric-impedance analyzer was used to study alternating current (AC) conductivities of the samples in the frequency range from 0.1 Hz to 3 MHz with respect to temperature. The samples were poured into the specially designed vacuum sealed liquid cell with parallel plate electrodes with a diameter of 20 mm and a thickness of approximately 4.5 mm and their conductivities were measured under an inert atmosphere.

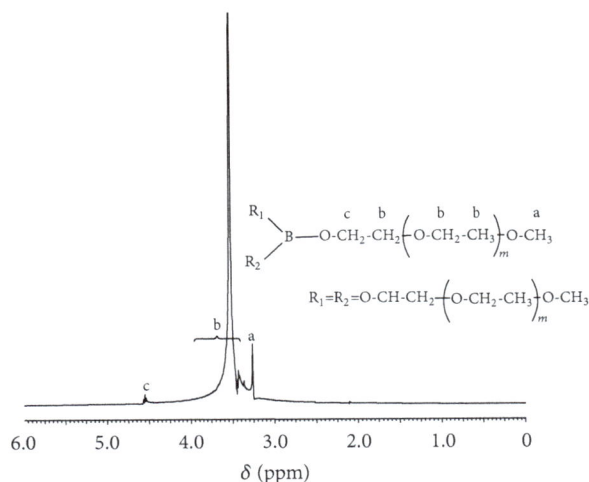

FIGURE 2: ^{1}H NMR spectra of PEGME coordinated boron.

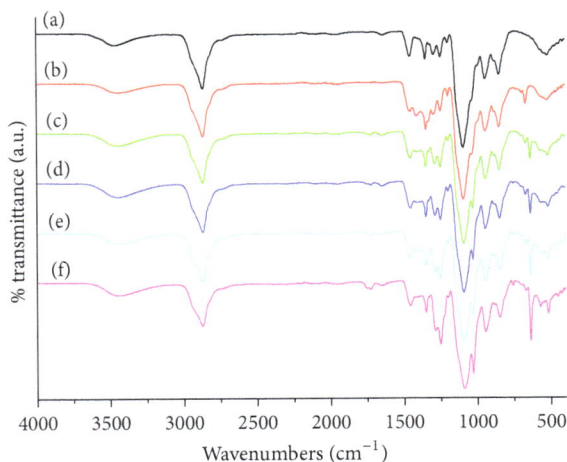

FIGURE 3: Comparison of FTIR spectra of PEGME350 (a); PEGME350-B (b); and PEGME350-B-Y, (Y = 100 (c), 75 (d), 50 (e), and 25 (f)).

3. Result and Discussion

3.1. NMR Study. The ^{1}H NMR spectrum of the final boron comprising branched polymer, PEGME550-B, is presented in Figure 2. The chemical shift at 3.26 ppm (a) clearly indicates the methoxy group at terminal position. The prominent peak at 3.42 and 3.53 ppm (b) can be attributed to ethylene oxide in the polymer backbone: esteric methyl groups, which were shifted to the downfield at 4.55 ppm (c) due to deshielding effect, which is a result of interactions between the boron and the ethylene oxide units.

3.2. FTIR Study. The FTIR spectrum of the branched polymer electrolyte PEGME350-B-Y is given in Figure 3. All the samples show a peak located at 3400 cm^{-1} which corresponds to the absorption of hydroxyl units. It is clearly shown that disappearance of this peak illustrates highest transformation of the hydroxyl groups into borate esters [21]. In addition, strong bands at 2910–2870 cm^{-1} due to C-H stretching of the polymer electrolytes and weaker peaks located around 667 cm^{-1} belong to B-O units [25, 26]. The peaks at ~1029 and ~1350 were characteristic band of B-O-C bond and also antisymmetric stretching vibration of B-O [17]. This confirms the reaction of BH$_3$ with hydroxyl units of PEGs. The broadening of the C-O-C peaks near 1150 cm^{-1}, especially at higher doping ratios, indicates a higher degree of interactions between Li$^+$ and EO [27]. The characteristic absorptions of the pure LiCF$_3$SO$_3$ are located at 1266 and 1033 cm^{-1}, corresponding to asymmetric SO$_3$ and symmetric SO$_3$ vibrations of LiCF$_3$SO$_3$, respectively [28]. Consequently, the FTIR spectra reveal that borate ester polymer electrolytes were successfully synthesized.

3.3. Thermal Properties. The thermal stability of polymer electrolytes is a critical issue for application in rechargeable lithium polymer batteries. The thermal stabilities of these novel electrolytes were studied by TGA at various Li-salt.

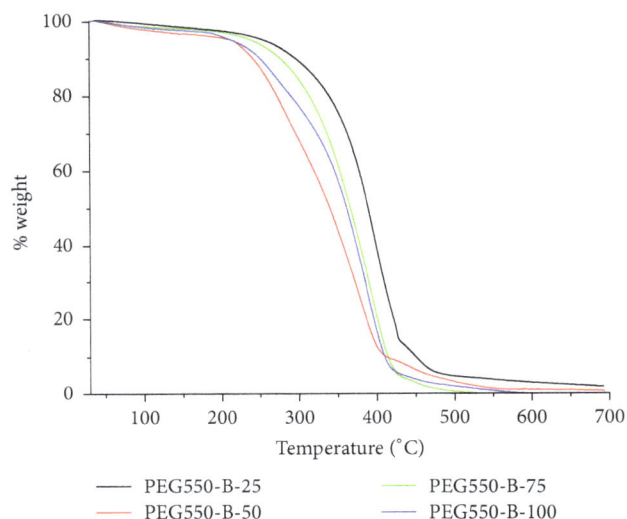

FIGURE 4: Thermogravimetric analysis of PEGME550-B-Y (Y = 25, 50, 75, and 100) polymer electrolytes under nitrogen atmosphere at a heating rate of 10°C·min^{-1}.

Figure 4 shows the TGA results of the polymer electrolytes PEGME550-B-Y. As seen in the figure, the onset of degradation temperature of LiCF$_3$SO$_3$ introduced PEGME-borate esters is approximately 180°C and degradation occurs via two steps. It was previously mentioned that this behavior may be due to the weakness of the C-O bond caused by a reduction in electronic density by Li/EO [29, 30]. All the samples can be said to be thermally stable up to 180°C which is far higher than the operating temperature of lithium ion batteries.

DSC was used to investigate the thermal transitions of the polymer electrolytes with various Li-salt contents. Figure 5 shows the melting endotherms of polymer electrolyte systems. Clearly, for all samples, there is no definite T_g within the given temperature range. The polymer electrolytes of

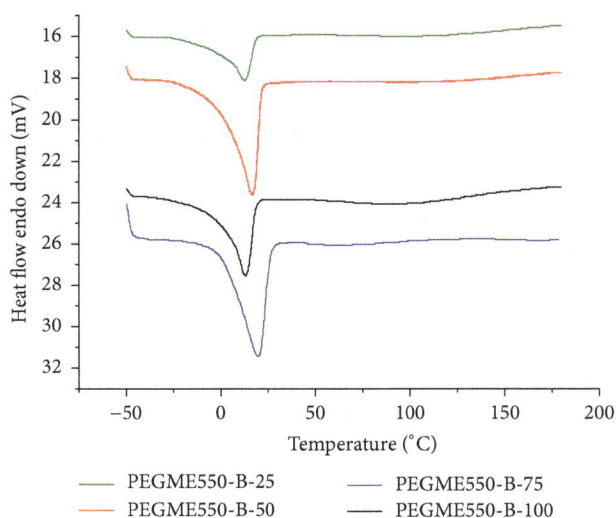

FIGURE 5: DSC curves of PEGME550-B-Y (Y = 25, 50, 75, and 100) polymer electrolytes. Second heating run endotherms are presented at a heating rate of 10°C/min.

PEGME550-B-Y (Y = 25, 50, 75, and 100) showed respective melting endotherms at 13.50, 16.40, 13.15, and 19.80°C.

3.4. *Ionic Conductivity*. Frequency dependence of the ac conductivity (σ_{ac}) of PEGME350 B-25 within 20–100°C is given in Figure 7. It can be observed that the ac conductivity versus frequency for various temperatures exhibits almost linear increasing at low frequencies which are assigned to the polarization of the electrodes; then frequency independent plateau regions which belong to σ_{dc} are well developed at higher frequencies ($F > 100$ Hz).

The σ_{dc} versus inverse temperature variations of the polymer electrolytes with different molecular weight of PEGME and doping ratios of $LiCF_3SO_3$ are shown in Figure 6. The bend in the curve has been observed in ion conducting polymer electrolytes and has been explained by the free volume concept [30, 31], which can be described by the Vogel–Tamman–Fulcher (VTF) equation (1). To have better insight into the temperature dependence of σ_{dc}, the conductivity data have been fitted to the equation.

$$\log \sigma = \log \sigma_0 - \frac{Ev}{[k(T - T_0)]}, \qquad (1)$$

where σ_0 (S/cm) is the conductivity at infinite temperature, Ev is the Vogel activation energy (eV), k is the Boltzmann constant, T is the absolute temperature, and T_0 is the Vogel temperature which is the onset of polymer segmental relaxations (°C).

Figure 7 shows the temperature dependence of ionic conductivity for the polymer electrolytes. The ionic conductivity increases with decreasing molecular weight of the starting PEGME.

Over the whole temperature range, the sample containing PEGME-borate ester whose molecular weight is $X = 350$ shows the highest ionic conductivity, and the values of the

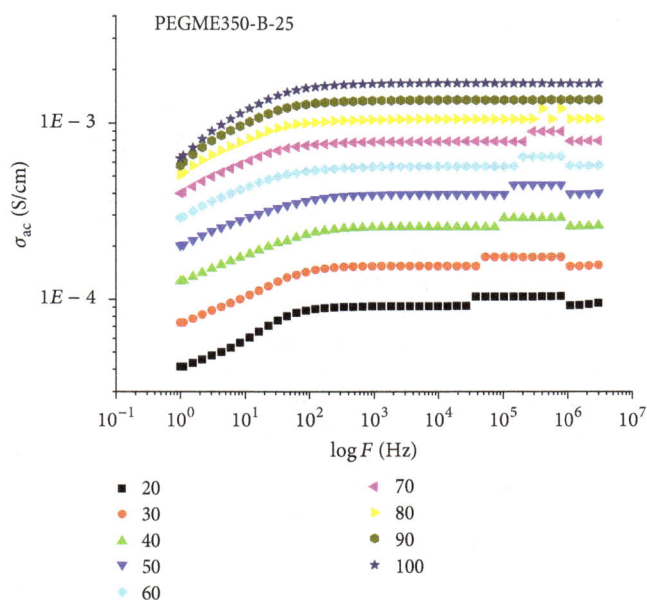

FIGURE 6: The ac conductivity versus frequency of anhydrous PEGME350-B-25 borate ester polymer electrolyte at various temperatures.

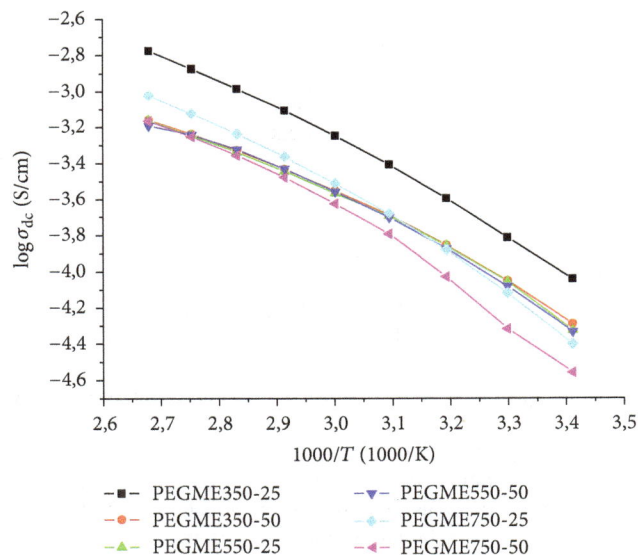

FIGURE 7: Temperature dependence of the DC conductivities of PEGMEX-B-Y polymer electrolytes.

ionic conductivity were 1.5×10^{-4} S cm^{-1} at 30°C and 1.7×10^{-3} S·cm^{-1} at 100°C. It is thought that the PEG-borate ester ($X = 350$) has the highest ability to dissolve and transfer Li ions which may be due to the shorter EO chain length. At higher chain lengths, it is clearly seen that there is almost no significant increase of the conductivity. In particular, PEGME 750-25 seems to be of optimum composition and shows slightly higher conductivity at higher temperatures.

4. Conclusions

This study presents a new approach to preparing polymer electrolytes having borate ester groups for lithium ion batteries. The PEGME-borate esters were synthesized by the reaction of poly(ethylene glycol) methyl ether and borane tetrahydrofuran complex solution, BH_3/THF. Then the electrolytes were successfully produced by doping of the polymer matrix with CF_3SO_3Li at different ratios which are abbreviated as PEGMEX-B-Y. NMR, FTIR, TGA, and dielectric-impedance spectroscopy methods were used for the characterization of these electrolytes. The FTIR and NMR spectra reveal that borate ester polymer electrolytes were successfully synthesized. TGA demonstrated that the samples are thermally stable up to 200°C. The conductivity results illustrated that the lithium ion conductivity of branched polymer electrolytes depends on chain length of PEGME as well as the CF_3SO_3Li content. The polymer electrolyte PEGME350-B-25 yielded a promising ionic conductivity of $1.5 \times 10^{-4}\,S\,cm^{-1}$ at 30°C and $1.7 \times 10^{-3}\,S{\cdot}cm^{-1}$ at 100°C. The curved lines at conductivity versus $1000/T$, especially at higher salt content, are typical for VTF behavior, indicating a positive contribution of segmental relaxation to the ion motion. These materials are nonvolatile, have promising ionic conductivity, and can be suggested for use in energy storage systems.

Conflicts of Interest

The authors declare that there are no conflicts of interest regarding the publication of this paper.

References

[1] M. Armand and J.-M. Tarascon, "Building better batteries," *Nature*, vol. 451, no. 7179, pp. 652–657, 2008.

[2] J.-M. Tarascon and M. Armand, "Issues and challenges facing rechargeable lithium batteries," *Nature*, vol. 414, no. 6861, pp. 359–367, 2001.

[3] P. Huang, C. Lethien, S. Pinaud et al., "On-chip and freestanding elastic carbon films for micro-supercapacitors," *Science*, vol. 351, no. 6274, pp. 691–695, 2016.

[4] G. Nagasubramanian and K. Fenton, "Reducing Li-ion safety hazards through use of non-flammable solvents and recent work at Sandia National Laboratories," *Electrochimica Acta*, vol. 101, pp. 3–10, 2013.

[5] W. Tang, Y. Zhu, Y. Hou et al., "Aqueous rechargeable lithium batteries as an energy storage system of superfast charging," *Energy & Environmental Science*, vol. 6, pp. 2093–2104, 2013.

[6] H. Kim, J. Hong, K.-Y. Park, H. Kim, S.-W. Kim, and K. Kang, "Aqueous rechargeable Li and Na ion batteries," *Chemical Reviews*, vol. 114, no. 23, pp. 11788–11827, 2014.

[7] L. Suo, O. Borodin, T. Gao et al., "'Water-in-salt' electrolyte enables high-voltage aqueous lithium-ion chemistries," *Science*, vol. 350, no. 6263, pp. 938–943, 2015.

[8] R.-S. Kühnel, D. Reber, and C. Battaglia, "A high-voltage aqueous electrolyte for sodium-ion batteries," *ACS Energy Letters*, vol. 2, no. 9, pp. 2005-2006, 2017.

[9] P. C. Barbosa, L. C. Rodrigues, M. M. Silva et al., "Solid-state electrochromic devices using pTMC/PEO blends as polymer electrolytes," *Electrochimica Acta*, vol. 55, no. 4, pp. 1495–1502, 2010.

[10] H. Gao, Q. Tian, and K. Lian, "Polyvinyl alcohol-heteropoly acid polymer electrolytes and their applications in electrochemical capacitors," *Solid State Ionics*, vol. 181, no. 19-20, pp. 874–876, 2010.

[11] D. K. Pradhan, R. N. P. Choudhary, and B. K. Samantaray, "Studies of dielectric and electrical properties of plasticized polymer nanocomposite electrolytes," *Materials Chemistry and Physics*, vol. 115, no. 2-3, pp. 557–561, 2009.

[12] J. R. Kim, S. W. Choi, S. M. Jo, W. S. Lee, and B. C. Kim, "Electrospun PVdF-based fibrous polymer electrolytes for lithium ion polymer batteries," *Electrochimica Acta*, vol. 50, no. 1, pp. 69–75, 2004.

[13] X. Li, G. Cheruvally, J.-K. Kim et al., "Polymer electrolytes based on an electrospun poly(vinylidene fluoride-co-hexafluoropropylene) membrane for lithium batteries," *Journal of Power Sources*, vol. 167, no. 2, pp. 491–498, 2007.

[14] O. Bohnke, G. Frand, M. Rezrazi, C. Rousselot, and C. Truche, "Fast ion transport in new lithium electrolytes gelled with PMMA. 1. Influence of polymer concentration," *Solid State Ionics*, vol. 66, no. 1-2, pp. 97–104, 1993.

[15] O. Bohnke, G. Frand, M. Rezrazi, C. Rousselot, and C. Truche, "Fast ion transport in new lithium electrolytes gelled with PMMA. 2. Influence of lithium salt concentration," *Solid State Ionics*, vol. 66, no. 1-2, pp. 105–112, 1993.

[16] D. Peramunage, D. M. Pasquariello, and K. M. Abraham, "Polyacrylonitrile-Based Electrolytes with Ternary Solvent Mixtures as Plasticizers," *Journal of the Electrochemical Society*, vol. 142, no. 6, pp. 1789–1798, 1995.

[17] P.-Y. Pennarun, P. Jannasch, S. Papaefthimiou, N. Skarpentzos, and P. Yianoulis, "High coloration performance of electrochromic devices assembled with electrolytes based on a branched boronate ester polymer and lithium perchlorate salt," *Thin Solid Films*, vol. 514, no. 1-2, pp. 258–266, 2006.

[18] F. Zhou, D. R. MacFarlane, and M. Forsyth, "Boroxine ring compounds as dissociation enhancers in gel polyelectrolytes," *Electrochimica Acta*, vol. 48, no. 12, pp. 1749–1758, 2003.

[19] B. S. Lalia, N. Yoshimoto, M. Egashira, and M. Morita, "Effects of Lewis-acid polymer on the electrochemical properties of alkylphosphate-based non-flammable gel electrolyte," *Journal of Power Sources*, vol. 194, no. 1, pp. 531–535, 2009.

[20] N.-S. Choi, S.-W. Ryu, and J.-K. Park, "Effect of tris(methoxy diethylene glycol) borate on ionic conductivity and electrochemical stability of ethylene carbonate-based electrolyte," *Electrochimica Acta*, vol. 53, no. 22, pp. 6575–6579, 2008.

[21] P.-Y. Pennarun and P. Jannasch, "Electrolytes based on LiClO4 and branched PEG-boronate ester polymers for electrochromics," *Solid State Ionics*, vol. 176, no. 11-12, pp. 1103–1112, 2005.

[22] J. McBreen, H. S. Lee, X. Q. Yang, and X. Sun, "New approaches to the design of polymer and liquid electrolytes for lithium batteries," *Journal of Power Sources*, vol. 89, no. 2, pp. 163–167, 2000.

[23] E. Zygadło-Monikowska, Z. Florjańczyk, A. Tomaszewska et al., "New boron compounds as additives for lithium polymer electrolytes," *Electrochimica Acta*, vol. 53, no. 4, pp. 1481–1489, 2007.

[24] Y. Kato, K. Hasumi, S. Yokoyama et al., "Polymer electrolyte plasticized with PEG-borate ester having high ionic conductivity and thermal stability," *Solid State Ionics*, vol. 150, no. 3-4, pp. 355–361, 2002.

[25] P. Chetri, N. N. Dass, and N. S. Sarma, "Conductivity measurement of poly(vinyl borate) and its lithium derivative in solid state," *Materials Science and Engineering B: Solid-State Materials for Advanced Technology*, vol. 139, no. 2-3, pp. 261–264, 2007.

[26] M. Şenel, A. Bozkurt, and A. Baykal, "An investigation of the proton conductivities of hydrated poly(vinyl alcohol)/boric acid complex electrolytes," *Ionics*, vol. 13, no. 4, pp. 263–266, 2007.

[27] P.-Y. Pennarun and P. Jannasch, "Influence of the alkali metal salt on the properties of solid electrolytes derived from a Lewis acidic polyether," *Solid State Ionics*, vol. 176, no. 23-24, pp. 1849–1859, 2005.

[28] H. Aydin and A. Bozkurt, "Synthesis, characterization, and ionic conductivity of novel crosslinked polymer electrolytes for Li-ion batteries," *Journal of Applied Polymer Science*, vol. 124, no. 2, pp. 1193–1199, 2012.

[29] Y.-H. Liang, C.-C. Wang, and C.-Y. Chen, "Synthesis and characterization of a new network polymer electrolyte containing polyether in the main chains and side chains," *European Polymer Journal*, vol. 44, no. 7, pp. 2376–2384, 2008.

[30] M. H. Cohen and D. Turnbull, "Molecular transport in liquids and glasses," *The Journal of Chemical Physics*, vol. 31, no. 5, pp. 1164–1169, 1959.

[31] G. S. Grest and M. H. Cohen, "Liquid-glass transition: dependence of the glass transition on heating and cooling rates," *Physical Review B*, vol. 21, no. 9, pp. 4113–4117, 1980.

14

Solid-Liquid Phase Equilibria of the Ternary System (NaCl + CH₃OH + H₂O) at 298.15, 308.15, 318.15 K, and 0.1 MPa

Jian Shi,[1] **Jiayin Hu** ⓘ,[1] **Long Li,**[1] **Yafei Guo** ⓘ,[1,2] **Xiaoping Yu** ⓘ,[1] **and Tianlong Deng** ⓘ[1]

[1]*Tianjin Key Laboratory of Marine Resources and Chemistry, College of Chemical Engineering and Materials Science, Tianjin University of Science and Technology, Tianjin 300457, China*
[2]*College of Chemistry and Materials Science, Northwest University, Xi'an 710127, China*

Correspondence should be addressed to Jiayin Hu; hujiayin@tust.edu.cn and Tianlong Deng; tldeng@tust.edu.cn

Academic Editor: Christophe Coquelet

The phase equilibrium of the ternary system (NaCl + CH₃OH + H₂O) at 298.15, 308.15, 318.15 K, and 0.1 MPa has been investigated by the isothermal dissolution equilibrium method. Solubilities and physicochemical properties including refractive index (n_D) and density (ρ) in the ternary system were determined experimentally. According to the experimental data, the phase diagrams and diagrams of physicochemical properties versus sodium chloride concentration in the solvents at 298.15, 308.15, and 318.15 K were plotted, respectively. The experimental results showed that the system did not cause stratification and the equilibrium solid phase was anhydrous sodium chloride. Neither double salt nor solid solution was found at the three temperatures. The physicochemical properties of the ternary system change regularly with the increase of sodium chloride concentration in the solution, and the solvation effect of CH₃OH on NaCl was significant. Moreover, the calculated values of NaCl solubility data based on the CNIBS/R-K equations agreed well with the experimental results, and the thermodynamic functions in the dissolution process of NaCl in the CH₃OH-H₂O binary system were also calculated for further investigation of NaCl dissolution process.

1. Introduction

A large number of mixture containing sodium chloride (NaCl), lithium chloride (LiCl), and organic pollution are obtained in association with the production of special plastics. For example, during the production of poly-phenylene sulfide (PPS), a large amount of by-product salt slurry containing NaCl, LiCl, and a small amount of oligomers were generated [1]. The green recycling of above high-value by-products is of great significance in the sustainable development of the plastics, salt chemical, and chloralkali industries [2]. For separating the inorganic salts, such as NaCl, from the by-product salt slurry, the phase equilibria of the related system are priorities to investigate.

Compared with vast data on the solubility and physicochemical properties of NaCl in aqueous electrolyte systems, the solubility and physicochemical properties data of NaCl in organic solvents are quite limited. Yang et al. [3] published the ternary system (NaCl + C₂H₅OH + H₂O)

equilibrium at 293.15 K. They found the equilibrium solid phase was anhydrous sodium chloride, and no double salt and solid solution formed, especially, ethanol had been found to have strong salting-out effect on NaCl.

In this work, the solubilities of NaCl in the ternary parameters of (NaCl + CH₃OH + H₂O) were studied with the isothermal dissolution method at 298.15, 308.15, and 318.15 K for the first time. The related physicochemical properties including density, refractive index, and salting-out rate were measured. Moreover, the solubilities of NaCl, as well as the thermodynamic functions in the dissolution process of NaCl in the CH₃OH-H₂O binary system were also calculated for further investigation.

2. Experimental

2.1. Apparatus and Reagents. The apparatus used for the isothermal equilibrium in this work was designed in our laboratory and is shown in Figure 1. The experiments were

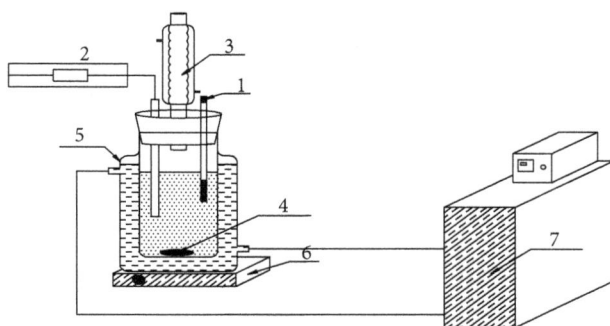

FIGURE 1: Apparatus used for the isothermal dissolution equilibrium: (1) sampler; (2) thermometer; (3) condenser; (4) magnetic stirring rod; (5) jacketed glass cell; (6) magnetic stirrer; and (7) temperature-controlled water bath.

carried out in a double-jacketed ground-glass cell with a volume of $100 \, cm^3$. The temperatures of the cell were controlled by an external water bath circulator (K20-cc-NR, Huber, Germany) with an uncertainty of ± 0.01 K [4]. To avoid the evaporation of methanol, a condensing unit was attached on the double-jacketed ground-glass cell.

The chemicals used in this study are shown in Table 1. It is worth mentioning that NaCl was recrystallized before used. Doubly deionized water (DDW) with pH 6.60 and conductivity less than $1 \times 10^{-4} \, S \cdot m^{-1}$ at room temperature (298.15 K) was used in this work.

2.2. Experimental Method. The phase equilibrium of the ternary system was studied with the isothermal dissolution method [5]. At the beginning, series of H_2O and methanol were exactly preweighted. The weight of H_2O was m_W, and the weight of methanol was m_M. Then, H_2O and methanol with known composition were added to the 100 mL jacketed glass cell and capped tightly. The jacketed glass cell was placed in the external water bath, whose temperature was set at a certain value (298.15 ± 0.01, 308.15 ± 0.01, and 318.15 ± 0.01 K). After stirring 1 h, a certain amount of NaCl was added into the jacketed glass cell and stirred for 36 h. A $0.5 \, cm^3$ sample of the clarified supernatant was taken from the liquid phase of the jacketed glass cell with a pipette at regular intervals for chemical analysis. It was worthy saying that the magnetic stirrer was allowed to rest for 1 h in order to ensure the separation of the solid and liquid phase before sampling. If the compositions of the liquid phase in the bottle became constant, it indicated that the equilibrium was achieved.

2.3. Analytical Method. Briefly, the solubilities of sodium chloride in the binary solvent mixtures were measured by the gravimetric method using high precision balance (OHAUS, USA, precision of 0.1 mg) with standard uncertainty of 0.2 mg at 0.681 level of confidence [6]. The weight of the empty bottle was m_0, and the weight of the bottle with solution (supernatant that has achieved equilibrium) was m_1. The bottle with liquid samples was

first dried in an oven at 388.15 K. After a fine grind of this NaCl solid, again the fine NaCl was dried at 388.15 K until the mass did not change. In addition, the dried NaCl solid was tested by infrared spectroscopy, and the result (Figure S1) showed the characteristic peaks of both of H_2O and CH_3OH were not observed. The result above indicated that the solid NaCl obtained did not trap or embed some of water and/or methanol molecules. At this time, the weight of the bottle was obtained and kept constant as m_2. All results were the average of 3 repeated tests. The content of each component in the equilibrium supernatant system is calculated by the following equations:

$$w_{NaCl} = \frac{m_2 - m_0}{m_1 - m_0},$$

$$w_{H_2O} = \frac{m_W (m_1 - m_2)}{(m_W + m_M)(m_1 - m_0)}, \quad (1)$$

$$w_{CH_3OH} = \frac{m_M (m_1 - m_2)}{(m_W + m_M)(m_1 - m_0)},$$

where w_{H_2O} and w_{CH_3OH} are the mass fractions of water and methanol in the equilibrium supernatant system, respectively. All experimental data containing m_0, m_1, m_2, m_w, and m_M are given in Table S1 in the supporting information.

The densities (ρ) were measured by the automatic oscillating U-tube densimeter (DMA 4500, Anton Paar, Austria, precision of $1.0 \times 10^{-5} \, g \cdot cm^{-3}$) with standard uncertainty of $0.5 \, mg \cdot cm^{-3}$ at 0.681 level of confidence [7]. The refractive indices (n_D) were measured by an Abbe refractometer (model WZS-1, Shanghai, precision ± 0.0001) with standard uncertainty of 0.001 at 0.681 level of confidence [8]. The solid phase minerals were identified by an X-ray diffractometer (MSAL XD-3, Beijing) and BX51 digital polarizing microscope (Olympus, Japan). All measurements of the above physicochemical properties were maintained in a supper thermostatic water bath that controlled at the desired temperature (298.15 ± 0.01, 308.15 ± 0.01, and 318.15 ± 0.01 K).

3. Results and Discussion

3.1. For the Ternary System NaCl + CH_3OH + H_2O. In order to verify the reliability of our experimental method, we compared the equilibrium solubility, refractive index, and density data with the reported binary system (NaCl + H_2O) at different temperatures, and the results are shown in Table 2. The confidence interval of NaCl solubility (mass fraction) in water was 26.43 ± 0.15, 26.67 ± 0.09, and 26.80 ± 0.11 at 298.15, 308.15, and 318.15 K, respectively. The results above proved our experimental method was reliable.

The solubility and physicochemical properties of the ternary system (NaCl + CH_3OH + H_2O) at 298.15, 308.15, and 318.15 K are shown in Table 3. On the basis of the experimental solubility data in Table 3, the equilibrium phase diagrams of the ternary system (NaCl + CH_3OH +

TABLE 1: Chemicals used in this study.

Chemical name	Source	Initial mass fraction purity	Purification method	Final mass fraction purity	Analysis method
NaCl	[a]A.R.	0.995	Recrystallization	0.998	Chemical analysis and ICP-OES[c]
CH$_3$OH	[b]A.R.	0.998	None	—	—

[a]A.R.: the Group of States Chemical Reagent Co., Ltd.; [b]A.R.: the Tianjin Fuyu Fine Chemical Co. Ltd.; [c]ICP-OES: inductively coupled plasma optical emission spectrometer.

TABLE 2: Solubilities, refractive index, and density of the binary system (NaCl + H$_2$O) at 298.15, 308.15, and 318.15 K and p = 0.1 MPa.[a]

T/K	Solubility, 100 w[b]	Refractive index, n_D	Density, ρ (g·cm^{-3})	Reference
298.15	26.30	1.3796	1.1978	[9]
	26.37	1.3810	1.1983	[10]
	26.45	—[c]	—	[11]
	26.47	1.3801	1.19755	This work
308.15	26.59			[11]
	26.66	1.3766	1.1936	[9]
	26.62	—	1.19350	[12]
	26.63	1.3789	1.19362	This work
318.15	26.69	1.3766	1.1889	[9]
	26.80	1.3776	1.18959	This work

[a]Standard uncertainties u are $u(T)$ = 0.1 K and $u(p)$ = 0.005 MPa [13]; [b]w, mass fraction, respectively. [c]Not detected.

H$_2$O) at 298.15, 308.15, and 318.15 K are shown in Figures 2–4.

Points A$_1$ in Figure 2, A$_2$ in Figure 3, and A$_3$ in Figure 4 represented the solubilities of NaCl in pure water in mass fraction (100 w) with 26.47, 26.63, and 26.80 at 298.15, 308.15, and 318.15 K, respectively. Similarly, Points B$_1$ in Figure 2, B$_2$ in Figure 3, and B$_3$ in Figure 4 represents the solubilities of NaCl in pure methanol in mass fraction (100w) with 1.36, 1.37, and 1.38 at 298.15, 308.15, and 318.15 K, respectively. There was no turning point and no double salt and solid solution in this system at different temperatures. The equilibrium solid phase was anhydrous sodium chloride. It can be seen from Figure 5 that the solubilities of NaCl increased with the increasing of the percentage of H$_2$O in the mixed solvent at three temperatures, i.e., the solubilities of NaCl decreased with the increase of the percentage of CH$_3$OH.

On the basis of the physicochemical property data (densities and refractive indices) in Table 3, the diagrams of physicochemical properties versus sodium chloride content in the ternary system at 298.15, 308.15, and 318.15 K were plotted in Figures 6(a) and (b). It was found that densities and refractive indices in this ternary system at three temperatures change regularly with the increasing of sodium chloride content. Generally, the densities and the refractive indices have the same varying trend. Both of them increased with the increasing of sodium chloride content at each temperature.

3.2. Salting-Out Rate.
According to the results above, the solubilities of NaCl decreased with the increase of the percentage of CH$_3$OH. It suggested that methanol had strong salting-out effect on NaCl. The salting-out effects of methanol on NaCl can be expressed by salting-out rate (SOR). SOR [14] is identified in the following equation:

$$SOR(\%) = \frac{w_{NaCl-W} - w_{NaCl-Mix}}{w_{NaCl-W}} \times 100\%, \quad (2)$$

where w_{NaCl-W} is the mass fraction of NaCl in saturated sodium chloride pure water solution and $w_{NaCl-Mix}$ expresses the mass fraction of NaCl in the equilibrium supernatant system.

According to the data in Table 3, the values of SOR were calculated by the above equation in accordance with the solubilities of NaCl, and the curve of SOR versus the mass fraction of methanol in mixture solvent at 298.15, 308.15, and 318.15 K is shown in Figure 7. The SOR was increased gradually with the increasing of methanol content in mass fraction.

3.3. Correlation of the Solubility of NaCl and the Composition of Mixed Solvents.
Acree [15] proposed a CNIBS/R-K equation, which was shown in the following equation (3). This equation could study the correlation of the solubility of a solute and the composition of the mixed solvent at isothermal temperature:

$$\ln x_A = x_B^0 \ln (x_A)_B + x_C^0 \ln (x_A)_C + x_B^0 x_C^0 \sum_{i=0}^{N} S_i \left(x_B^0 - x_C^0 \right)^i, \quad (3)$$

where x_A is the molar solubility of the solute in the mixed solvent; x_B^0 and x_C^0, respectively, represent the molar ratio of the solvents B and C in the mixed solvent in the absence of solute; $(x_A)_B$ and $(x_A)_C$ represent the saturated molar solubility of solute in the pure solvent B and C, respectively; N represents the number of the component; S_i is the parameter of the model.

Our system was the two-component mixed solvent system, and then N equals 2 and x_C^0 can be replaced by

TABLE 3: The experimental solubilities and the physicochemical properties in the ternary system (NaCl + CH$_3$OH + H$_2$O) at 298.15, 308.15, and 318.15 K, and p = 0.1 MPa.[a]

No.	Composition of liquid phase, 100 w[b]			Physicochemical properties		Equilibrium solid phase
	CH$_3$OH	H$_2$O	NaCl	Refractive index, n_D	Density, ρ (g/cm^3)	
				298.15 K		
1, A$_1$	0.00	73.53	26.47	1.3801	1.19755	NaCl
2	4.74	70.98	24.28	1.3773	1.16931	NaCl
3	8.00	69.13	22.87	1.3755	1.15040	NaCl
4	16.13	64.34	19.53	1.3717	1.10901	NaCl
5	24.84	58.85	16.31	1.3678	1.06641	NaCl
6	35.03	52.09	12.88	1.3623	1.02088	NaCl
7	44.53	45.35	10.12	1.3579	0.98190	NaCl
8	54.38	38.05	7.57	1.3531	0.94401	NaCl
9	65.97	28.98	5.05	1.3475	0.90274	NaCl
10	77.05	19.70	3.25	1.3426	0.86571	NaCl
11	87.86	10.07	2.07	1.3374	0.83531	NaCl
12, B$_1$	98.64	0.00	1.36	1.3301	0.79767	NaCl
				308.15 K		
1, A$_2$	0.00	73.37	26.63	1.3789	1.19362	NaCl
2	4.04	71.17	24.79	1.3762	1.16933	NaCl
3	7.80	68.99	23.21	1.3748	1.14854	NaCl
4	15.87	64.23	19.90	1.3701	1.10590	NaCl
5	25.19	58.31	16.50	1.3658	1.06084	NaCl
6	34.53	52.06	13.41	1.3608	1.01920	NaCl
7	44.73	44.89	10.37	1.3560	0.97650	NaCl
8	55.97	36.55	7.48	1.3500	0.93320	NaCl
9	66.13	28.60	5.27	1.3451	0.89573	NaCl
10	76.97	19.62	3.41	1.3399	0.85862	NaCl
11	87.66	10.19	2.15	1.3338	0.82366	NaCl
12, B$_2$	98.63	0.00	1.37	1.3270	0.78797	NaCl
				318.15 K		
1, A$_3$	0.00	73.20	26.80	1.3776	1.18959	NaCl
2	3.90	71.04	25.06	1.3752	1.16633	NaCl
3	7.69	68.79	23.51	1.3732	1.14571	NaCl
4	15.97	63.84	20.19	1.3689	1.10219	NaCl
5	24.84	58.19	16.97	1.3649	1.05944	NaCl
6	34.55	51.75	13.70	1.3594	1.01414	NaCl
7	44.55	44.71	10.74	1.3547	0.97288	NaCl
8	55.09	36.98	7.93	1.3481	0.92982	NaCl
9	65.92	28.59	5.49	1.3430	0.88964	NaCl
10	76.18	20.18	3.64	1.3373	0.85332	NaCl
11	87.64	10.18	2.18	1.3310	0.81509	NaCl
12, B$_3$	98.62	0.00	1.38	1.3244	0.77871	NaCl

[a]Standard uncertainties u are $u(T)$ = 0.1 K and $u(p)$ = 0.005 MPa; $u(w)$ for NaCl, CH$_3$OH, and H$_2$O are 0.0049, 0.0005, and 0.0005 in mass fraction, respectively; $u(x)$ for n_D and ρ are 0.001 and 0.5 mg·cm^{-3}, respectively; [b]w, mass fraction.

$(1 - x_B^0)$. x_A is the molar solubility of the NaCl in the mixed methanol + water system. x_B^0 and x_C^0 are referred to the molar fraction of methanol and water in the binary methanol + water system recalculated considering a NaCl-free supernatant, respectively. $(x_A)_B$ and $(x_A)_C$ represent the saturated molar solubility of NaCl in the pure methanol and water, respectively. In this way, Equation (3) can be changed to Equation (4):

$$\ln x_A - x_B^0 \ln (x_A)_B - (1 - x_B^0) \ln (x_A)_C$$
$$= x_B^0 (1 - x_B^0) \left[S_0 + S_1 (2x_B^0 - 1) + S_2 (2x_B^0 - 1)^2 \right]. \quad (4)$$

According to Equation (4), the calculated parameters and multiple correlation coefficient R^2 at different temperatures

are presented in Table 4. All the R^2 values are distributed in the range of 0.9983–0.9986. The results inferred the calculated values agreed well with the experimental data, and this agreement showed that the parameter S_i obtained in this work were reliable and can be used to calculate any solubilities of NaCl in the mixed solvent (CH$_3$OH + H$_2$O) in the corresponding temperatures.

3.4. Calculated Thermodynamic Functions. To exploit the valuable NaCl from organic solvents, the thermodynamic functions for the dissolution process of NaCl in the mixed solvent (CH$_3$OH + H$_2$O) are also essential. Hence, we calculated the relevant thermodynamic functions [16]

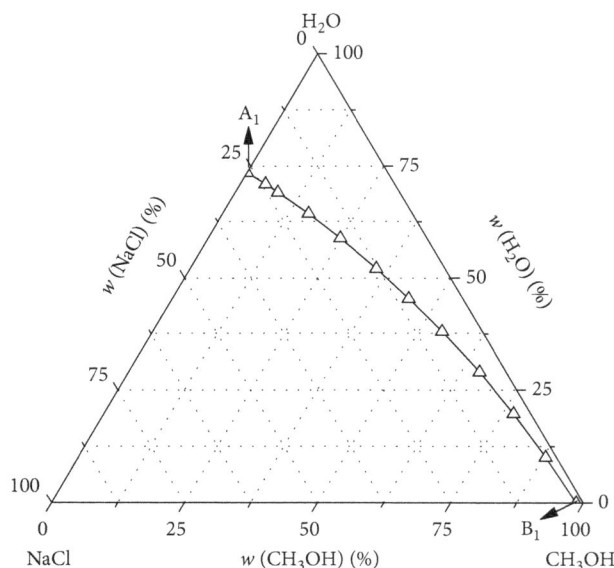

FIGURE 2: Phase diagram of the ternary system (NaCl + CH$_3$OH + H$_2$O) at 298.15 K.

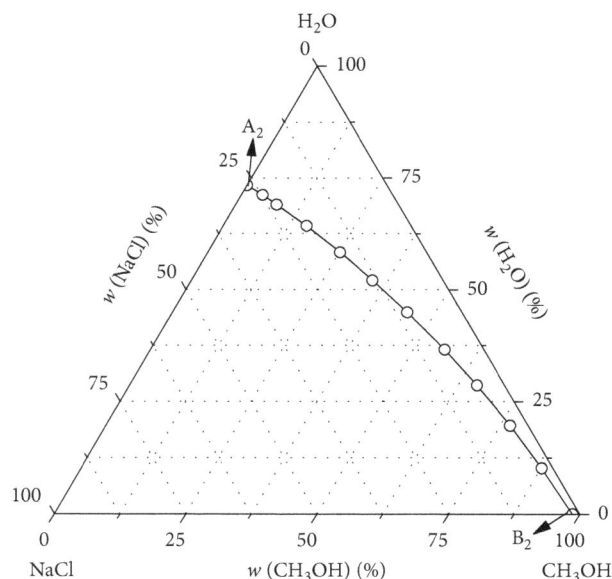

FIGURE 3: Phase diagram of the ternary system (NaCl + CH$_3$OH + H$_2$O) at 308.15 K.

during the dissolution process of NaCl in the mixed solvent, such as Gibbs free energy ($\Delta_{sol}G$), entropy ($\Delta_{sol}S$), and enthalpy ($\Delta_{sol}H$), and the following correlation equations were used [17]:

$$T_{mean} = \frac{n}{\sum_{i=1}^{n}(1/T)},$$

$$x_A = \frac{(m_A/M_A)}{(m_A/M_A + m_W/M_W + m_M/M_M)},$$

$$\Delta_{sol}H = -R\left(\frac{\partial \ln x_A}{\partial(1/T - 1/T_{mean})}\right),$$

$$\Delta_{sol}G = -RT_{mean} \times \text{intercept}, \tag{5}$$

$$\Delta_{sol}S = \frac{\Delta_{sol}H - \Delta_{sol}G}{T_{mean}},$$

$$\zeta_H = \frac{|\Delta_{sol}H|}{|\Delta_{sol}H| + |T_{mean} \times \Delta_{sol}S|},$$

$$\zeta_{TS} = \frac{|T_{mean} \times \Delta_{sol}H|}{|\Delta_{sol}H| + |T_{mean} \times \Delta_{sol}S|},$$

where m_A, m_W, and m_M are the mass (g) of sodium chloride, water, and methanol from Table 3, respectively; M_A, M_W, and M_M are the relative molar mass of sodium chloride, water, and methanol, respectively; ζ_H and ζ_{TS} present the enthalpy compensation and entropy compensation of NaCl dissolved in the CH$_3$OH-H$_2$O binary solvents, respectively; n is the number of studied temperature points. Intercept is obtained from the linear fit between $\ln x_A$ and $(1/T - 1/T_{mean})$. The temperature was selected in the range of 298.15–318.15 K; as a result, the value of T_{mean} was 307.93 K.

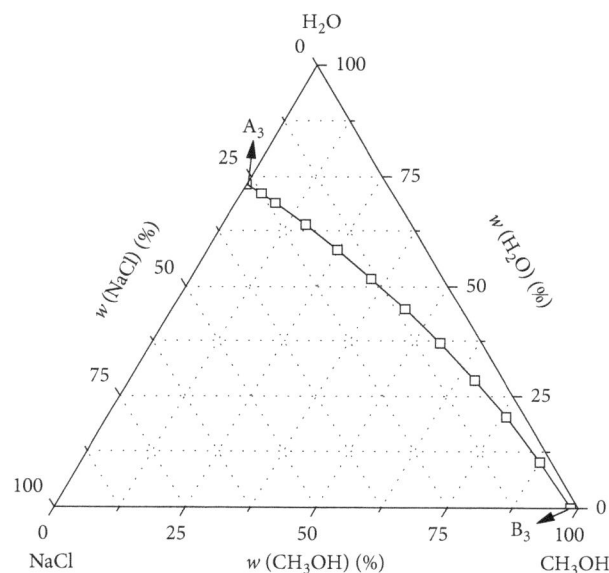

FIGURE 4: Phase diagram of the ternary system (NaCl + CH$_3$OH + H$_2$O) at 318.15 K.

The calculated thermodynamic functions of NaCl dissolved in the CH$_3$OH-H$_2$O binary system are presented in Table 5. No matter how the mixture ratios of CH$_3$OH-H$_2$O binary system changed, the values of $\Delta_{sol}H$ were all greater than zero, suggesting that the dissolution of NaCl in CH$_3$OH-H$_2$O binary system was an endothermal process. In addition, the maximum ζ_H was 0.400, which was much lower than the minimum ζ_{TS} (0.600). This result inferred that the contribution of ζ_{TS} to $\Delta_{sol}G$ is more significant than that of ζ_H, and enthalpy has limited effect on $\Delta_{sol}G$ while entropy is the major contributor.

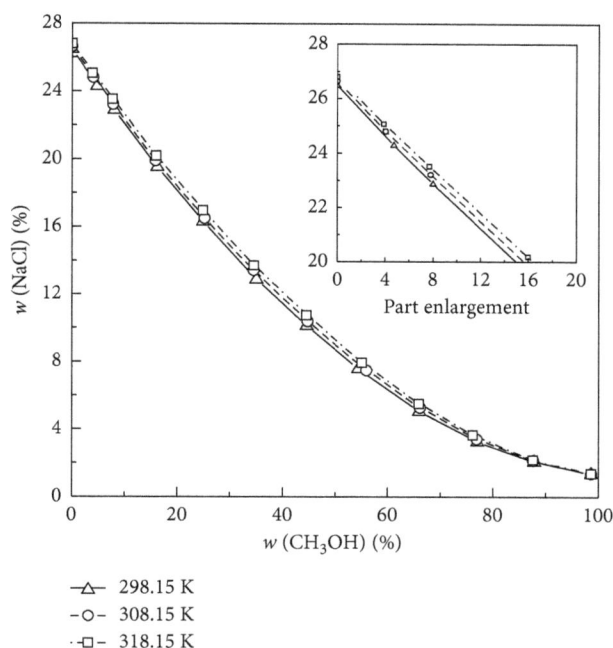

FIGURE 5: Comparison of the sodium chloride content of the ternary system(NaCl + CH₃OH + H₂O) at 298.15, 308.15, and 318.15 K; (△), experimental point at 298.15 K; (—), solubility curve at 298.15 K; (○), experimental point at 308.15 K; (- -), solubility curve at 308.15 K; (□), experimental point at 318.15 K; (⋯), solubility curve at 318.15 K.

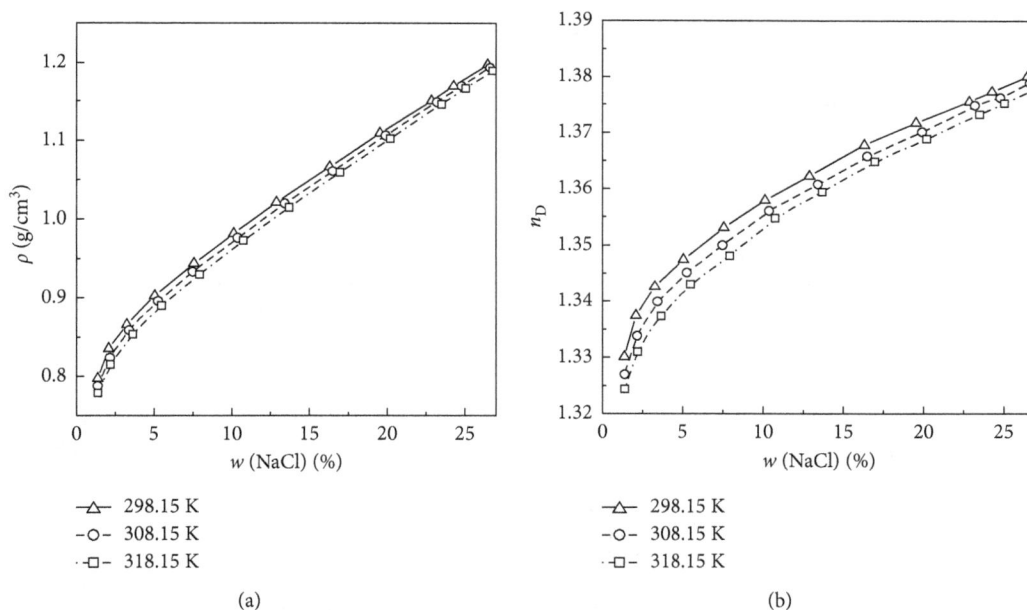

(a) (b)

FIGURE 6: Comparison of the physicochemical properties versus sodium chloride content for the ternary system (NaCl + CH₃OH + H₂O) at 298.15, 308.15, and 318.15 K; (△), experimental point at 298.15 K; (○), experimental point at 308.15 K; (□), experimental point at 318.15 K: (a) density versus NaCl composition; (b) refractive index versus NaCl composition.

4. Conclusion

The solubilities and physicochemical properties including refractive index and density of the ternary mixture solvent system (NaCl + CH₃OH + H₂O) at 298.15, 308.15, and 318.15 K were investigated using the isothermal dissolution equilibrium method. Based on the experimental data, the equilibrium phase diagrams and diagrams of physicochemical properties versus composition were plotted. It was found that there was no turning point and no double salt and solid solution, and the equilibrium solid phase is anhydrous sodium chloride. The physicochemical properties of the ternary system at three temperatures show regular change with the increase of sodium chloride concentration in

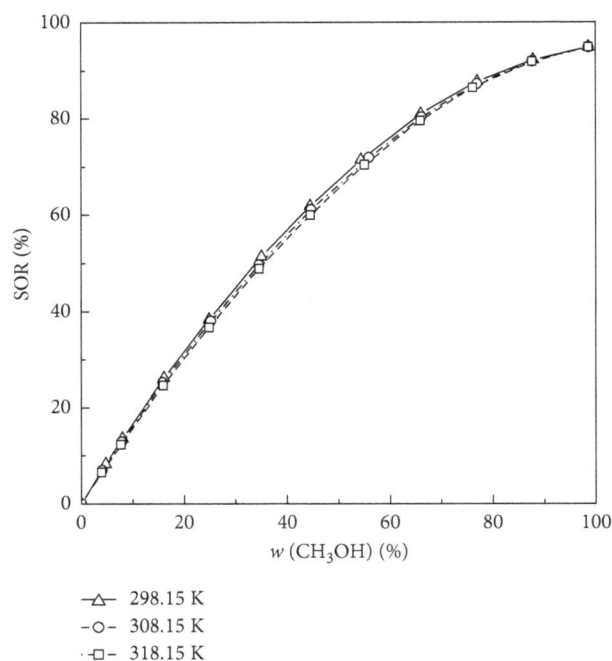

FIGURE 7: Salting-out rate (SOR) versus methanol content for the ternary system (NaCl + CH$_3$OH + H$_2$O) at 298.15, 308.15, and 318.15 K; (△), experimental point at 298.15 K; (O), experimental point at 308.15 K; (□), experimental point at 318.15 K.

TABLE 4: Parameters of CNIBS/R-K model for solubility in solvent mixtures.

T/K	S_0	S_1	S_2	R^2
298.15	0.065	−0.810	−0.464	0.9984
308.15	0.236	−0.751	−0.505	0.9986
318.15	0.363	−0.785	−0.585	0.9983

TABLE 5: Calculated various thermodynamic functions of NaCl dissolved in different mixture ratios of CH$_3$OH-H$_2$O binary system.

$x^0_{CH_3OH}$	$\Delta_{sol}H/(kJ \cdot mol^{-1})$	$\Delta_{sol}G/(kJ \cdot mol^{-1})$	$\Delta_{sol}S/(J \cdot mol^{-1} \cdot K^{-1})$	ζ_H	ζ_{TS}
0.00	0.599	5.879	−17.148	0.102	0.898
0.10	1.310	6.423	−16.605	0.204	0.796
0.20	2.107	6.998	−15.885	0.301	0.699
0.30	2.819	7.625	−15.607	0.370	0.630
0.40	3.320	8.311	−16.207	0.400	0.600
0.50	3.531	9.052	−17.928	0.390	0.610
0.60	3.416	9.829	−20.827	0.348	0.653
0.70	2.984	10.613	−24.773	0.281	0.719
0.80	2.293	11.360	−29.444	0.202	0.798
0.90	1.443	12.014	−34.331	0.120	0.880
1.00	0.579	12.508	−38.737	0.046	0.954

solution. Methanol had strong salting-out effect on NaCl, and the SOR was increased gradually with the increasing of methanol content in the mixture solvent ternary system. The calculated values of NaCl solubility data based on the CNIBS/R-K equations agreed well with the experimental results. In addition, the thermodynamic functions in the dissolution process of NaCl in the CH$_3$OH-H$_2$O binary system were also calculated for further investigation of NaCl dissolution process. Generally, studies on phase equilibria

and phase diagrams of the ternary systems (NaCl + CH$_3$OH + H$_2$O) at 298.15, 308.15, 318.15 K, and 0.1 MPa could provide the based thermodynamic data to exploit the valuable inorganic salts from organic solvents.

Conflicts of Interest

The authors declare that they have no conflicts of interest.

Acknowledgments

The authors thank the National Natural Science Foundation of China (U1607123, U1607129, and 21773170), the Chinese Postdoctoral Science Foundation (2016M592827 and 2016M592828), and the Yangtze Scholars and Innovative Research Team of the Chinese University (IRT_17R81).

References

[1] Z. J. Shao, W. W. Jiang, J. R. Wang et al., "LiFePO$_4$ synthesized with PPS lithium-containing filtrate," *China Measurement and Test*, vol. 43, pp. 42–45, 2017.

[2] Z. H. Cheng, X. D. Li, H. Y. Wang et al., "Research progress of high concentration organic wastewater treatment," *Advances in Environmental Protection*, vol. 6, pp. 130–136, 2016.

[3] J. M. Yang, J. J. Chen, R. Z. Zhang, and M. Ping, "Study on the phase equilibrium of (NaCl + C$_2$H$_5$OH + H$_2$O) ternary system at 20°C," *Henan Chemical Industry*, vol. 24, pp. 20–22, 2007.

[4] T. J. Fortin, A. Laesecke, M. Freund, and S. Outcalt, "Advanced calibration, adjustment, and operation of a density and sound speed analyzer," *Journal of Chemical Thermodynamics*, vol. 57, pp. 276–285, 2013.

[5] Y. Zhong, H. Yang, H. Wang, H. Ge, and M. Wang, "Solid–liquid phase equilibrium in the ternary system MgSO$_4$ + MgCl$_2$ + H$_2$O at 263.15 K," *Journal of Chemical and Engineering Data*, vol. 63, no. 5, pp. 1300–1303, 2018.

[6] D. W. Wei and X. R. Zhang, "Solubility of puerarin in the binary system of methanol and acetic acid solvent mixtures," *Fluid Phase Equilibria*, vol. 339, pp. 67–71, 2013.

[7] T. L. Deng and D. C. Li, "Solid–liquid metastable equilibria in the quaternary system (NaCl–KCl–CaCl$_2$–H$_2$O) at 288 K," *Fluid Phase Equilibria*, vol. 269, no. 1-2, pp. 98–103, 2008.

[8] B. H. Bu, L. Li, N. Zhang et al., "Solid–liquid metastable phase equilibria for the ternary system (Li$_2$SO$_4$ + K$_2$SO$_4$ + H$_2$O) at 288 and 323 K, p = 0.1 MPa," *Fluid Phase Equilibria*, vol. 402, pp. 78–82, 2015.

[9] Y. X. Gao, S. Li, Q. G. Zhai, Y. C. Jiang, and M. C. Hu, "Phase diagrams and physicochemical properties for the ternary system (CsCl + NaCl + H$_2$O) at T = (298.15, 308.15, and 318.15) K," *Journal of Chemical and Engineering Data*, vol. 62, no. 9, pp. 2533–2540, 2017.

[10] X. P. Ding, B. Sun, L. J. Shi, H. T. Yang, and P. S. Song, "Study on phase equilibria in (NaCl + SrCl$_2$ + H$_2$O) ternary system at 25°C," *Inorganic Chemicals Industry*, vol. 42, pp. 9–11, 2010.

[11] A. Apelblat and E. Korin, "The vapour pressures of saturated aqueous solutions of sodium chloride, sodium bromide, sodium nitrate, sodium nitrite, potassium iodate, and rubidium chloride at temperatures from 227 to 323 K," *Journal of Chemical Thermodynamics*, vol. 30, no. 1, pp. 59–71, 1998.

[12] B. J. Zhang, *Metastable Phase Equilibrium of Quaternary System Li$^+$, Na$^+$, Mg^{2+}//Cl$^-$ – H$_2$O and Ternary System Na$^+$(K$^+$), Mg^{2+}//Cl$^-$ – H$_2$O at 35°C*, Cheng Du University of Technology, Chengdu, China, 2007.

[13] W. J. Cui, X. B. Fan, Y. Liao et al., "Experimental and thermodynamic modeling study of the solid-liquid equilibrium in the ternary system (NaCl + NaClO$_3$ + H$_2$O) at 293.15 and 333.15 K and 0.1 MPa," *Journal of Chemical Thermodynamics*, vol. 126, pp. 99–104, 2018.

[14] H. Z. Lu, J. J. Lu, and J. M. Yang, "Study on phase equilibrium of (lithium sulfate + ethanol + water) ternary system at 25°C," *Inorganic Chemicals Industry*, vol. 41, pp. 21–23, 2009.

[15] W. E. J. Acree, "Mathematical representation of thermodynamic properties: Part 2. Derivation of the combined nearly ideal binary solvent (NIBS)/Redlich-Kister mathematical representation from a two-body and three-body interactional mixing model," *Thermochimica Acta*, vol. 198, no. 1, pp. 71–79, 1992.

[16] S. Y. Li, L. J. Jiang, J. X. Qiu, and P. Wang, "Solubility and solution thermodynamics of the δ form of L-citrulline in water + ethanol binary solvent mixtures," *Journal of Chemical and Engineering Data*, vol. 61, no. 1, pp. 264–271, 2015.

[17] S. Y. Li, *Dissolution Equilibrium of Myo-inositol and L-Citrulline in Binary Solvent Mixtures*, Changchun University of Technology, Changchun, China, 2016.

15

Molecular Dynamics of POPC Phospholipid Bilayers through the Gel to Fluid Phase Transition: An Incoherent Quasi-Elastic Neutron Scattering Study

U. Wanderlingh,[1] C. Branca,[1] C. Crupi,[1] V. Conti Nibali,[2] G. La Rosa,[3] S. Rifici,[1] J. Ollivier,[4] and G. D'Angelo[1]

[1]Dipartimento di Scienze Matematiche e Informatiche, Scienze Fisiche e Scienze della Terra, Università di Messina, Via F. Stagno d'Alcontres 31, 98166 Messina, Italy
[2]Institute for Physical Chemistry II, Ruhr-University Bochum, Bochum, Germany
[3]Physik-Department, Technische Universität München, Munich, Germany
[4]Institut Laue-Langevin, 6 rue J. Horowitz, BP 156, 38042 Grenoble, France

Correspondence should be addressed to U. Wanderlingh; uwanderlingh@unime.it and G. D'Angelo; gdangelo@unime.it

Academic Editor: José L. Arias Mediano

The microscopic dynamics for the gel and liquid-crystalline phase of highly aligned D_2O-hydrated bilayers of 1-palmitoyl-oleoyl-sn-glycero-phosphocholine (POPC) were investigated in the temperature range from 248 to 273 K by using incoherent quasi-elastic neutrons scattering (QENS). We develop a model for describing the molecular motions of the liquid phase occurring in the 0.3 to 350 ps time range. Accordingly, the complex dynamics of hydrogen are described in terms of simple dynamical processes involving different parts of the phospholipid chain. The analysis of the data evidences the existence of three different motions: the fast motion of hydrogen vibrating around the carbon atoms, the intermediate motion of carbon atoms in the acyl chains, and the slower translational motion of the entire phospholipid molecule. The influence of the temperature on these dynamical processes is investigated. In particular, by going from gel to liquid-crystalline phase, we reveal an increase of the segmental motion mainly affecting the terminal part of the acyl chains and a change of the diffusional dynamics from a localized rattling-like motion to a confined diffusion.

1. Introduction

The lipid bilayer is one of the most important self-assembled structures in nature. It regulates the flow of nutrients signaling events between cells and provides a selective barrier for cells and subcellular structures. It is also the structural host for functional proteins that control molecular traffic and cellular communication across the membrane. Membrane phospholipid bilayers are extensively used as model membranes for investigating the biological processes that occur at the cellular level [1]. In the last years, they received an increasing interest owing to a growing number of applications in biophysical and biomedical research, which include the passive transport through biomembranes [2], the pharmacokinetics of drugs [3, 4] or anaesthetics [5–8], and

the lodging of membrane proteins [9]. Despite the efforts in studying these properties, many challenges still remain for the proper design of highly performing systems. To successfully apply the biomembrane technologies to real problems, it is necessary to deepen the understanding of the mechanisms that regulate the relative mobility (fluidity) of the individual lipid molecules in the bilayer [10] and to clarify how this mobility is influenced by changes in the environmental conditions (e.g., hydration and temperature) [11–13]. In particular, the fluidity of the membrane lipid bilayer is believed to play a main role in controlling the cell growth [14] and in modulating the function of transmembrane proteins and membrane enzymes [15], just to cite some notable examples.

As well known, lipid membranes have a rich phase diagram that is dominated by a main transition temperature, T_m, between the gel phase (L_β) and the liquid-crystalline phase (L_α) [16]. For most phospholipids of practical interest, the phase transition temperature ranges from 270 to 310 Kelvin; it is higher for those phospholipids with longer tails, whereas the presence of unsaturated carbon bonds reduces T_m. The membrane fluidity critically depends on the structural phase state of lipids. In the gel phase, van der Waals interactions cause highly ordered packing of the acyl chains, which can also be tilted with respect to the lipid bilayer normal, resulting in limited rotational and translational motions. The gel phase is characterized by very low lipid mobility in the bilayer plane [17]. At the gel-liquid transition, the hydrocarbon chains change from an all-trans to a trans-gauche conformation. In the fluid phase, the lipid hydrocarbon chains experience higher translational and rotational mobility compared to the gel phase, resulting in enhanced membrane fluidity [18].

Several studies have demonstrated the coexistence of gel and liquid phases in living cell membranes [19] and in model membrane as well [20, 21]. Other recent studies have highlighted the existence of a complex collective THz dynamics of phospholipid membrane [22, 23] and of an anomalous trend of the in-plane collective dynamics in crossing the main phase transition [24]. These anomalous vibrational dynamics were proven to be correlated with the formation of short-lived nanometer-scale lipid clusters that play a role in facilitating the passive molecular transport across the bilayer plane [24].

Consequently, studies addressing the polymorphic phase transitions of phospholipids, the dynamics and conformation of phospholipids in both gel and liquid-crystalline states, and the phase coexistence and phase interconversion in membrane model systems are of utmost importance to better understand the biological function of cell membranes.

In this regard, deeper knowledge of the single-particle dynamics of lipid molecules in bilayers above, at, and below the phase transition is highly desirable.

Recently, we have studied the liquid phase of hydrated DMPC and POPC bilayers [25] by means of quasi-elastic incoherent neutron scattering measurements and have identified three dominant dynamical processes on the ns-ps time scale, ascribed to three different motions of phospholipid molecules: (i) a fast uniaxial rotational diffusion, (ii) conformational, jump-like dynamics of the acyl chain along with threefold rotation of methyl groups, and (iii) overall diffusion dynamics confined within the first-neighbours cages. Here we present an extension of our previous investigation [25]. In this study, we examine the changes in the dynamics of hydrated POPC bilayers in the gel phase and near the gel-liquid phase transition temperature.

2. Experimental Details and Data Analysis

POPC is a synthetic monounsaturated phospholipid with a phosphocholine head group and one saturated 16:0 fatty acid chain and one unsaturated 18:1 fatty acid chain with two ester carbonyls (see Figure 1). Powder POPC was purchased from ROF Corporation.

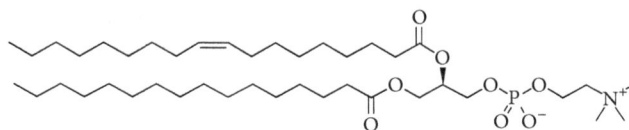

FIGURE 1: Chemical structure of POPC.

Mica V-1 grade Muscovite $KAl_2(Si_3Al)O_{10}(OH,F)_2$ sheets were used as substrate to support highly oriented and hydrated multilayer stacks.

The samples were hydrated with 28 D_2O molecules per lipid. Details of sample preparation are reported elsewhere [25].

QENS experiments were performed using the time-of-flight spectrometer IN5 at Institut Laue-Langevin, Grenoble, France. The incident wavelength was 8 Å and the **Q**-range covered 0.2–1.3 Å^{-1} with an energy resolution of 10 μeV. All spectra were recorded for the 135° orientation of planar membrane stacks relative to the incident beam at the temperatures of 248 K and 273 K, respectively, below and close to the gel-liquid phase transition temperature of POPC ($T_m = 271$ K).

The spectra were corrected for absorption and self-shielding and then converted to dynamic structure factor $S(\mathbf{Q}, \hbar\omega)$ and interpolated at constant **Q** values. Correction for multiple scattering was estimated to be negligible because the sample transmission was ≥90%.

Quasi-elastic neutron scattering can provide valuable information about microscopic dynamics of atoms or molecules on time scale ranging from picoseconds to nanoseconds [26]. It is widely employed in the investigation of biological systems, including phospholipids bilayers [11, 27–29]. When applied to organic or biological systems, which are hydrogen-rich, neutron scattering spectra are dominated by the high incoherent cross section of H and essentially provide information on single-particle dynamics. In the case of phospholipids, hydrogen atoms are evenly distributed over the molecule and they can act as a marker for average phospholipid diffusion and lipid chain dynamics.

The incoherent scattering function $S_{inc}(\mathbf{Q}, \omega)$ in the elastic region is usually separated into an elastic "delta" component, $A_0(\mathbf{Q})\delta(\omega)$, and a quasi-elastic component, $A_1(\mathbf{Q})L(\omega)$, where $L(\omega)$ is a Lorentzian function, centered on $\omega = 0$. The "delta" term and the Lorentzian term convey information on the geometry of the diffusion and on the time scale of diffusion, respectively. The fall in the elastic intensity as a function of **Q** is described by a form factor called the elastic incoherent structure factor (EISF), which is defined as

$$\text{EISF} = \frac{A_0\left(\vec{Q}\right)}{A_0\left(\vec{Q}\right) + A_1\left(\vec{Q}\right)} = \frac{\text{elastic intensity}}{\text{total intensity}}. \quad (1)$$

The exact shape of EISF is indicative of the geometry of the diffusional process and, by a careful line shape analysis of the spectra, it is possible to distinguish different dynamics that can be correlated to specific molecular groups in the system.

The bound cross sections for the whole POPC molecule for the component groups and for 28 D_2O molecules were

TABLE 1: Total cross section (barn) for a single phospholipid (POPC) with 28 hydration water molecules. Percentage values for constituent groups are given relative to the total cross section for a whole phospholipid.

		σ_{inc}	σ_{coh}	$\% \sigma$
Water	$28D_2O$	114.8	431.7	7.2
Phospholipid	POPC	6581.8	421.	92.8
Head methyls	$3CH_3$	722.3	32.5	10.8
Head linkers	$4CH_2 + CH + 2C$	722.3	54.7	11.1
Head phosphate	$N + 8O + P$	0.5	48.2	0.7
Tail acyl chains	$29CH_2$	4655.1	265.4	70.2
Tail methyls	$2CH_3$	481.6	21.6	7.2

calculated and listed in Table 1. To a good approximation, the D_2O-hydrated sample exhibits only scattering from the phospholipids. In fact, the incoherent scattering cross section for a POPC molecule is about 95% of the total cross section and is due to the predominant incoherent scattering of hydrogen in the biomolecules. Nevertheless, water contribution was explicitly taken into account as described in the supplementary material of our previous study [25].

3. Results and Discussion

Similar to what was observed for the liquid phase [25] and in general for all biological macromolecules, the QENS spectral line of the gel phase of POPC/D_2O system appears to be broadened over a very wide energy range with respect to the resolution function (see Figure 2). This feature is a signature of the existence of structural fluctuations on the picosecond time scale. The notable extension of the broadening implies that a single Lorentzian cannot accurately model the spectrum and suggests that more than one dynamic contributes to the quasi-elastic scattering. Thus we analyzed the neutron data following the same procedure and applying the same model used for the POPC liquid phase [25].

The incoherent contribution from the phospholipid dynamics, $S_{lip}(\mathbf{Q}, \hbar\omega)$, was calculated as the sum of a delta function ($\delta(\hbar\omega)$) and a set of Lorentzians [11, 30, 31]:

$$S_{lip}(\mathbf{Q}, \hbar\omega) = A_0 \delta(\hbar\omega) + \sum_{i=1}^{3} \frac{A_i}{\pi} \frac{\Gamma_i}{\Gamma_i^2 + \hbar\omega^2}. \quad (2)$$

The delta function represents the elastic response, while the Lorentzian contributions account for the relaxational motions that cause the quasi-elastic broadening of the elastic peak. The HWHMs (Half Widths at Half Maximums) of the Lorentzian contributions, Γ_i, are directly related to the characteristic times associated with the motions of the scattering nuclei involved: $\tau_i = \hbar/\Gamma_i$.

Notably, as for the liquid phase [25], three Lorentzians were required to adequately fit the gel phase of POPC bilayers. An example of the quality of the fit of the gel phase of POPC ($T = 248$ K) at $\mathbf{Q} = 0.7$ Å^{-1} is shown in Figure 2. The fitting procedure shows/suggests the presence of a very broad contribution of about 1800–2500 μeV, a second contribution of about 80–120 μeV, and a narrow contribution ranging from 5 to 15 μeV. As already observed in the liquid phase, the values

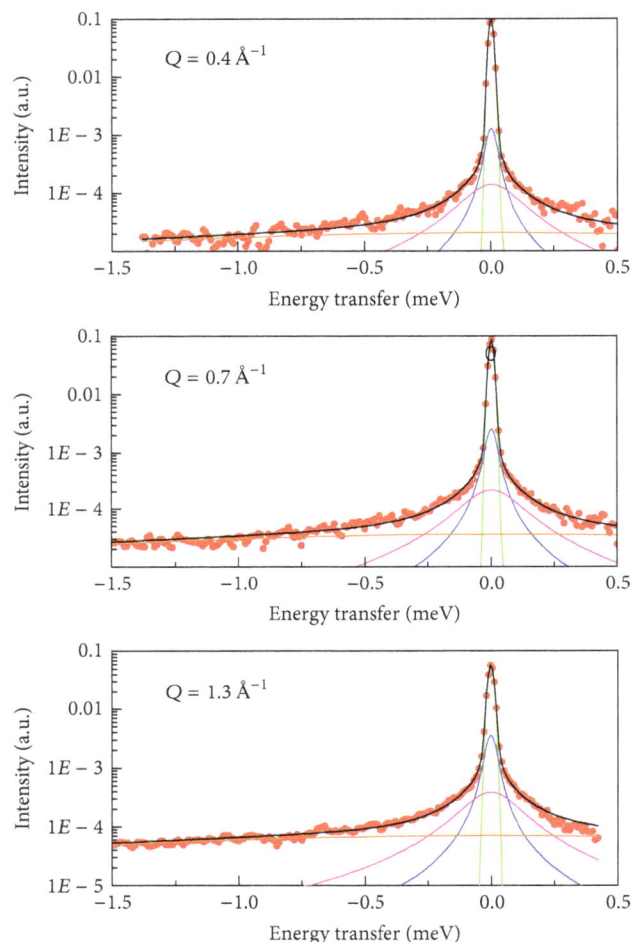

FIGURE 2: Semilog plot of the QENS intensity for POPC at 248 K at three different **Q** values. The best fit of equation (2) is displayed as a black continuous line.

of Γ_i differ nearly by an order of magnitude over the entire **Q**-range (data not shown). More precisely, these findings confirm the existence of three dominant dynamical processes taking place on different time scales in both structural phases.

To properly describe each of these dynamics and analyze the effect of temperature on the conformational motions of lipid bilayers, the same microscopic model proposed to

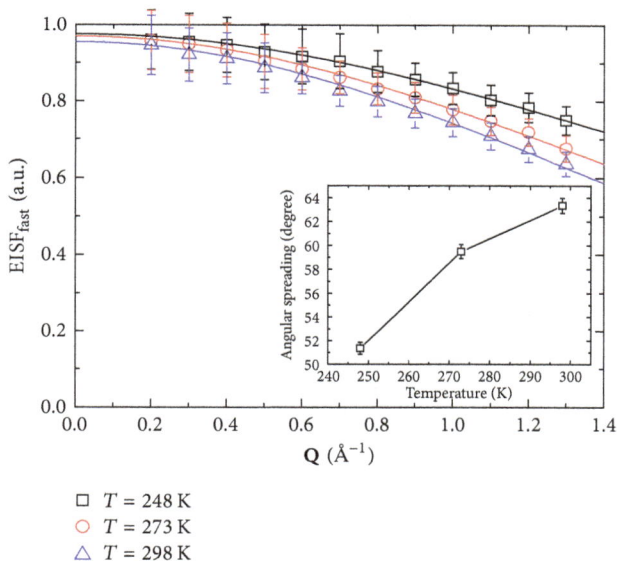

□ $T = 248$ K
○ $T = 273$ K
△ $T = 298$ K

FIGURE 3: **Q**-dependence of the experimental $EISF_{fast}$ at $T = 248$ K and 273 K compared with the data at $T = 298$ K taken from [25]. Solid lines are fits to the uniaxial rotation model. In the inset, the temperature dependence of the angular spreads is shown.

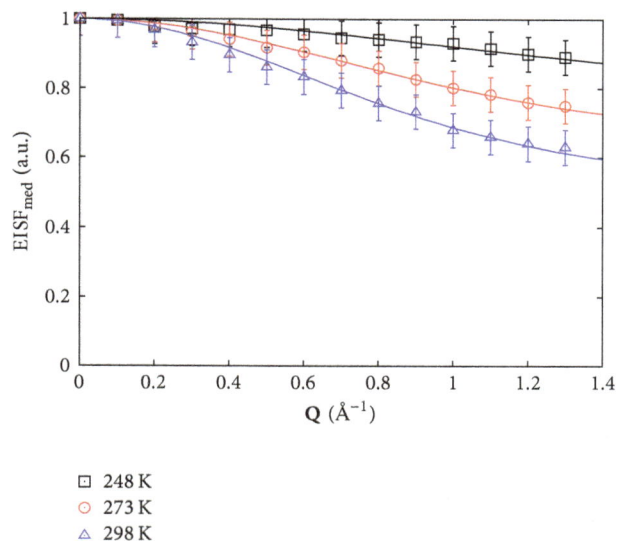

□ 248 K
○ 273 K
△ 298 K

FIGURE 4: **Q**-dependence of the experimental $EISF_{med}$ for POPC at three temperatures. The solid lines are fits obtained with the bead model, described in the text.

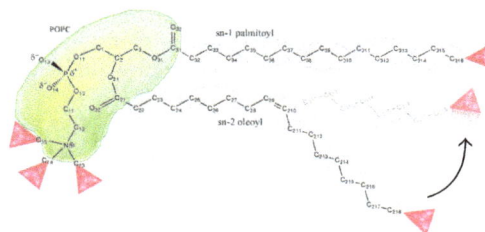

FIGURE 5: Sketch of the bead model for POPC. See text for details.

describe the hydrogen dynamics for the liquid phase of phospholipid bilayers was considered.

In this model, the dynamical structure factor $S(\mathbf{Q}, \hbar\omega)$ is described by a convolution of three types of motions (see [25] and supplementary materials therein for more details). Each contribution describes the dynamics for the protons in a specific confined geometry and is represented as the sum of an EISF and an elastic term. The three dynamics evolve on three different time scales: fast, intermediate, and slow.

The fast process occurs on the picoseconds time scale ($\hbar/\Gamma \approx 0.33$ ps) and is believed to describe the hydrogen motion with respect to the acyl carbon atoms that act like a center of mass for a restricted librational motion about the C–H bond direction [25]. Similar to what was observed for liquid crystals [32, 33], this motion can be described by the uniaxial rotational diffusion model [34]. The corresponding EISF, averaged over all the possible orientations of the bond direction with respect to **Q**, can be written as follows:

$$\text{EISF}_{fast}(\mathbf{Q}) = \sum_{l=0}^{\infty} (2l + 1)\, j_l^2(QR)\, S_l^2(\mu), \qquad (3)$$

where R is the C–H bond length and $S_l(\mu)$ is an orientational order parameter dependent on the width of the angular distribution, μ, which is the only fitting parameter [25].

The experimental EISF for this component is shown in Figure 3 for POPC at the two investigated temperatures, $T = 248$ and 273 K, and compared with the results obtained at $T = 298$ K for the liquid phase [25]. The best-fit curves are also shown. As can be inferred from the figure, the model satisfactorily reproduces the experimental data, suggesting that the fast, small amplitude fluctuation of the bond angle between H and C atoms in the CH_2 groups is responsible for these dynamics. Nevertheless, the possibility that a similar

hindered libration of the bond angle takes place also in the methyl groups cannot be a priori excluded.

The values of angular spreads, $\Delta\alpha = \cos^{-1}(S_l(\mu))$, obtained from the fitting procedure, are shown in the inset of Figure 3 as a function of temperature. The increase of $\Delta\alpha$ value with temperature describes wider spreading of the hydrogen position and can be explained due to the softening of the interchain interactions occurring during the passage from the gel to the liquid-crystalline phase.

The intermediate dynamics are characterized by a time ($\tau = \hbar/\Gamma$) of 6-7 ps, which is the typical time of both the conformational dynamics in the lipid chains and the rotational diffusion of methyl groups [11, 35, 36]. In Figure 4, the experimental $EISF_{med}$, derived by the line shape analysis, for the POPC at $T = 248$ and 273 K, are shown and compared with the results obtained at $T = 298$ K for the liquid phase. These data were fitted by means of a simple "bead model": the phospholipid molecule was built up with several linked beads representing head groups, chain segments, head methyls, and tail methyls (see Figure 5).

In this scheme, the intermediate time scale motion is described in terms of two processes: a two-site jump transition between trans and gauche conformation of carbon atoms in the CH_2 groups and a rotation of the CH_3 methyl groups

[25]. It is worth noting that in the present paper the "bead model" was improved by increasing the number of beads modelling the chains. Increasing the number of beads results in the increase of spatial resolution and in the possibility to consider in a more realistic way the chain dynamics of POPC, which shows a rigid cis double bond in the middle of one of the two hydrocarbon chains.

More specifically, the first bead, termed as 0, accounts for the CH_2 groups in the polar head. The other beads, six for the unsaturated and longer (oleoyl) chain and five for the saturated (palmitoyl) chain, account for the segmental dynamics of the chain. Each bead includes two single C–C bonds and can perform a two-site jump motion. Of course each bead displacement is convoluted with that of the preceding beads. The rotational motions of the head and terminal methyl groups, also modelled as small beads, are included in the model.

For the two-site jump dynamics, EISF is given as follows [37]: $EISF_i(Q) = 1 - 2P1P2(1 - j_0(Qd))$, where $P1$ and $P2$ are the probabilities of finding a bead in a trans or gauche conformation, respectively, with $(P1 + P2) = 1$. The value of the jump length was chosen to be equal to the distance between the carbon positions in the two conformations: $d = 2.49$ Å.

As far as the methyl group dynamics are concerned, they are described by a three-site jump rotational diffusion model, and the corresponding EISF is given by [26]

$$EISF_m = \frac{\left(1 + J_0\left(QR_m\sqrt{3}\right)\right)}{3}. \tag{4}$$

The contributions of the two dynamics were conveniently weighted taking into account the sample composition. Furthermore, since the motion of each segment has to be convoluted with the motion of the previous one, the dynamic structure factor (relative to a center of mass, corresponding to the head-tail contact) was obtained as the sum of products of the single contributions. Consequently, the resulting $EISF_{bead}$ for our model has been written as

$$EISF_{bead} = A\, Eisf_{meth} Eisf_0 + B\, Eisf_0 + \frac{C_u}{6} \sum_{i=1}^{6} \prod_{j=1}^{i} Eisf_i$$

$$+ D\, Eisf_6 Eisf_{meth} + \frac{C_s}{5} \sum_{i=1}^{5} \prod_{j=1}^{i} Eisf_i \tag{5}$$

$$+ D\, Eisf_5 Eisf_{meth},$$

where A, B, C, and D are the fractional cross sections for the CH_3 groups in the polar head, the CH_2 groups in the polar head, the CH_2 groups in the chains, and the terminal CH_3 groups, respectively. Note that C_u and C_s indicate the contribution for the unsaturated and saturated chains. The corresponding fitted curves are shown in Figure 4 and the best-fit parameters of the probability $P1_i$ for each bead are given in Table 2, where they are compared with the data obtained for the POPC liquid phase.

It is worth noting that the quality of the fit for the POPC liquid phase obtained by increasing the number of beads of

TABLE 2: Values for the probability of finding a bead in the lower energy site.

	248 K	273 K	298 K
$P1_0$	1.00	1.00	1.00
$P1_1$	1.00	1.00	1.00
$P1_2$	1.00	1.00	1.00
$P1_3$	1.00	1.00	0.93
$P1_4$	0.99	0.95	0.67
$P1_5$	0.98	0.77	0.49
$P1_6$	0.91	0.49	0.50

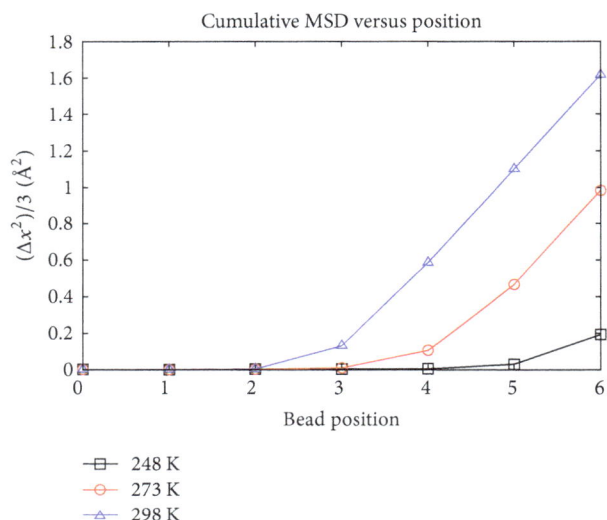

FIGURE 6: Mean square displacement as function of bead position for POPC at the investigated temperatures.

the model to seven was improved compared to the previous analysis [25]. Furthermore, we observe that the values of $P1$ decrease nonlinearly as the position of carbon atoms in alkyl chain increases, showing a substantial reduction starting from the fourth bead. This indicates higher motional freedom of the acyl chain in the region below the double-bond position of the unsaturated chain of POPC. This finding suggests that the conformational sampling of the molecules is correctly represented in the level of resolution of the seven-bead model.

The ability of this model to describe the experimental data of both liquid and gel phases of phospholipid bilayers proves that the transition between trans and gauche conformation is a relevant motion for the lipid chain dynamics.

Within the used model, it is possible to evaluate the mean square displacement (MSD) from the jump probabilities $P1$ and $P2$: $\langle \Delta x^2 \rangle = P1P2d^2/3$. In Figure 6, a plot of $\langle \Delta x^2 \rangle$ as a function of bead position is shown. We observed that in the gel phase a residual motion associated with anti-gauche transition is only observed in correspondence of the terminal part of the chain. Conversely, rising the temperature, these dynamics become even more relevant and involve about half of chain at T_m, engaging up to 2/3 of the chain when bilayers are in the liquid phase. The results for the jump length/rates

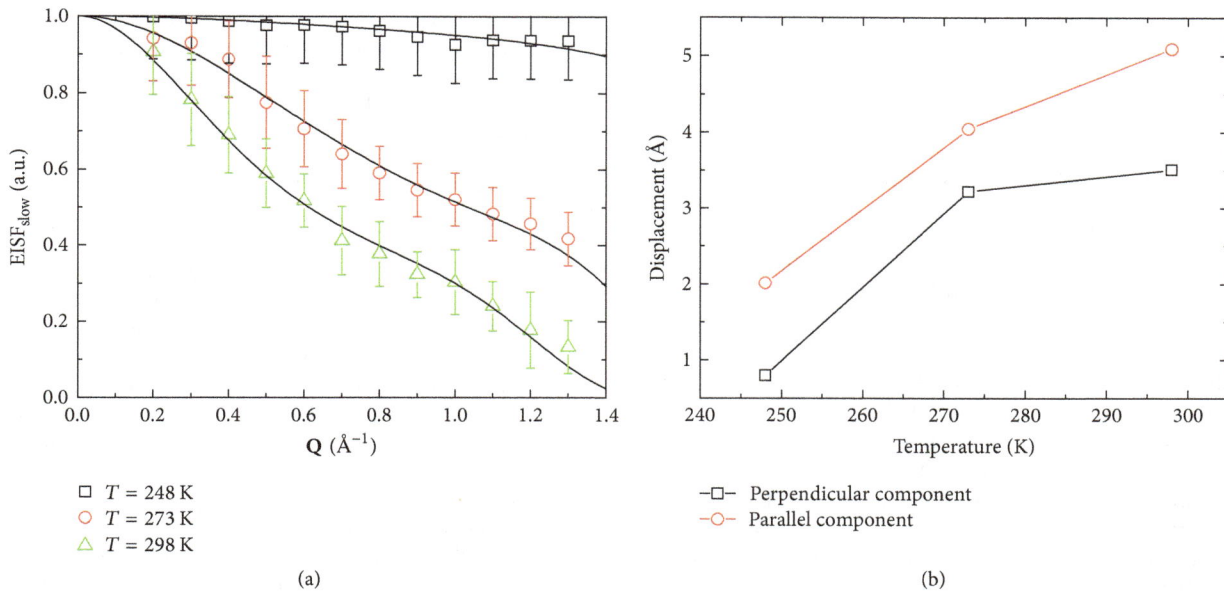

FIGURE 7: (a) **Q**-dependence of the experimental EISF$_{slow}$ for POPC at three temperatures. The solid lines represent the fitted model described in the text (see equation (6)). (b) Parallel (in-plane) displacement and perpendicular (off-plane) displacement versus temperature.

and the MSD are consistent with those previously obtained on similar systems by neutron scattering techniques [31, 38, 39], even if the details of the chain motion are modelled by different approaches. The nonlinear increase of the motion's amplitude when moving down the chain can be associated with the order parameter obtained by NMR measurements in similar systems [40, 41].

Finally, we have examined the EISFs relative to the slow dynamics of the POPC gel phase ($\hbar/\Gamma$ ranging from 40 to 350 ps). Analogously to what was reported for the liquid phase [25], we assume that the slow component is associated with the translational mobility of the whole phospholipid and we describe it as a center-of-mass motion.

As first step, we adopted the same model used for the liquid phase [25], that is, a continuum diffusion inside a cylindrical shape. Anyway, this model resulted to be unsuitable to describe the experimental EISF at and below the main transition temperature (data not shown).

Conversely, we obtained a fairly good description using a simpler model similar to that described by (4) and based on the convolution of two jump diffusions for the in-plane and off-plane displacements, respectively. Within this framework, assuming that, in long times, a two-state equilibrium with $P1$ equal to $P2$ is reached, EISF of the slow motion can be written as follows:

$$\text{EISF}_{\text{slow}}(\mathbf{Q})$$
$$= \frac{1}{4}\left(1 + j_0\left(\vec{\mathbf{Q}} \cdot \vec{d}_\parallel\right)\right)\left(1 + j_0\left(\vec{\mathbf{Q}} \cdot \vec{d}_\perp\right)\right). \tag{6}$$

Here the fitting parameters $d_\parallel$ and $d_\perp$ are the jump distances for the in-plane and off-plane displacements, whereas the angle between the exchanged wave-vector $\vec{\mathbf{Q}}$ and the

displacement $\vec{d}$ is defined by the detector positions and the sample orientation.

In Figure 7(a) the experimental EISF and the fitting curves for the phase at and below the main transition are reported and compared with the results for the liquid phase obtained by the continuous diffusion model inside a cylinder [25]. In Figure 7(b) the best-fit values for $d_\parallel$ and $d_\perp$ as a function of temperature are shown. Note that the displacements for the liquid phase reported here are the diameter and the height of the confining cylindrical region.

It is worth observing that the structure factor increases as the temperature is decreased, indicating that by going from the liquid to the gel phase the motion becomes more and more restricted. Furthermore, while the in-plane displacement exhibits a nearly linear increase with temperature, the perpendicular component shows saturation. This is consistent with the fact that the transition from the liquid into the gel phase involves a drastic decrease in volume and area of the lipid molecules, which results in significant strengthening of the van der Waals attractive forces in the lipid bilayer. The increase in the packing density of acyl chains imposes geometrical constraints that restrict the local diffusion of lipid molecules, affecting mainly the out-of-plane direction. It is also reasonable to consider the observed saturation as a consequence of the limited protrusion allowed for phospholipids packed in a bilayer. We infer that the inadequacy of the model of diffusion into a cylinder in describing the slow dynamics at and below the main transition should indicate that the phospholipid diffusion in the gel phase is rattling-like with short jumps around a quasi-equilibrium position [42], while in the liquid phase the dynamics, although rather confined, evolve into a continuous diffusion. However, because of the limited instrumental resolution, we are not able to identify with certainty a diffusive motion outside the confinement region for both liquid and gel phases.

4. Conclusion

The microscopic dynamics of lipid multibilayer POPC system in the gel phase and at the gel-liquid phase transition were investigated by means of incoherent quasi-elastic neutron scattering and compared with the dynamics previously observed in the liquid phase. In both phases, the line shape analysis of the QENS spectra revealed the presence of three contributions to the global dynamics of phospholipid. Such contributions take place on well-separated time scales and were ascribed to different parts of the phospholipids, that is, local hydrogen libration, segmental acyl tails motion, and overall diffusion. The present analysis is based on the refinement of a model, recently developed to describe the ps time scales dynamics for the liquid phase of bilayers. This model was revealed to be more accurate and suitable for describing the dynamics of more complex phospholipids characterized by unsaturated and asymmetric chains.

Through the analysis of the temperature dependence of EISF, it was possible to infer that the gel to liquid phase transition entails (i) softening of the interchains interaction, (ii) an increase of the segmental motion occurring in the final part of the acyl chain, and (iii) a change of the diffusional dynamics from a very localized rattling-like motion to a confined diffusion.

This study provides valuable data for characterizing the dynamical behaviour of lipid bilayers. The simplicity of the model allows for its application in studying various subjects such as lipid raft and more complicated cases where functional components are included in the membrane bilayer.

Conflicts of Interest

The authors declare that there are no conflicts of interest regarding the publication of this paper.

References

[1] T. Yang, O. K. Baryshnikova, H. Mao, M. A. Holden, and P. S. Cremer, "Investigations of bivalent antibody binding on fluid-supported phospholipid membranes: The effect of hapten density," *Journal of the American Chemical Society*, vol. 125, no. 16, pp. 4779–4784, 2003.

[2] S. Paula, A. G. Volkov, A. N. Van Hoek, T. H. Haines, and D. W. Deamer, "Permeation of protons, potassium ions, and small polar molecules through phospholipid bilayers as a function of membrane thickness," *Biophysical Journal*, vol. 70, no. 1, pp. 339–348, 1996.

[3] J. L. Arias, "Liposomes in drug delivery: a patent review (2007 present)," *Expert Opinion on Therapeutic Patents*, vol. 23, pp. 1399–1414, 2013, Arias, J. L. *Expert Opin. Ther. Pat.* **2013,** *23,* 1399-1414.

[4] M. A. Kiselev and D. Lombardo, "Structural characterization in mixed lipid membrane systems by neutron and X-ray scattering," *Biochimica et Biophysica Acta - General Subjects*, vol. 1861, no. 1, pp. 3700–3717, 2017.

[5] U. Wanderlingh, G. D'Angelo, V. Conti Nibali et al., "Interaction of alcohol with phospholipid membrane: NMR and XRD investigations on DPPC-hexanol system," *Spectroscopy*, vol. 24, no. 3-4, pp. 375–380, 2010.

[6] S. Rifici, C. Corsaro, C. Crupi et al., "Lipid diffusion in alcoholic environment," *Journal of Physical Chemistry B*, vol. 118, no. 31, pp. 9349–9355, 2014.

[7] S. Rifici, G. D'Angelo, C. Crupi et al., "Influence of alcohols on the lateral diffusion in phospholipid membranes," *Journal of Physical Chemistry B*, vol. 120, no. 7, pp. 1285–1290, 2016.

[8] H. A. Pillman and G. J. Blanchard, "Effects of ethanol on the organization of phosphocholine lipid bilayers," *Journal of Physical Chemistry B*, vol. 114, no. 11, pp. 3840–3846, 2010.

[9] G. D'angelo, U. Wanderlingh, V. C. Nibali, C. Crupi, C. Corsaro, and G. Di Marco, "Physical study of dynamics in fully hydrated phospholipid bilayers," *Philosophical Magazine*, vol. 88, no. 33-35, pp. 4033–4046, 2008.

[10] H. K. Kimelberg, "Alterations in phospholipid-dependent (Na++K+)-ATPase activity due to lipid fluidity. Effects of cholesterol and Mg2+," *BBA - Biomembranes*, vol. 413, no. 1, pp. 143–156, 1975.

[11] J. Fitter, R. Lechner, G. Buldt, and N. Dencher, "Internal molecular motions of bacteriorhodopsin: hydration-induced flexibility studied by quasielastic incoherent neutron scattering using oriented purple membranes," *National Academy of Sciences of the United States of America*, vol. 93, pp. 7600–7605, 1996.

[12] J. Fitter, R. Lechner, and N. Dencher, "Interactions of hydration water and biological membranes studied by neutron scattering," *The Journal of Physical Chemistry B*, vol. 103, pp. 8036–8050, 1999.

[13] S. Rifici, C. Crupi, G. D'Angelo et al., "Effects of a short length alcohol on the dimyristoylphosphatidylcholine system," *Philosophical Magazine*, vol. 91, no. 13-15, pp. 2014–2020, 2011.

[14] R. McElhaney, *Biomembranes*, vol. 12, Plenum Press, New York, NY, USA, 1984.

[15] R. McElhaney, *Current Topics in Membranes and Transport*, vol. 17, Academic Press, New York, NY, USA, 1982.

[16] J. F. Nagle and S. Tristram-Nagle, "Structure of lipid bilayers," *Biochimica et Biophysica Acta—Reviews on Biomembranes*, vol. 1469, no. 3, pp. 159–195, 2000.

[17] R. N. A. H. Lewis and R. N. McElhaney, *The Structure of Biological Membranes*, vol. 2, CRC press, 3rd edition, 2005.

[18] G. Pabst, H. Amenitsch, D. P. Kharakoz, P. Laggner, and M. Rappolt, "Structure and fluctuations of phosphatidylcholines in the vicinity of the main phase transition," *Physical Review E - Statistical, Nonlinear, and Soft Matter Physics*, vol. 70, no. 2, Article ID 021908, pp. 1–21908, 2004.

[19] G. van Meer, D. R. Voelker, and G. W. Feigenson, "Membrane lipids: where they are and how they behave," *Nature Reviews Molecular Cell Biology*, vol. 9, no. 2, pp. 112–124, 2008.

[20] A. F. Xie, R. Yamada, A. A. Gewirth, and S. Granick, "Materials Science of the Gel to Fluid Phase Transition in a Supported Phospholipid Bilayer," *Physical Review Letters*, vol. 89, no. 24, Article ID 246103, 2002.

[21] M. C. Rheinstädter, C. Ollinger, G. Fragneto, F. Demmel, and T. Salditt, "Collective dynamics of lipid membranes studied by inelastic neutron scattering," *Physical Review Letters*, vol. 93, no. 10, pp. 1–108107, 2004.

[22] V. C. Nibali, G. D'Angelo, and M. Tarek, "Molecular dynamics simulation of short-wavelength collective dynamics of phospholipid membranes," *Physical Review E*, vol. 89, no. 5, Article ID 050301, 2014.

[23] D. Angelo, G. C. Nibali, V. Crupi et al., "Probing intermolecular interactions in phospholipid bilayers by far-infrared spectroscopy," *The Journal of Physical Chemistry B*, vol. 121, no. 6, pp. 1204–1210, 2017.

[24] M. Zhernenkov, D. Bolmatov, D. Soloviov et al., "Revealing the mechanism of passive transport in lipid bilayers via phonon-mediated nanometre-scale density fluctuations," *Nature Communications*, vol. 7, Article ID 11575, 2016.

[25] U. Wanderlingh, G. D'Angelo, C. Branca et al., "Multi-component modeling of quasielastic neutron scattering from phospholipid membranes," *Journal of Chemical Physics*, vol. 140, no. 17, Article ID 174901, 2014.

[26] M. Bee, *Quasielastic Neutron Scattering*, Adam Hilger, Bristol, 1988.

[27] W. Pfeiffer, T. H. Henkel, E. Sackmann, W. Knoll, and W. Knoll, "Local dynamics of lipid bilayers studied by incoherent quasi-elastic neutron scattering," *EPL*, vol. 8, no. 2, pp. 201–206, 1989.

[28] E. Endress, H. Heller, H. Casalta, M. F. Brown, and T. M. Bayerl, "Anisotropic motion and molecular dynamics of cholesterol, lanosterol, and ergosterol in lecithin bilayers studied by quasi-elastic neutron scattering," *Biochemistry*, vol. 41, no. 43, pp. 13078–13086, 2002.

[29] G. Schiro, M. Sclafani, C. Caronna, F. Natali, M. Plazanet, and A. Cupane, "Dynamics of myoglobin in confinement: an elastic and quasi-elastic neutron scattering study," *Chemical Physics*, vol. 345, no. 2-3, pp. 259–266, 2008.

[30] S. König, W. Pfeiffer, T. Bayerl, D. Richter, and E. J. Sackmann, *Journal de Physique II*, vol. 2, pp. 1589–1615, 1992.

[31] Y. Gerelli, V. G. Sakai, J. Ollivier, and A. Deriu, "Conformational and segmental dynamics in lipid-based vesicles," *Soft Matter*, vol. 7, no. 8, pp. 3929–3935, 2011.

[32] F. Volino, A. J. Dianoux, and H. Hervet, "Neutron quasi-elastic scattering study of rotational motions in the smectic C, H and VI phases of terephtal-bis-butyl-aniline (TBBA)," *Le Journal de Physique Colloques*, vol. 37, no. C3, pp. C3-55–C3-64, 1976.

[33] A. J. Dianoux, F. Volino, and H. Hervet, "Incoherent scattering law for neutron quasi-elastic scattering in liquid crystals," *Molecular Physics*, vol. 30, no. 4, pp. 1181–1194, 1975.

[34] M. Bée, "Localized and long-range diffusion in condensed matter: State of the art of QENS studies and future prospects," *Chemical Physics*, vol. 292, no. 2-3, pp. 121–141, 2003.

[35] T. Pradeep, S. Mitra, A. Sreekumaran Nair, and R. Mukhopadhyay, "Dynamics of alkyl chains in monolayer-protected au and ag clusters and silver thiolates: A comprehensive quasielastic neutron scattering investigation," *Journal of Physical Chemistry B*, vol. 108, no. 22, pp. 7012–7020, 2004.

[36] S. König and E. Sackmann, "Molecular and collective dynamics of lipid bilayers," *Current Opinion in Colloid and Interface Science*, vol. 1, no. 1, pp. 78–82, 1996.

[37] W. Doster, S. Cusack, and W. Petry, "Dynamical transition of myoglobin revealed by inelastic neutron scattering," *Nature*, vol. 337, no. 6209, pp. 754–756, 1989.

[38] V. K. Sharma, E. Mamontov, D. B. Anunciado, H. Oneill, and V. Urban, "Nanoscopic dynamics of phospholipid in unilamellar vesicles: Effect of gel to fluid phase transition," *Journal of Physical Chemistry B*, vol. 119, no. 12, pp. 4460–4470, 2015.

[39] C. Castellano, F. Natali, D. Pozzi, G. Caracciolo, and A. Congiu, "Dynamical properties of oriented lipid membranes studied by elastic incoherent neutron scattering," *Physica B: Condensed Matter*, vol. 350, no. 1-3, pp. e955–e958, 2004.

[40] M. Lafleur, B. Fine, E. Sternin, P. Cullis, and M. Bloom, "Smoothed orientational order profile of lipid bilayers by 2H-nuclear magnetic resonance," *Biophysical Journal*, vol. 56, pp. 1037–1041, 1989.

[41] L. S. Vermeer, B. L. De Groot, V. Réat, A. Milon, and J. Czaplicki, "Acyl chain order parameter profiles in phospholipid bilayers: computation from molecular dynamics simulations and comparison with 2H NMR experiments," *European Biophysics Journal*, vol. 36, no. 8, pp. 919–931, 2007.

[42] C. L. Armstrong, M. Trapp, J. Peters, T. Seydel, and M. C. Rheinstädter, "Short range ballistic motion in fluid lipid bilayers studied by quasi-elastic neutron scattering," *Soft Matter*, vol. 7, no. 18, pp. 8358–8362, 2011.

16

Correlation between Formic Acid Oxidation and Oxide Species on Pt(Bi)/GC and Pt/GC Electrode through the Effect of Forward Potential Scan Limit

Jelena D. Lović

ICTM, Institute of Electrochemistry, University of Belgrade, Njegoševa 12, Belgrade, Serbia

Correspondence should be addressed to Jelena D. Lović; jlovic@tmf.bg.ac.rs

Academic Editor: Gonzalo Garcia

Following earlier works from our laboratory, further experiments on electrochemical behavior in formic acid oxidation at electrodeposited Pt(Bi)/GC and Pt/GC electrode were performed in order to examine the effect of successive increase of the forward potential scan limit. Correlation between formic acid oxidation and oxide species on Pt(Bi)/GC electrode with increases of forward potential scan limit is based on the dependency of the backward peak potential from backward peak current. The obtained dependency reveals Bi influence for the scan limits up to 0.8 V. Since the Pt(Bi)/GC electrode is composed of Bi core occluded by Pt and Bi-oxide surface layer, the observed behavior is explained through the influence of surface metal oxide on easier formation of OH_{ad} species. Nevertheless, the influence of electronic modification of Pt surface atoms by underlying Bi is present and leads to the stronger adsorption of OH on Pt. At higher forward potential scan limits (from 0.8 V), Pt has a dominant role in HCOOH oxidation.

1. Introduction

Formic acid oxidation (FAO) reaction represents an important reaction in electrocatalysis, since it can be used in fundamental studies and, moreover, formic acid has been proposed as a fuel for direct liquid fuel cells (DLFCs), which can be used as small power supply in electronic devices [1, 2].

FAO on Pt electrodes has been widely studied over the last decades due to the high activity of this metal for the oxidation of different small organic molecules (SOMs) [1, 3, 4]. Since FAO probably has the simplest oxidation mechanism among all different SOMs, understanding of the FAO mechanism on Pt should be very useful for other electrocatalytic oxidation reactions. It is well accepted that FAO on Pt electrodes follows two different reaction pathways [5–7]. The presence of the parallel paths in formic acid oxidation has been confirmed by DEMS, ATR-FTIR, and SEIRAS measurements [8–10]. One of them is known as the direct pathway and implies the formation of an active intermediate, which is immediately oxidized into CO_2. The other so-called indirect pathway involves the formation of CO on the electrode surface, which acts as a poison intermediate. Spectroscopic analysis clearly demonstrated that the poisoning species in the indirect path is adsorbed CO, while the reactive intermediate is a formate species [8–10].

Further improving of Pt electrochemical activity for FAO with addition of metals, such as Pb, Cu, Au, Ni, Ag, Sn, and Co, has been performed [11–15], and, among bimetallic catalysts, Pt-Bi is the most extensively studied. Different types of Pt-Bi catalysts obtained by various methods have been reported, such as the synthesis method capable of producing intermetallic Pt-Bi nanoparticles 1–3 nm in diameter [16] or Pt-Bi intermetallics with a particle size of 20–100 nm with certain preferential surface structures [17], Pt-Bi alloy nanoparticles following the cathodic corrosion method [18], electrochemically codeposited Pt-Bi [19], or Pt modified by Bi either by UPD or by irreversible adsorption [20, 21]. Enhanced activity of Pt in combination with Bi for FAO reaction was referred to as an electronic effect where the modification of the Pt electronic structure due to the presence of Bi enhances the activity of the surface [22], improving tolerance to CO poisoning by steric interference of Bi [23] or to ensemble effect [24, 25]. It is generally considered that the presence of a second metal has an impact on catalyst activity and stability but can also greatly reduce Pt loading so as to lower the usage of precious metals.

For direct formic acid fuel cell, dehydrogenation is the desired reaction pathway to enhance overall cell efficiency [2]. Bimetallic Pt-Bi catalysts selectively enhance the dehydrogenation reaction rate, thus offering the possibility of overcoming the problem of poisoning the electrode, such as Pt with the intermediate CO [1, 2]. Also, onset potential of FAO is shifted toward less positive potentials on bimetallic Pt-Bi catalysts compared to Pt so that, by reducing the overpotential, a better performance of direct formic acid fuel cell could be fulfilled [2].

Our studies of Pt_2Bi electrode and Pt-Bi clusters showed high activity and stability for FAO [24, 26]. The main reason for their high stability is the inhibition of dehydration path in the reaction, as well as suppression of Bi leaching in the presence of formic acid. Comparing the results obtained for two types of Pt-Bi catalysts, polycrystalline Pt modified by irreversible adsorbed Bi, and Pt_2Bi catalyst, the role of the ensemble effect and electronic effect in the oxidation of formic acid was distinguished [27]. The electronic effect contributes to lower onset potential of the reaction, while the high current comes from the ensemble effect.

The aim of the work is to investigate the effect of successive increase of forward potential scan limit on electrochemical behavior of electrodeposited Pt(Bi)/GC and Pt/GC electrodes in formic acid oxidation. In order to reveal the correlation between FAO and oxide species, the dependency of the backward peak potential from backward peak current is performed. The intention is to get some additional insight in examination of the role of Bi on formic acid oxidation.

2. Experimental

Platinum-bismuth deposits on glassy carbon (GC) substrate were prepared by a two-step process as described previously [28]. To recapitulate, Bi was deposited onto the GC from 2 mM Bi perchlorate in 0.1 M H_2SO_4 using chronocoulometry at −0.1 V versus SCE. After the Bi deposition, the electrode was rinsed and transferred to the electrochemical cell containing 1 mM H_2PtCl_6 solution in 0.1 M H_2SO_4. Pt was deposited under the same deposition condition as Bi and the amount of both metals was the same. The bimetallic electrode was denoted as Pt(Bi)/GC. For the sake of comparison, Pt/GC electrode was prepared using the same electrochemical procedure and quantity corresponding to one for bimetallic electrode. Pt(Bi)/GC electrodes were activated by cycling potential with scan rate of 50 mV s^{-1} between hydrogen and oxygen evolution in 0.1 M H_2SO_4 solution 50 times leading to quasi-steady-state cyclic voltammogram (CV). Formic acid oxidation (0.125 M) was investigated in 0.1 M H_2SO_4 solution after holding the potential for 2 min at −0.2 V by potentiodynamic (scan rate 50 mV s^{-1}) measurements. The rotating speed of the disc electrode (ω) was 1500 rpm in all CV experiments. The results are compared to Pt/GC treated in the same manner. Real surface area of the investigated electrodes was calculated from the charge of CO stripping, assuming 420 μCcm^{-2} for the CO monolayer [28].

All the experiments were performed in standard three-compartment electrochemical glass cells with Pt wire as the counterelectrode and saturated calomel electrode as the reference electrode at room temperature. The potentials reported in the paper are expressed on the scale of the saturated calomel electrode (SCE). The electrolytes were prepared with high purity water (Millipore, 18 MΩ cm resistivity) and the p.a. chemicals provided by Merck. The experiments were conducted at 295 ± 0.5 K. A VoltaLab PGZ 402 (Radiometer Analytical, Lyon, France) was employed.

3. Results and Discussion

3.1. Effect of Successive Increase of the Forward Potential Scan Limit in Base Voltammograms. The bimetallic catalyst was prepared by a two-step process using chronocoulometry in a way that controlled the amount of Bi that was electrodeposited onto glassy carbon followed by electrodeposition of Pt layer [28]. Cyclic voltammetry revealed Bi leaching from the electrode surface indicating that Bi was not completely occluded by Pt. Continuous cycling after 50 cycles up to 1.2 V versus SCE in supporting electrolyte assures quasi-steady-state voltammogram. On this way the bimetallic electrode composed of Bi core occluded by Pt and Bi-oxide surface layer was obtained and denoted as Pt(Bi)/GC. For the sake of comparison, Pt/GC electrode was prepared using the same quantity corresponding to one for bimetallic electrode. The base voltammetric profiles of the Pt(Bi)/GC and Pt/GC electrodes in 0.1 M H_2SO_4 solution are shown is Figure 1. As one can see, the region between −0.25 and 0.2 V versus SCE is associated with the hydrogen adsorption/desorption process [26]. This region is followed by the region of oxide formation/reduction. In the case of the bimetallic catalyst (Figure 1(b)), the base voltammetric profile shows a current contribution in the region of oxide formation/reduction, associated with the redox behavior of Bi at the surface which is superimposed to Pt oxide formation/reduction [18].

In order to investigate the origin of the oxide formation/reduction in the potential region 0.4–1.2 V on Pt(Bi)/GC electrodes, a series of cyclic voltammograms for successive increases of the forward potential scan limit was recorded on this electrodeposited bimetallic electrode and compared with voltammograms obtained on Pt/GC electrode as it was shown in Figure 1(a). Previous analysis concerning the reversibility of various stages of surface oxidation on polycrystalline Pt reveals the presence of irreversible oxygen species below the monolayer of electrosorbed OH [29]. Concerning Pt/GC, the increase of the forward potential scan limit causes the enhancement of charge for the oxide reduction, indicating that a more extended or thicker oxide film is formed. This finding was proven by nanogravimetry measurements [30] and also by using time-resolved energy dispersive X-ray absorption spectroscopy and time-resolved X-ray diffraction [31]. Besides, it was shown that Pt oxide formation/reduction depends on particle size and film roughness, electrolyte composition, as well as electrode scan rate [32].

Further analysis of base voltammograms obtained by a continuous increase of the forward potential scan limit expressed through the dependency of the charge associated

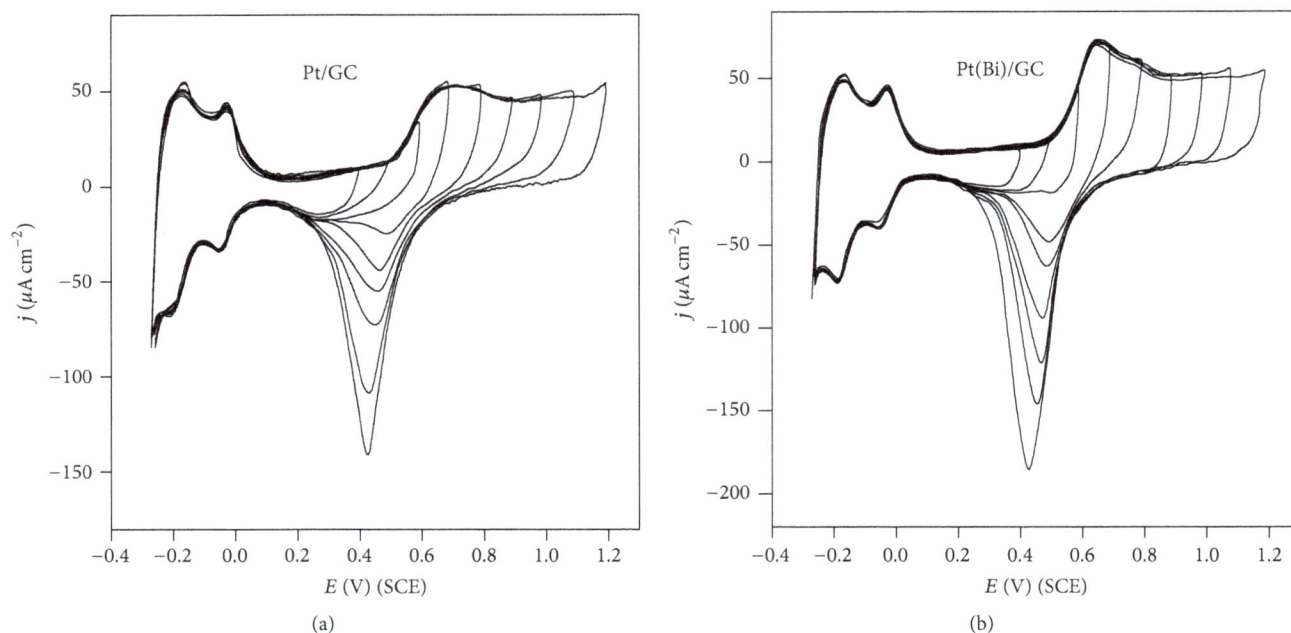

FIGURE 1: CVs on Pt/C (a) and Pt(Bi)/GC (b) electrodes in 0.1 M H$_2$SO$_4$ at 50 mV s^{-1} and ϖ = 1500 rpm for successive increases of the forward potential scan limit.

with anodic film formation from the electrode potential is presented in Figure 2. Existences of two slopes on both investigated electrodes signify the presence of different forms of surface oxides. Transition in the region with higher slope (II), that is, the region of irreversibly adsorbed oxygenated species, occurred ~100 mV earlier on Pt(Bi)/GC electrode indicating a faster transformation of oxygenated species from reversible to irreversible state. Similar slope in the first region (I) is expected since there is an overlap in CVs up to 0.6 V [28]. For higher potentials, larger content of charge on Pt(Bi)/GC in relation to Pt/GC electrode brings about a steeper slope.

The observed behavior could be explained through the electronic model based on the modification of Pt electronic structure in the presence of some other metal [33]. According to the theoretical DFT calculations, Pt overlayer on Bi is under a tensile strain since Bi has a larger lattice constant than Pt. This results in an increase in d-band center of Pt atoms in the out-layer, leading to the stronger adsorption of OH on Pt [33, 34]. In addition, the influence of surface metal oxide is possible since the Pt(Bi)/GC electrode is composed of Bi core occluded by Pt and Bi-oxide surface layer. It is well documented in literature that metal oxide causes an increase in adsorption ability of the hydroxyl ion onto catalyst surface [35]. Therefore, in the case of electrodeposited bimetallic electrode, Bi influences two modes of action contributing to a stronger and enhanced adsorption of OH species.

3.2. Formic Acid Oxidation with the Increase of the Forward Potential Scan Limit. Before analyzing the results obtained with the increase of the forward potential scan limit during FOA, it is important to recall some characteristics of Pt electrode in formic acid electrooxidation. In the forward scan direction, taking the Pt/GC electrode into

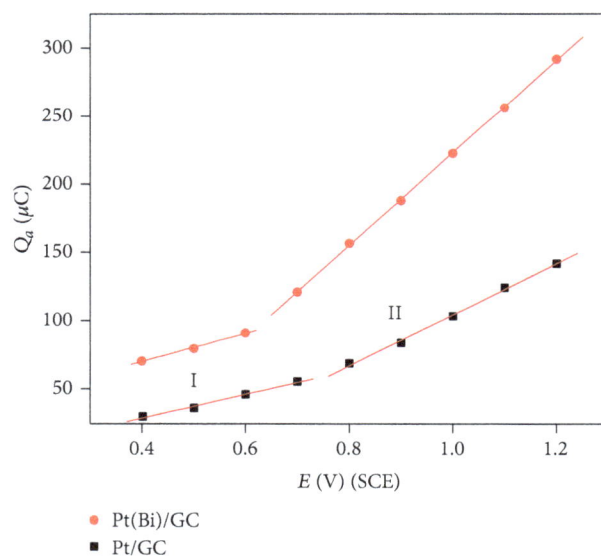

FIGURE 2: Oxide formation charge as a function of the forward potential scan limit (data obtained from Figure 1).

consideration (Figure 3), reaction current slowly increases reaching a plateau at ~0.25 V followed by an ascending current starting at 0.5 V, attaining a maximum at ~0.6 V. Bearing in mind the dual path mechanism, at low potentials, HCOOH oxidizes through the direct path with the simultaneous formation of CO$_{ad}$ through the indirect path. Increased coverage by CO$_{ad}$ reduces the Pt sites available for the direct path and, because of that, current slowly increases reaching a plateau. Subsequent formation of oxygenated species on Pt

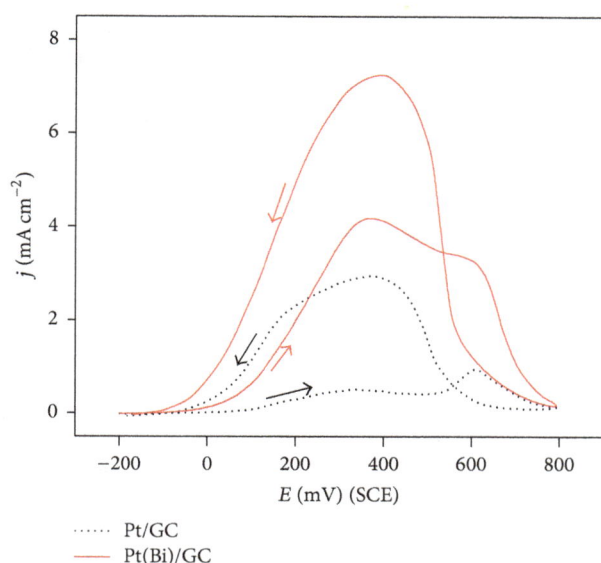

FIGURE 3: CVs for oxidation of 0.125 M HCOOH in 0.1 M H_2SO_4 on Pt/GC and Pt(Bi)/GC electrode at 50 mV s^{-1} and ω = 1500 rpm.

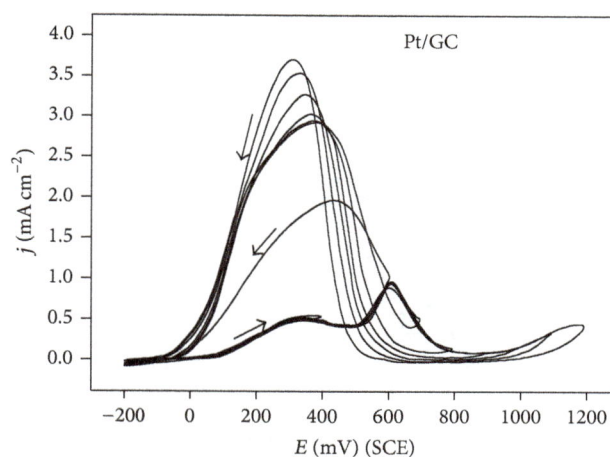

FIGURE 4: CVs for oxidation of 0.125 M HCOOH in 0.1 M H_2SO_4 on Pt/GC electrode at 50 mV s^{-1} and ω = 1500 rpm for the successive increases of the forward potential scan limit.

enables the oxidative removal of CO_{ad} and as a result more Pt sites are disposed of for FAO. A reaction current increase until Pt–oxide, inactive for HCOOH oxidation, is formed inducing a peak. In the backward scan after the surface oxide reduction, more Pt sites are released for FAO so the reaction current increases reaching a peak. At more negative potentials, current decreases due to a lack of oxygenated species necessary for CO_{ad} oxidation. The currents are much higher than, in the forward scan, because Pt surface is freed of CO_{ad}. The polarization curve for electrodeposited bimetallic Pt(Bi)/GC surface indicates different behavior in comparison to Pt/GC. Figure 3 shows that the onset potential for the reaction on Pt(Bi)/GC electrode is about 100 mV less positive in comparison with Pt/GC. In the case of Pt(Bi)/GC electrode, the current rises up to 0.3 V and reaches a peak about 8 times higher than the plateau on Pt/GC. This peak indicates predominant direct oxidation path of HCOOH to CO_2, occurring on Pt sites that are not blocked by the poisoning CO_{ad} species, while the appearance of a poorly defined shoulder on the descending part of the curve signifies some participation of indirect path in the reaction as well. The currents recorded in the backward direction are higher, but the difference between forward and backward scan is not as large as on Pt/GC electrode indicating lower surface poisoning of bimetallic electrodes [36]. It has been shown that a number of neighboring Pt sites on the electrode surface determine the reaction path in the sense that the dehydration path needs an ensemble of at least three adjacent Pt atoms, while the dehydrogenation path is possible on a smaller atomic ensemble [37]. The obtained behavior is explained primarily by the ensemble effect induced by surface Bi oxides interrupting Pt domains but to some extent could also be attributed to the influence of the underlying Bi onto the Pt surface layer, affecting the extent of poison adsorption on the Pt [28].

The results of the potentiodynamic measurements for HCOOH oxidation support the finding of the quasi-steady-state measurements [28]. Namely, the Tafel slope on Pt(Bi)/GC electrodes is about 120 mV dec^{-1}, indicating that the HCOOH oxidation takes place on a surface almost free of CO_{ad} through dehydrogenation path. Therefore, the first electron transfer is the rate-determining step, and the slow step that determines the rate of formic acid oxidation on Pt(Bi)/GC electrodes is the C–H bond cleavage and formation of $COOH_{ad}$. The Tafel slope of about 150 mV dec^{-1} obtained during formic acid oxidation on Pt/GC electrode indicates that CO was formed and adsorbed on the surface slowing down the reaction rate.

CVs of the FAO on Pt/GC electrode recorded for the successive increases of the forward potential scan limit are shown in Figure 4.

Extension of the forward potential scan limit slightly influences the reaction currents in the forward scan, while the currents recorded in the backward direction become higher and the peak is shifted to a more negative potential. For higher forward potential scan limits, greater amount of surface oxide is formed [30, 31]. Consequently in the backward direction more surface oxide is reduced, releasing more Pt sites for HCOOH oxidation, thus increasing reaction currents.

The polarization curves for FAO obtained for successive increases of the forward potential scan limit from 0.4 V up to 1.2 V on electrodeposited bimetallic Pt(Bi)/GC surface (Figure 5) indicates quite different behavior in comparison to Pt/GC. For the forward potential scan limit 0.4 V there is almost an overlap of reaction currents in forward and backward scan direction suggesting an absence of surface poisoning. It can be noticed that forward reaction currents decrease almost two times at peak potential with further extension of forward potential scan limits. Other contribution of forward potential scan limit extension is obtained in the backward scan direction and it is presented as a dependency of the backward peak potential from backward

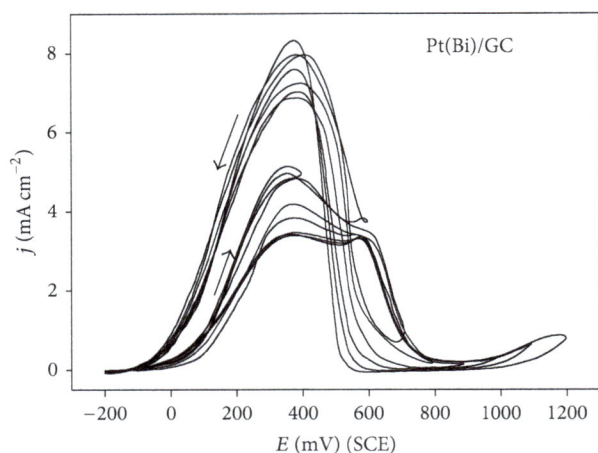

FIGURE 5: CVs for oxidation of 0.125 M HCOOH in 0.1 M H_2SO_4 on Pt(Bi)/GC electrode at 50 mV s^{-1} and $\omega = 1500$ rpm for successive increases of the forward potential scan limit.

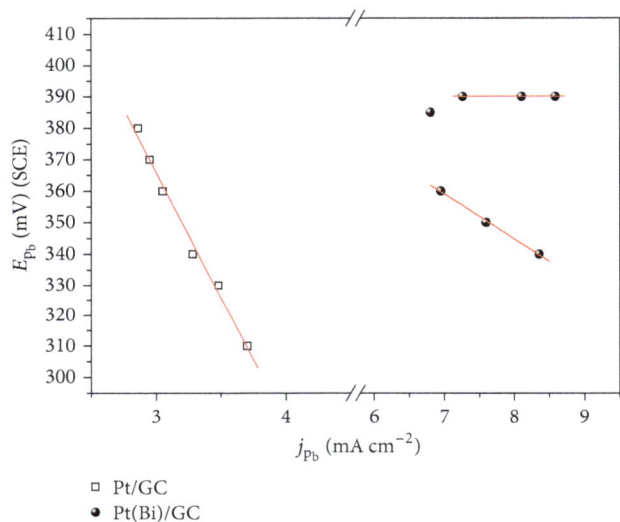

□ Pt/GC
● Pt(Bi)/GC

FIGURE 6: Backward peak potential versus backward peak current (data obtained from Figures 4 and 5).

peak current (Figure 6). This dependency on Pt/GC gave one negative slope [30, 38] which means that the increase of the forward potential scan limit enables current enhancement in the backward scan since the reduction of more Pt oxide releases more sites for FAO. However on electrodeposited bimetallic Pt(Bi)/GC surface the dependency of the backward peak potential from backward peak current depicts zero slope for forward potential scan limit 0.6–0.8 V and a small negative slope in potential scan limit 1.0–1.2 V. Zero slope obtained for specified scan limits means that peak potential remained the same while reaction currents decrease indicating the surface blockade. Nevertheless, further increases of forward potential scan limit (1.0–1.2 V) enable some current enhancement since after the reduction of surface oxides, the oxidation of formic acid occurs.

This finding is in accordance with the results presented in Figure 2. So, for the forward potential scan limit 0.4 V,

enhanced adsorption of OH species participates in the oxidation of CO_{ad} species, suggesting an absence of surface poisoning. Stronger adsorption of OH species for forward potential scan limit 0.6–0.8 V causes surface blockade. At higher forward potential scan limits (from 0.8 V) Pt has a dominant role in HCOOH oxidation. Besides, the overlap of base voltammograms confirms that observation [28]. To summarize, in the case of electrodeposited bimetallic electrode, it was shown that Bi influences two modes of action affecting the oxidation of surface poisoning (forward potential scan limit 0.4 V) and surface blockade (forward potential scan limit 0.6–0.8 V), whereby potential region of Bi influence in FAO is determined.

It has been reported in literature that the beneficial effect of Bi on Pt on bimetallic Pt-Bi electrodes for HCOOH oxidation reaction followed several different mechanisms as is stated in the introduction part [30–33]. Depending on the fact that the preparation method of the catalysts and their resulting surface composition the contribution of the effects may vary, our previously published results on bimetallic Pt-Bi electrodes confirm the ensemble effect as dominant in HCOOH oxidation [26–28]. Based on the presented electrochemical results, it was possible to offer same additional insight into examination of the role of Bi in formic acid oxidation.

4. Conclusion

Successive increase of the forward potential scan limit in base voltammograms of Pt(Bi)/GC and Pt/GC electrodes is expressed through the dependency of the charge associated with anodic film formation from the electrode potential. This dependency shows two slopes on both investigated electrodes due to the presence of different forms of surface oxides. Transition in the region with higher slope occurred ~100 mV earlier on Pt(Bi)/GC electrode indicating the faster transformation of oxygenated species from reversible to irreversible state. The observed behavior is explained through the electronic modification of Pt surface atoms by Bi leading to the stronger adsorption of OH on Pt. Also, the influence of surface metal oxide on easier formation of OH_{ad} species is possible. With the increase of the forward potential scan limit, correlation between formic acid oxidation and oxide species on Pt(Bi)/GC electrode reveals Bi influence for scan limits up to 0.8 V. At higher forward potential scan limits Pt has a dominant role in HCOOH oxidation. In this way, potential region of Bi influence in FAO at electrodeposited bimetallic Pt(Bi)/GC electrode is determined.

Conflicts of Interest

The author declares that there are no conflicts of interest regarding the publication of this paper.

Acknowledgments

This work was financially supported by the Ministry of Education, Science and Technological Development, Republic of Serbia, Contract no. H-172060.

References

[1] J. M. Feliu, E. Herrero W, H. A. Gasteiger, and A. Lamm, Eds., *Handbook of fuel cells – fundamentals, technology and applications, Electrocatalysis*, vol. 2, John Wiley Sons, New York, NY, USA, 2003.

[2] C. Rice, S. Ha, R. Masel, P. Waszczuk, A. Wieckowski, and T. Barnard, "Direct formic acid fuel cells," *Journal of Power Sources*, vol. 111, no. 1, pp. 83–89, 2002.

[3] N. M. Marković and P. N. Ross Jr., "Surface science studies of model fuel cell electrocatalysts," *Surface Science Reports*, vol. 45, no. 4–6, pp. 117–229, 2002.

[4] M. T. M. Koper, S. C. S. Lai, and E. Herrero, "Fuel Cell Catalysis: A Surface Science Approach," in *Fuel Cell Catalysis: A Surface Science Approach*, M. T. M. Koper, Ed., p. 183, John Wiley Sons Hoboken, New Jersey, NJ, USA, 2009.

[5] A. Capon and R. Parsons, "The oxidation of formic acid at noble metal electrodes Part III. Intermediates and mechanism on platinum electrodes," *Journal of Electroanalytical Chemistry*, vol. 45, no. 2, pp. 205–231, 1973.

[6] Y.-X. Chen, M. Heinen, Z. Jusys, and R. J. Behm, "Bridge-bonded formate: Active intermediate or spectator species in formic acid oxidation on a Pt film electrode?" *Langmuir*, vol. 22, no. 25, pp. 10399–10408, 2006.

[7] S. G. Sun, J. Clavilier, and A. Bewick, "The mechanism of electrocatalytic oxidation of formic acid on Pt (100) and Pt (111) in sulphuric acid solution: an emirs study," *Journal of Electroanalytical Chemistry*, vol. 240, no. 1-2, pp. 147–159, 1988.

[8] G. Samjeské and M. Osawa, "Current oscillations during formic acid oxidation on a Pt electrode: Insight into the mechanism by time-resolved IR spectroscopy," *Angewandte Chemie - International Edition*, vol. 44, no. 35, pp. 5694–5698, 2005.

[9] Y. X. Chen, S. Ye, M. Heinen, Z. Jusys, M. Osawa, and R. J. Behm, "Application of in-situ attenuated total reflection-Fourier transform infrared spectroscopy for the understanding of complex reaction mechanism and kinetics: Formic acid oxidation on a pt film electrode at elevated temperatures," *Journal of Physical Chemistry B*, vol. 110, no. 19, pp. 9534–9544, 2006.

[10] R. Gómez, J. M. Orts, J. M. Feliu, J. Clavilier, and L. H. Klein, "The role of surface crystalline heterogeneities in the electrooxidation of carbon monoxide adsorbed on Rh(111) electrodes in sulphuric acid solutions," *Journal of Electroanalytical Chemistry*, vol. 432, no. 1-2, pp. 1–5, 1997.

[11] V. Perales-Rondóna, J. Solla-Gullóna, E. Herreroa, and C. M. Sánchez-Sánchez, "Enhanced catalytic activity and stability for the electrooxidation of formic acid on lead modified shape controlled platinum nanoparticles," *Applied Catalysis B: Environmental*, vol. 201, pp. 48–57, 2017.

[12] L. J. Zhang, Z. Y. Wang, and D. G. Xia, "Bimetallic PtPb for formic acid electro-oxidation," *Journal of Alloys and Compounds*, vol. 426, no. 1-2, pp. 268–271, 2006.

[13] W. Liu and J. Huang, "Electro-oxidation of formic acid on carbon supported Pt-Os catalyst," *Journal of Power Sources*, vol. 189, no. 2, pp. 1012–1015, 2009.

[14] Z. Awaludin, T. Okajima, and T. Ohsaka, "Formation of Pt-Li alloy and its activity towards formic acid oxidation," *Electrochemistry Communications*, vol. 31, pp. 100–103, 2013.

[15] O. Winjobi, Z. Zhang, C. Liang, and W. Li, "Carbon nanotube supported platinum-palladium nanoparticles for formic acid oxidation," *Electrochimica Acta*, vol. 55, no. 13, pp. 4217–4221, 2010.

[16] X. Ji, K. T. Lee, R. Holden et al., "Nanocrystalline intermetallics on mesoporous carbon for direct formic acid fuel cell anodes," *Nature Chemistry*, vol. 2, no. 4, pp. 286–293, 2010.

[17] H. Liao, J. Zhu, and Y. Hou, "Synthesis and electrocatalytic properties of PtBi nanoplatelets and PdBi nanowires," *Nanoscale*, vol. 6, no. 2, pp. 1049–1055, 2014.

[18] E. Bennett, J. Monzó, J. Humphrey et al., "A Synthetic Route for the Effective Preparation of Metal Alloy Nanoparticles and Their Use as Active Electrocatalysts," *ACS Catalysis*, vol. 6, no. 3, pp. 1533–1539, 2016.

[19] X. Yu and P. G. Pickup, "Carbon supported PtBi catalysts for direct formic acid fuel cells," *Electrochimica Acta*, vol. 56, no. 11, pp. 4037–4043, 2011.

[20] S. Daniele and S. Bergamin, "Preparation and voltammetric characterisation of bismuth-modified mesoporous platinum microelectrodes. Application to the electrooxidation of formic acid," *Electrochemistry Communications*, vol. 9, no. 6, pp. 1388–1393, 2007.

[21] B.-J. Kim, K. Kwon, C. K. Rhee, J. Han, and T.-H. Lim, "Modification of Pt nanoelectrodes dispersed on carbon support using irreversible adsorption of Bi to enhance formic acid oxidation," *Electrochimica Acta*, vol. 53, no. 26, pp. 7744–7750, 2008.

[22] N. de-los-Santos-Álvarez, L. R. Alden, E. Rus, H. Wang, F. J. DiSalvo, and H. D. Abruña, "CO tolerance of ordered intermetallic phases," *Journal of Electroanalytical Chemistry*, vol. 626, no. 1-2, pp. 14–22, 2009.

[23] E. Casado-Rivera, Z. Gál, A. C. D. Angelo, C. Lind, F. J. DiSalvo, and H. D. Abruña, "Electrocatalytic oxidation of formic acid at an ordered intermetallic PtBi surface," *ChemPhysChem*, vol. 4, no. 2, pp. 193–199, 2003.

[24] J. D. Lović, M. D. Obradović, D. V. Tripković et al., "High Activity and Stability of Pt$_2$Bi Catalyst in Formic Acid Oxidation," *Electrocatalysis*, vol. 3, no. 3-4, pp. 346–352, 2012.

[25] A. López-Cudero, F. J. Vidal-Iglesias, J. Solla-Gullón, E. Herrero, A. Aldaz, and J. M. Feliu, "Formic acid electrooxidation on Bi-modified polyoriented and preferential (111) Pt nanoparticles," *Physical Chemistry Chemical Physics*, vol. 11, no. 2, pp. 416–424, 2009.

[26] J. D. Lović, S. I. Stevanović, D. V. Tripković et al., "Formic acid oxidation at platinum-bismuth clusters," *J Electrochem Soc*, vol. 16, no. 8, pp. 2223–2233, 2014.

[27] J. Lovic, D. Tripkovic, K. Popovic, V. Jovanovic, and A. Tripkovic, "Electrocatalytic properties of Pt–Bi electrodes towards the electro-oxidation of formic acid," *Journal of the Serbian Chemical Society*, vol. 78, no. 8, pp. 1189–1202, 2013.

[28] J. Lović, S. Stevanović, D. Tripković et al., "Catalytic activities of Pt thin films electrodeposited onto Bi coated glassy carbon substrate toward formic acid electrooxidation," *Journal of Electroanalytical Chemistry*, vol. 735, pp. 1–9, 2014.

[29] H. Angerstein-Kozlowska, B. E. Conway, and W. B. A. Sharp, "The real condition of electrochemically oxidized platinum surfaces. Part I. Resolution of component processes," *Journal of Electroanalytical Chemistry*, vol. 43, no. 1, pp. 9–36, 1973.

[30] M. Tian and B. E. Conway, "Electrocatalysis in oscillatory kinetics of anodic oxidation of formic acid: At Pt; Nanogravimetry and voltammetry studies on the role of reactive surface oxide," *Journal of Electroanalytical Chemistry*, vol. 616, no. 1-2, pp. 45–56, 2008.

[31] H. Imai, K. Izumi, M. Matsumoto, Y. Kubo, K. Kato, and Y. Imai, "In Situ and Real-Time Monitoring of Oxide Growth in a Few Monolayers at Surfaces of Platinum Nanoparticles in Aqueous

Media," *Journal of the American Chemical Society*, vol. 131, no. 17, pp. 6293–6300, 2009.

[32] D. Chen, Q. Tao, L. W. Liao, S. X. Liu, Y. X. Chen, and S. Ye, "Determining the Active Surface Area for Various Platinum Electrodes," *Electrocatalysis*, vol. 2, no. 3, pp. 207–219, 2011.

[33] J. R. Kitchin, J. K. Nørskov, M. A. Barteau, and J. G. Chen, "Modification of the surface electronic and chemical properties of Pt(111) by subsurface 3d transition metals," *Journal of Chemical Physics*, vol. 120, no. 21, pp. 10240–10246, 2004.

[34] M. T. Paffett, C. T. Campbell, R. G. Windham, and B. E. Koel, "A multitechnique surface analysis study of the adsorption of H2, CO and O2 on Bi Pt(111) surfaces," *Surface Science*, vol. 207, no. 2-3, pp. 274–296, 1989.

[35] C. Xu, P. k. Shen, and Y. Liu, "Ethanol electrooxidation on Pt/C and Pd/C catalysts promoted with oxide," *Journal of Power Sources*, vol. 164, no. 2, pp. 527–531, 2007.

[36] A. López-Cudero, F. J. Vidal-Iglesias, J. Solla-Gullón, E. Herrero, A. Aldaz, and J. M. Feliu, "Formic acid electrooxidation on Bi-modified Pt(1 1 0) single crystal electrodes," *Journal of Electroanalytical Chemistry*, vol. 637, no. 1-2, pp. 63–71, 2009.

[37] A. Cuesta, M. Escudero, B. Lanova, and H. Baltruschat, "Cyclic voltammetry, FTIRS, and DEMS study of the electrooxidation of carbon monoxide, formic acid, and methanol on cyanide-modified Pt(111) electrodes," *Langmuir*, vol. 25, no. 11, pp. 6500–6507, 2009.

[38] Z. Liu and L. Hong, "Electrochemical characterization of the electrooxidation of methanol, ethanol and formic acid on Pt/C and PtRu/C electrodes," *Journal of Applied Electrochemistry*, vol. 37, no. 4, pp. 505–510, 2007.

17

Research on Crude Oil Demulsification Using the Combined Method of Ultrasound and Chemical Demulsifier

Mingxu Yi, Jun Huang, and Lifeng Wang

School of Aeronautic Science and Engineering, Beihang University, Beijing 100191, China

Correspondence should be addressed to Lifeng Wang; wanglf1972@163.com

Academic Editor: Davide Vione

In this paper, experiments of crude oil demulsification using ultrasound, chemical demulsifier, and the combined method of ultrasound and chemical demulsifier, respectively, at different temperatures (40°C, 60°C, and 70°C) are carried out. The photos of water droplet distribution in crude oil, taken with microscopic imaging system, before and after demulsification using the above methods at 70°C are given. Research results show that the combined method of ultrasound and chemical demulsifier has the best demulsification effect, followed by chemical demulsifier. Ultrasound without using chemical demulsifier has the least demulsification effect. Furthermore, the impact of ultrasonic power, treatment time, and temperature on crude oil demulsification using the combined method of ultrasound and chemical demulsifier is studied. Results indicate that the final dehydration rate increases with the increase of temperatures and ultrasonic power and almost does not change with the increase of ultrasonic treatment time. These important conclusions will provide the foundation for an extensive application of the combined method of ultrasound and chemical demulsifier.

1. Introduction

The formation of water-in-oil emulsions during crude oil production is a problem for crude oil industry. Until now, the most common way of separating water from crude oil is by using chemical demulsifiers [1]. Although chemical demulsifiers are widely used in crude oil industry, they are not very efficient [2].

Until recently, ultrasonic irradiation was considered as an efficient method to desalt heavy crude oil [3]. As an easier, simpler, and more efficient method, ultrasonic technology has recently been applied to water-oil emulsion separation on the laboratory scale in many literature papers [3–11]. Experimental results in these papers suggest that the separation is enhanced under ultrasonic irradiation. And then, many kinds of ultrasonic dehydration apparatus [12–15] on the large scale have been designed successfully, which enabled the scale-up of ultrasound-assisted dehydration. This method was first used in refinery oil dehydration by Yu et al. [16] to remove water in crude oil emulsion, which is mostly made by deliberate injection of water into the crude oil in order to dissolve soluble salts especially NaCl [3]. Their experimental

study on the influence of sound field parameters on the water in crude oil emulsion behavior indicated that the response efficiency of the ultrasonically irradiated dispersed water phase to drive, coalesce, and segregate was similar to the behavior of suspended particles [17], droplets [18], and bubbles [19, 20] in sound field.

Ultrasonic demulsification technique mainly uses the agglomeration effect and the viscosity reducing effect of ultrasonic wave on crude oil to make water droplets in the crude oil gather and accelerate oil and water two-phase separation [21, 22]. References [2, 23] prove that lower ultrasonic frequency is good for ultrasonic demulsification when power of transducer remained unchanged.

Higher efficiency of demulsification could be achieved by combining demulsifier and electrostatic treatment [24]. However, in order to achieve suitable demulsification efficiency, desalted water (in case of desalting water) and more demulsifiers must be added. Other methods, such as the combination of demulsifier and hydrocyclones [25] and the combination of demulsifier and ultrasonic radiation [4, 10, 23, 26], have been evaluated, but these techniques are still not used in industrial crude oil demulsification processes

[2]. In a word, the influence of parameters of transducer (frequency, power, and treatment time) on the effect of crude oil demulsification is investigated in these references, and the effect of the primary factors on crude oil demulsification using chemical demulsifier is also studied. Results show that the rate of droplet coalescence at higher ultrasonic frequency is significantly lower than that at lower ultrasonic frequency at earlier time when the droplets are small; bigger values of ultrasonic power are not always better, as they should be controlled in an optimal range; bigger values of ultrasonic treatment time are not always better.

However, the systematic comparison of crude oil demulsification using ultrasonic treatment, chemical agent alone, and the combination of ultrasonic and chemical demulsification methods, respectively, was not fully investigated.

In this paper, experiments of crude oil demulsification using ultrasound, natural sedimentation, chemical demulsifier, and the combined method of ultrasound and chemical demulsifier, respectively, at different temperatures are investigated. It is shown here that higher efficiency of crude oil demulsification can be achieved by the combined method of ultrasound and chemical demulsifier. The purpose of this paper is to provide the foundation for an extensive application of the combined method of ultrasound and chemical demulsifier.

2. Experimental Apparatus, Materials, and Procedures

Experiments of demulsification of produced liquid containing heavy oil by sonochemical treatment were carried out by Zhang in 2008 in China University of Petroleum (East China) [27]. On the basis of these experiments, the comparison of crude oil demulsification using ultrasonic wave, chemical demulsifier, and the combined method of ultrasound and chemical demulsifier is further investigated in this section.

2.1. Experimental Apparatus and Materials. The main experimental apparatuses are explosion-proof pneumatic mixer, stopwatch, HH-8 digital resistance heating furnace, densitometer, constant temperature waters, and WL-8XZ temperature controller. Explosion-proof pneumatic mixer, HH-8 digital resistance heating furnace, and WL-8XZ temperature controller are shown in Figures 1, 2, and 3, respectively. Three NC-M type transducers, produced in Wuxinni Ultrasonic Equipment Co., Ltd., are used to study crude oil demulsification using the combined method of ultrasound and chemical demulsifier. Transducers and their parameters are shown in Figure 4 and Table 1, respectively.

Explosion-proof pneumatic mixer is produced by DE-BAO pneumatic company in Zhejiang Province in China. The advantages of this mixer are stepless speed regulation, elevated torque, explosion-proof, operating intelligence, automation, and suitability for factory operations. HH-8 digital resistance heating furnace is produced by Zheng Kai Precision Instrument Co., Ltd., in the city of Dongguan in China. Its power is 1000 W, the temperature control range is room temperature ~100°C, and advantages are accurate

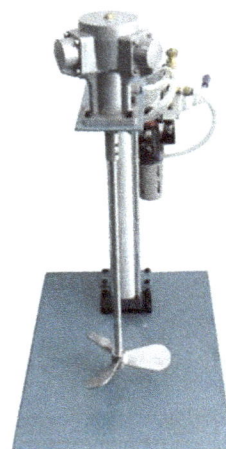

FIGURE 1: Explosion-proof pneumatic mixer.

FIGURE 2: HH-8 digital resistance heating furnace.

temperature control, digital display, and automatic temperature control. WL-8XZ temperature controller is produced by Tai Xi Electronics, Ltd., in the city of Guangzhou in China. Setting temperature is −30°C~120°C. Digital temperature showing in real time can be achieved using this temperature controller.

In order to make the experimental results more representative of reality, watery crude oils from Daqing Oilfield (the freezing point is 67.2°C) were selected as the samples to study the effects of crude oil demulsification using ultrasound, chemical demulsifier, and the combined method of ultrasound and chemical demulsifier, respectively.

The initial water content, density, viscosity, and salt content of crude oil are 25.23%, 0.9247 g·cm^{-3}, 145000 mP·s, and 95367 mg/L, respectively. The concentration of Type SP demulsifier is 250 mg/L. The container of crude oil was a mixing cylinder with stopper, with the following features: total volume 100 ml, outside diameter 30 mm, and height 260 mm. The amount of crude oil sample was 70 ml. After stirring crude oil well using pneumatic mixer, the experiment of ultrasonic demulsification was carried out immediately.

FIGURE 3: WL-8XZ temperature controller.

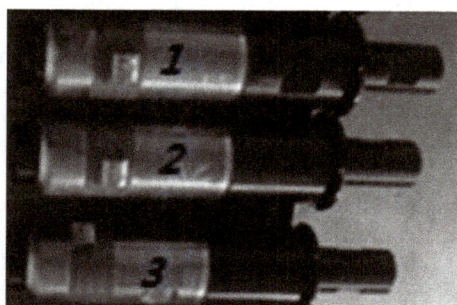

FIGURE 4: NC-M ultrasound transducer.

TABLE 1: Parameters of ultrasonic transducers.

Transducer number	Transducer frequency (kHz)	Transducer rated power (W)
1	20	150
2	30	100
3	40	50

2.2. Experimental Procedures

(1) We heated the constant temperature water to the required temperature.

(2) We added the SP demulsifier to the prepared watery crude oil and stirred the mixture of crude oil and demulsifier using high speed mixer.

(3) We processed the mixed sample using ultrasonic transducer.

(4) When the dewatering amount did not appear to add up, we recorded the total volumes of the dehydrated water and the whole mixed sample.

3. Experimental Results and Discussion

3.1. The Comparison of Crude Oil Demulsification Using Ultrasound, Natural Sedimentation, Chemical Demulsifier, and the Combined Method of Ultrasound and Chemical Demulsifier. Under the same ultrasonic power (150 W) and ultrasonic frequency (20 kHz), experiments of crude oil demulsification using ultrasound, natural sedimentation, chemical demulsifier, and the combined method of ultrasound and chemical

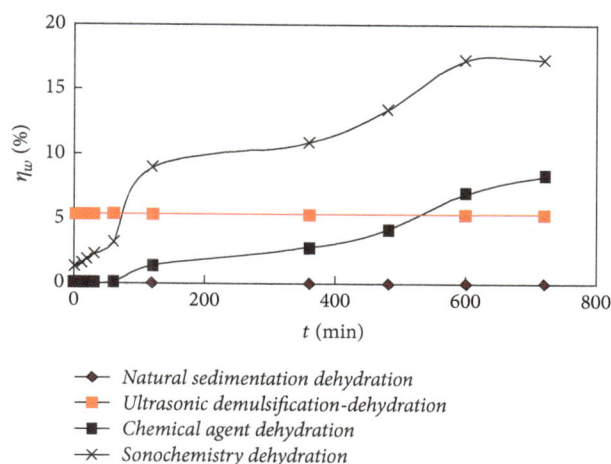

— ◆ — Natural sedimentation dehydration
— ■ — Ultrasonic demulsification-dehydration
— ■ — Chemical agent dehydration
— × — Sonochemistry dehydration

FIGURE 5: The comparison of demulsifying and dehydrating crude oil using ultrasonic excitation, natural sedimentation, chemical demulsifier, and sonochemistry method, respectively, at 40°C.

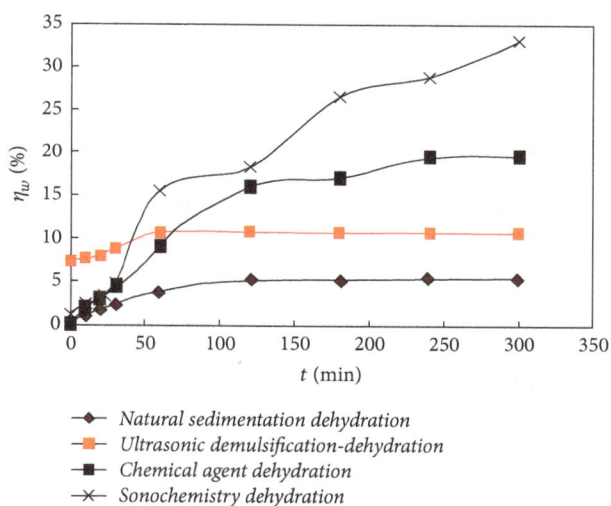

— ◆ — Natural sedimentation dehydration
— ■ — Ultrasonic demulsification-dehydration
— ■ — Chemical agent dehydration
— × — Sonochemistry dehydration

FIGURE 6: The comparison of demulsifying and dehydrating crude oil using ultrasonic excitation, natural sedimentation, chemical demulsifier, and sonochemistry method, respectively, at 60°C.

demulsifier, respectively, at different temperatures (40°C, 60°C, and 70°C) were carried out. Results are shown in Figures 5, 6, and 7 respectively.

As can be seen in Figure 5, ultrasound can dehydrate crude oil instantaneously, but the dehydration rate is nearly invariable and natural sedimentation has the lowest final dehydration rate; as time goes on, the combined method of ultrasound and chemical demulsifier has the best demulsification effect compared to that of using ultrasound and chemical demulsifier separately at 40°C.

As can be seen in Figure 6, the combined method of ultrasonic irradiation and chemical demulsifier has the best demulsification effect, followed by chemical demulsifier, and ultrasound without using chemical demulsifiers has the least demulsification effect at 60°C.

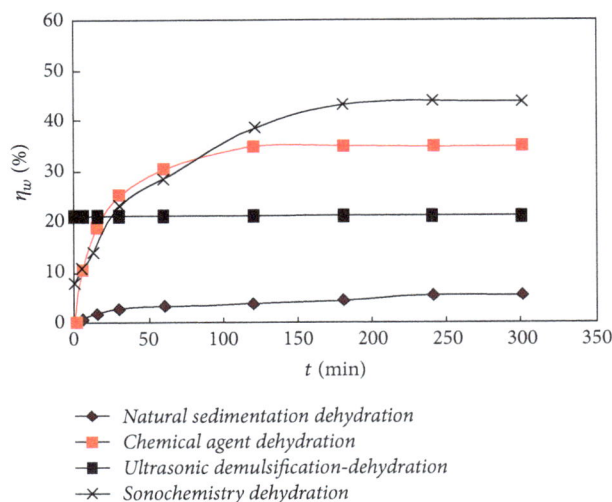

FIGURE 7: The comparison of demulsifying and dehydrating crude oil using ultrasonic excitation, natural sedimentation, demulsifier chemical, and sonochemistry method, respectively, at 70°C.

It can be seen from Figure 7 that the dehydration rate using the combined method of ultrasound and chemical demulsifier is double than that using ultrasonic treatment, and it is nearly higher compared to the use of chemical demulsifier at 70°C over time. The above three figures allow for a similar conclusion: the combined method of ultrasound and chemical demulsifier has the best demulsification effect, followed by chemical demulsifier and ultrasound without using chemical demulsifiers. Furthermore, they also indicate that the final dehydration rates when using the above four demulsification methods all increase as the temperatures rise, which indicates that temperature is an important factor that affects the demulsification effect.

The photos of water droplet distribution in crude oil using the above demulsification methods are taken with microscopic imaging system at 70°C. Figure 8(a) is the water droplet distribution in crude oil before demulsification, Figure 8(b) is the water droplet distribution in crude oil using chemical demulsifier, Figure 8(c) is the water droplet distribution in crude oil using ultrasonic excitation without using chemical demulsifiers, and Figure 8(d) is the water droplet distribution in crude oil using the combined method of ultrasound and chemical demulsifier.

It can be seen from Figure 8(a) that there are many water droplets with different diameters in crude oil before demulsification. From Figure 8(b), the number and diameters of water droplets in crude oil have all been reduced with the chemical demulsifier. From Figure 8(c), the number of water droplets and their diameters have been somewhat reduced after ultrasonic excitation. However, the diameters of water droplets are bigger than that in Figure 8(b). Moreover, the number of water droplets with big diameter is still higher in Figure 8(c) than in Figure 8(b), which indicates that the demulsification effect using chemical demulsifier is better than that using ultrasound. It can be seen from Figure 8(d) that the number of water droplets has been greatly reduced

and the water droplets with big diameter has become fewer compared with Figures 8(b) and 8(c) after demulsification using the combined method of ultrasound and chemical demulsifier. It can be seen in Figure 8(a) that there are many big and smaller water drops before sonochemistry treatment. The water content is 48.5%. It can be seen in Figure 8(d) that the number of big water droplets in crude oil has all been greatly reduced after sonochemistry treatment. The water content was 18.5% after sonochemistry treatment. Briefly, by comparing Figures 8(a), 8(b), 8(c), and 8(d), the combined method of ultrasound and chemical demulsifier has the best demulsification effect, followed by chemical demulsifier and ultrasonic irradiation without using chemical demulsifiers; besides, it is proved once again from the above four figures that the smaller the drops, the more difficult the demulsification.

Natural sedimentation dehydration is a time-consuming process and has no good dehydration effect. That is why the other three dehydrating methods are significantly better than natural sedimentation dehydration. Extraheavy crude oil demulsification using ultrasound is achieved mainly using mechanical fluctuation and thermal effects. Mechanical fluctuation can promote water droplets in crude oil to gather together so that they move downwards in the direction of gravity. Furthermore, mechanical fluctuation can also improve the solubility of natural emulsifiers in crude oil, such as paraffin, pectin, and asphalt, so that the mechanical strength of the oil-water interfacial film can be decreased. The decrease of the mechanical strength of the oil-water interfacial film is conducive to oil-water settling separation.

Thermal effects produced by ultrasound can decrease the mechanical strength of the oil-water interfacial film and the viscosity of crude oil. On the one hand, the increase of temperature at the oil-water interface is conducive to breaking the oil-water interfacial film; on the other hand, thermal energy converted from part of the sound energy absorbed by crude oil can reduce the viscosity of crude oil, which is conducive to water droplets moving downwards in the direction of gravity.

But why does the combined method of ultrasound and chemical demulsifier have the best demulsification effect? That is because ultrasound can promote water droplets in crude oil to gather together; chemical demulsifier can greatly decrease the surface tension of water droplets so that the mechanical strength of the oil-water interfacial film can be degraded. Due to the mechanical strength decrease of oil-water interfacial film, big water droplets merge into larger ones, and then they will move downwards in the direction of gravity.

4. The Impact of Ultrasonic Parameters on Crude Oil Demulsification Using the Combined Method of Ultrasound and Chemical Demulsifier

4.1. The Impact of Ultrasonic Power on Crude Oil Demulsification Using the Combined Method of Ultrasound and Chemical Demulsifier. This experiment is carried out under constant temperature, constant chemical demulsifier concentration,

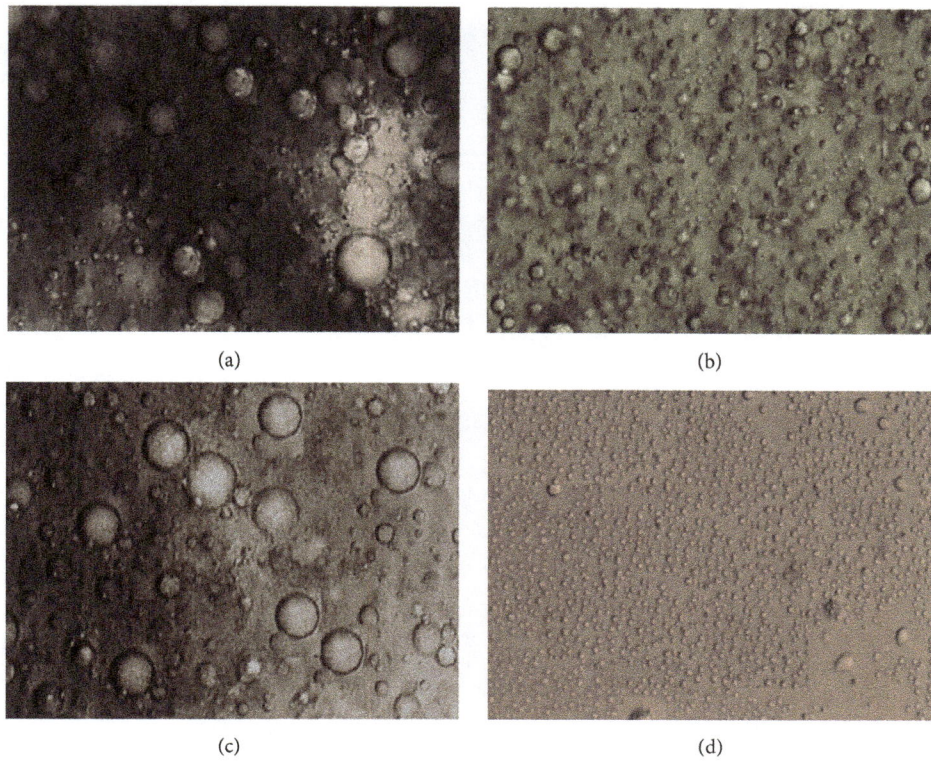

(a)

(b)

(c)

(d)

FIGURE 8: The effect of crude oil demulsification using different methods. (a) Before demulsification; (b) the effect of crude oil demulsification using chemical demulsifier; (c) the effect of crude oil demulsification using ultrasonic treatment; (d) the effect of crude oil demulsification using the combined method of ultrasound and chemical demulsifier.

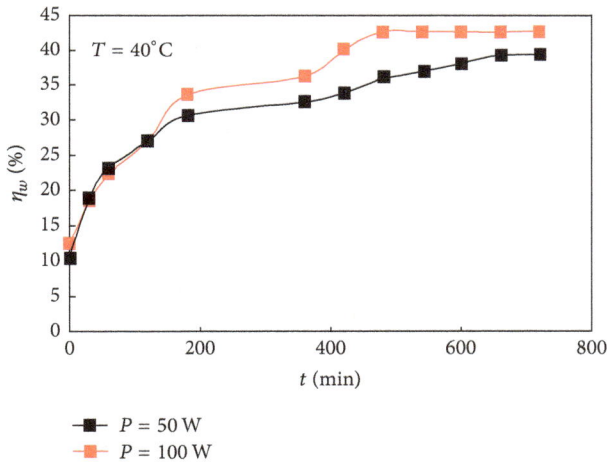

FIGURE 9: The impact of ultrasonic powers (50 W and 100 W) on sonochemistry dehydration.

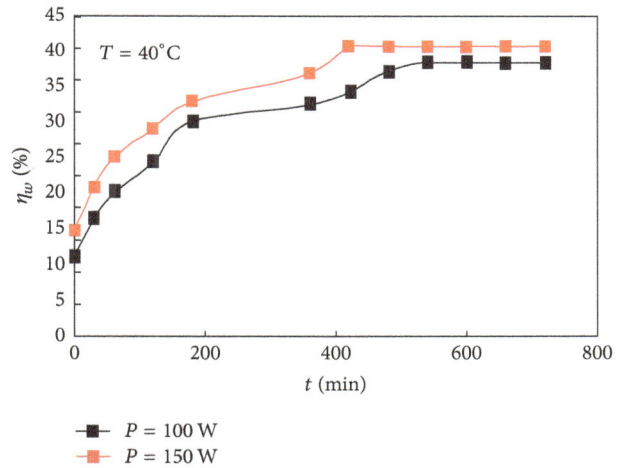

FIGURE 10: The impact of ultrasonic power (100 W and 150 W) on sonochemistry dehydration.

and constant ultrasonic treatment time. The impact of ultrasonic power on sonochemistry dehydration is shown in Figures 9 and 10.

As can be seen in Figures 9 and 10, the final dehydration rate by using the combined method of ultrasound and chemical demulsifier increases with the increase of ultrasonic power. That is because ultrasound with higher power can

cause more intensive mechanical vibrations so that the effect of ultrasonic dehydration can be improved significantly.

4.2. The Impact of Ultrasonic Treatment Time on Crude Oil Demulsification Using the Combined Method of Ultrasound and Chemical Demulsifier. This experiment is carried out under constant temperature, constant chemical demulsifier

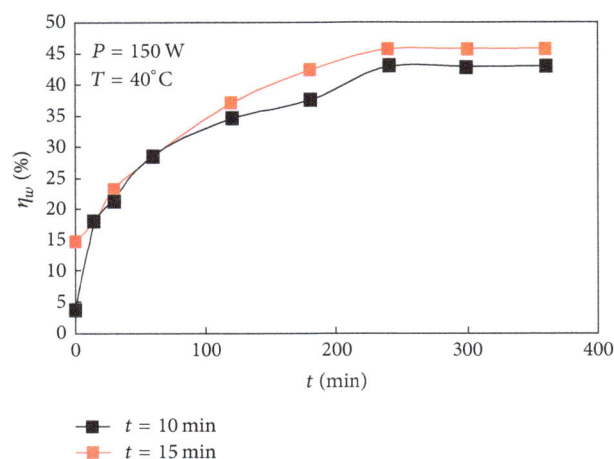

FIGURE 11: The impact of different ultrasonic treatment time (10 min and 15 min) on sonochemistry dehydration.

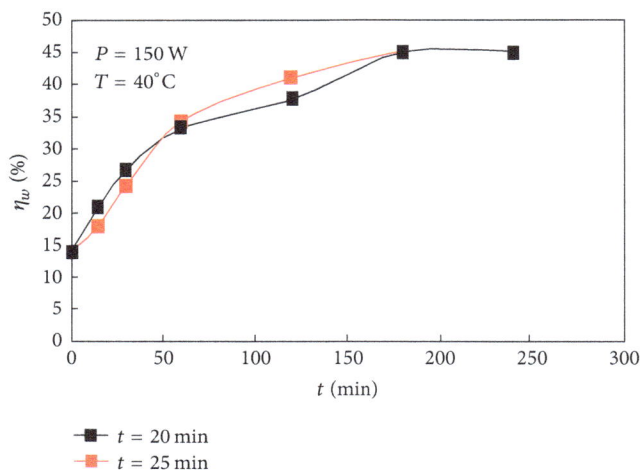

FIGURE 12: The impact of different ultrasonic treatment time (20 min and 25 min) on sonochemistry dehydration.

concentration, and constant ultrasonic power. The impact of ultrasonic treatment time on sonochemistry dehydration is shown in Figures 11 and 12.

As can be seen in Figures 11 and 12, this important conclusion indicates that there is no need to spend more time to dehydrate watery crude oil using the combined method of ultrasound and chemical demulsifier.

5. Conclusion

The novelty of this manuscript is that the comparison of crude oil demulsification using ultrasonic treatment, chemical agent alone, and the combination of ultrasonic and chemical demulsification methods, respectively, was investigated. Results prove that the combined method of ultrasound and chemical demulsifier has the best demulsification effect, followed by chemical demulsifier and ultrasonic treatment. Experiments also indicate that the combined method of

ultrasound and chemical demulsifier not only can achieve the purpose of heavy crude oil demulsification-dehydration but also can shorten demulsification -dehydration time. Furthermore, it is proven that the final dehydration rate by using the combined method of ultrasound and chemical demulsifier increases with the increase of ultrasonic power and temperature and is almost not changed with the increase of ultrasonic treatment time. These conclusions provide important reference for on-site application of sonochemistry dehydration technique.

Conflicts of Interest

The authors declare that they have no conflicts of interest.

References

[1] J. Wu, Y. Xu, T. Dabros, and H. Hamza, "Effect of demulsifier properties on destabilization of water-in-oil emulsion," *Energy and Fuels*, vol. 17, no. 6, pp. 1554–1559, 2003.

[2] H. Ming, *High-Temperature Heavy Oil Dehydration Static Experimental Study of Ultrasonic Demulsificatian*, Daqing Petroleum Institute, 2010.

[3] G. R. Check, "Two-stage ultrasonic irradiation for dehydration and desalting of crude oil: a novel method," *Chemical Engineering and Processing: Process Intensification*, vol. 81, pp. 72–78, 2014.

[4] W. Xie, R. Li, and X. Lu, "Pulsed ultrasound assisted dehydration of waste oil," *Ultrasonics Sonochemistry*, vol. 26, pp. 136–141, 2015.

[5] L. J. Stack, P. A. Carney, H. B. Malone, and T. K. Wessels, "Factors influencing the ultrasonic separation of oil-in-water emulsions," *Ultrasonics Sonochemistry*, vol. 12, no. 3, pp. 153–160, 2005.

[6] G. D. Pangu and D. L. Feke, "Acoustically aided separation of oil droplets from aqueous emulsions," *Chemical Engineering Science*, vol. 59, no. 15, pp. 3183–3193, 2004.

[7] J. Jiao, Y. He, T. Leong et al., "Experimental and theoretical studies on the movements of two bubbles in an acoustic standing wave field," *Journal of Physical Chemistry B*, vol. 117, no. 41, pp. 12549–12555, 2013.

[8] C. Browne, R. F. Tabor, D. Y. C. Chan, R. R. Dagastine, M. Ashokkumar, and F. Grieser, "Bubble coalescence during acoustic cavitation in aqueous electrolyte solutions," *Langmuir*, vol. 27, no. 19, pp. 12025–12032, 2011.

[9] W. Xie, R. Li, X. Lu, P. Han, and S. Gu, "Acoustically aided coalescence of water droplets and dehydration of crude oil emulsion," *Korean Journal of Chemical Engineering*, vol. 32, no. 4, pp. 643–649, 2015.

[10] G. R. Check and D. Mowla, "Theoretical and experimental investigation of desalting and dehydration of crude oil by assistance of ultrasonic irradiation," *Ultrasonics Sonochemistry*, vol. 20, no. 1, pp. 378–385, 2013.

[11] S. Nii, S. Kikumoto, and H. Tokuyama, "Quantitative approach to ultrasonic emulsion separation," *Ultrasonics Sonochemistry*, vol. 16, no. 1, pp. 145–149, 2009.

[12] F. J. Trujillo, P. Juliano, G. Barbosa-Cánovas, and K. Knoerzer, "Separation of suspensions and emulsions via ultrasonic standing waves—a review," *Ultrasonics Sonochemistry*, vol. 21, no. 6, pp. 2151–2164, 2014.

[13] G. Ye, X. Lü, F. Peng, P. Han, and X. Shen, "Pretreatment of crude oil by ultrasonic-electric united desalting and dewatering," *Chinese Journal of Chemical Engineering*, vol. 16, no. 4, pp. 564–569, 2008.

[14] M. Mohsin and M. Meribout, "Oil-water de-emulsification using ultrasonic technology," *Ultrasonics Sonochemistry*, vol. 22, pp. 573–579, 2015.

[15] S. Gou, J. Da, Y. Zhang, P. Hang, and J. Zhang, "A method and apparatus for demulsifying an oil-water emulsion via ultrasonic effect," Patent no. WO/2005/030360, 2005.

[16] J. Y. Yu, P. Yuan, and L. Yu, "Experiment study on the sound-field parameters for crude-oil dehydration by ultrasonic demulsification," *Applied Acoustics*, vol. 20, no. 3, pp. 27–30, 2001 (Chinese).

[17] N. Aboobaker, D. Blackmore, and J. Meegoda, "Mathematical modeling of the movement of suspended particles subjected to acoustic and flow fields," *Applied Mathematical Modelling*, vol. 29, no. 6, pp. 515–532, 2005.

[18] G. D. Pangu, *Acoustically aided coalescence of droplets in aqueous emulsions [Ph.D. thesis]*, Case Western Reserve University, Cleveland, Ohio, USA, 2006.

[19] M. Postema, P. Marmottant, C. T. Lancée, S. Hilgenfeldt, and N. D. Jong, "Ultrasound-induced microbubble coalescence," *Ultrasound in Medicine and Biology*, vol. 30, no. 10, pp. 1337–1344, 2004.

[20] X. Xi, F. B. Cegla, M. Lowe et al., "Study on the bubble transport mechanism in an acoustic standing wave field," *Ultrasonics*, vol. 51, no. 8, pp. 1014–1025, 2011.

[21] L. Hongwei, "High show in the army. Guo Limei. Ground engineering model of crude oil dehydration method," *Oil and Gas Field*, vol. 26, no. 3, pp. 32–33, 2007.

[22] H. Tongliang and Y. K. M. Liangjun, "Progress in research of crude oil dewatering and desalting," *Journal of Fushun Petroleum Institute*, vol. 23, no. 3, pp. 1–5, 2003.

[23] T. J. Mason, A. J. Cobley, J. E. Graves, and D. Morgan, "New evidence for the inverse dependence of mechanical and chemical effects on the frequency of ultrasound," *Ultrasonics Sonochemistry*, vol. 18, no. 1, pp. 226–230, 2011.

[24] J. S. Eow and M. Ghadiri, "Electrostatic enhancement of coalescence of water droplets in oil: a review of the technology," *Chemical Engineering Journal*, vol. 85, no. 2-3, pp. 357–368, 2002.

[25] Z.-S. Bai and H.-L. Wang, "Crude oil desalting using hydrocyclones," *Chemical Engineering Research and Design*, vol. 85, no. 12, pp. 1586–1590, 2007.

[26] R. Zolfaghari, A. Fakhru'l-Razi, L. C. Abdullah, S. S. E. H. Elnashaie, and A. Pendashteh, "Demulsification techniques of water-in-oil and oil-in-water emulsions in petroleum industry," *Separation and Purification Technology*, vol. 170, pp. 377–407, 2016.

[27] X. Zhang, *Experimental studies on demulsification of produced liquid containing heavy oil by sonochemical treatment [M.S. thesis]*, School of Petroleum Engineering, China University of Petroleum (East China), 2008.

18

First-Principles Study of Properties of Alpha Uranium Crystal and Seven Alpha Uranium Surfaces

Shan-Qisong Huang and Xue-Hai Ju

Key Laboratory of Soft Chemistry and Functional Materials of MOE, School of Chemical Engineering,
Nanjing University of Science and Technology, Nanjing 210094, China

Correspondence should be addressed to Shan-Qisong Huang; cqhchsqs@163.com

Academic Editor: Teodorico C. Ramalho

First-principles calculation based on the GGA methods has been applied to the prediction of the properties of bulk α-uranium and seven α-uranium surfaces. The number of layers in the slab has great effects on the simulated surface properties. The predicted surface properties are trustworthy when the slab number is nine or more. The surface energies of the seven low index uranium surfaces are in the range from 1.756 to 2.151 J/m^2. The hybrid between the $5f$ orbital and $6d$ orbital also has somewhat impacts on the surface energies of uranium.

1. Introduction

Uranium is a very typical early actinide metal. It has a very wide range of applications in the field of aerospace and military industry [1, 2]. Uranium exists in alpha phase under normal pressure [3]. Recently, the behavior of U at elevated pressure and temperature was studied experimentally [4, 5], and it was shown that alpha phase is stable up to at least 1 Mbar at ambient temperature with a bcc phase developing [5] at higher temperatures. Modern reviews of the physical properties of uranium were given by Fisher [6] and Lander [7] et al.

Uranium easily reacts with hydrogen [8], oxygen [9], and water [10] due to its lively chemical nature. These reactions take place on the surface and are mainly determined by the nature of the 5f electrons of the surface atoms. The effect of 5f electrons of early actinide metals (Th–Np) has attracted considerable attention for years [11–14]. There are a lot of cases about the surface properties of uranium and uranium compound in experiments [15, 16]. Most of the experimental surface energies stem from surface tension measurements in the liquid phase extrapolated to zero temperature. These experimental data of surface energies include uncertainties of unknown magnitudes and correspond to an isotropic crystal. Hence, they do not yield information about the surface energy of a particular surface facet. There are no exhaustive experimental determinations of the anisotropy in the surface energy of the alpha uranium in solid. Therefore, theoretical calculations play important roles in predicting the surface properties of alpha uranium. With the development in electronic structure simulations theory, in particular in density function theory (DFT), the available computing capacities deliver unprecedented power to compute various properties at an atomistic level. The DFT methods have been used with great success to predict an accurate surface energy. During the last decade there have been many calculations of the surface energy of metals from either the first-principles or the semiempirical methods [4, 17–19]. For example, Taylor investigated the properties of the (001)-oriented α-uranium single-crystal surface in particular by the projector-augmented wave potential method (PAW) [18]. Recently, Söderlind investigated the actinide metals to describe primarily phase stability, bonding, and electronic structure by density-functional theory (DFT) calculations. He found that the early actinides are governed predominantly by fully active 5f bonding [11]. The DFT-PAW method was actually able to reproduce the bonding and electronic structure of alpha uranium [12]. Zhang [13] and Michael [14] et al. also investigated the actinide metals by computer simulations. Their results indicated that the 5f electrons of uranium are similar to the d

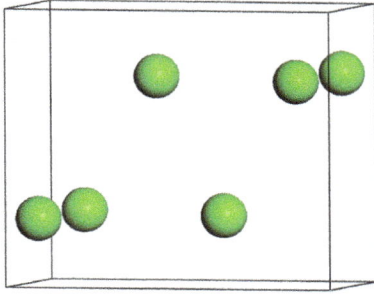

FIGURE 1: Alpha uranium unit cell.

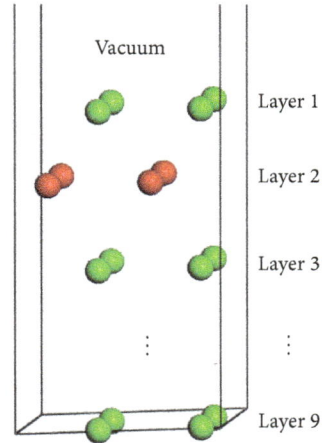

FIGURE 2: Side view of U(001) nine-layer slabs.

orbital electrons in the transition elements mainly as parade electron. Compared with the bulk atoms, the deficit of atoms near the uranium surface will break the balance between the delocalization and localization of 5f electrons leading to the electrons changing from delocalization to localization. As the electrons change from delocalization to localization, the arrangement of the surface atoms is interrupted and the total energy of surface system increases. In order to reduce the surface system energy, the geometry adjustment will lead to surface relaxation and reconstruction.

Although there are lots of surface property calculations of uranium and uranium compound, the comprehensive and accurate surface properties for alpha uranium are lacked. In this paper we investigated the calculation accuracy of alpha uranium bulk by using different GGA methods (PAW, PBE, RPBE, and PW91) at first. Hereafter, in this paper, we focused on the investigation of different surfaces of alpha uranium bulk using the most precise GGA methods. These GGA methods have been previously shown to treat the ground state properties of the light actinides [12]. The surface energies, work function, bulk modulus, elastic constants, and electronic state density of seven basic alpha uranium surfaces were calculated and the results are presented in Section 3.

2. Computational Details

The calculations of surface properties of alpha uranium (Figure 1) have been performed with both the Vienna ab initio simulation package (VASP) [20] and the CASTEP [21] package. All the calculations of VASP are performed within the framework of DFT using electron-ion interaction with the projector-augmented wave (PAW) method. Whereas, all the calculations of CASTEP are done with the GGA method using PBE, RPBE, and PW91 functionals. The electronic wave functions were obtained by a density-mixing scheme, and the structures were relaxed using the BFGS method. The cutoff energy of the plane waves was set to be 500.0 eV. Brillouin-zone sampling was performed using the Monkhorst-Pack scheme. The values of the kinetic energy cutoff and the k-point grid were determined to ensure the convergence of total energies.

The periodic boundary system was used to simulate seven basic surfaces of α-uranium. In order to make the surfaces reasonable, it must be sure that the Z orientation has no interaction with the atomic slabs. It is generally acknowledged that

the vacuum layers thicker than 16 angstroms can meet the requirements [18] (Figure 2). All the seven uranium surfaces were represented by periodic slabs of nine stacked layers with a $1 * 1$ surface unit cell and a large vacuum region of 16 Angstrom to avoid any interaction between the faces of the slab. The slab incremental energy of U(001), ΔE_n, which is depicted in Figure 3, is defined as follows:

$$\Delta E_n = E_{\text{tot}}(n) - E_{\text{tot}}(n-1), \tag{1}$$

where ΔE_n is interpreted as the change in total energy as more "layer" is added and n is the number of layers for the selected model. ΔE_n should converge exactly to a constant when n is large enough.

In our research, five layers of the surface models are reliable for the calculations of surface energies. After being fully relaxed these supercells and the surface energy (E_{surf}) can be calculated by

$$E_{\text{surf}} = \frac{E_{\text{slab}} - nE_{\text{bulk}}}{2A}. \tag{2}$$

Here the E_{slab} is total energy of the selected supercell, the E_{bulk} is the energy for per atom in the primitive cell of the bulk, n is the number of atoms for the selected supercell, and A is the area of the slab. Therefore, nine layers of the surface models are reliable for work function calculations as being demonstrated hereafter.

3. Results and Discussion

3.1. Bulk Properties. Uranium is an orthorhombic crystal structure. Before we studied the surface properties of uranium, we first benchmarked our calculation of the ground state properties of the alpha uranium crystal by comparing with previous first-principles calculations and experimental values (Table 1) [4, 11, 18, 22–24]. The calculated lattice parameters and total energies of alpha uranium bulk were listed in Table 1. All the lattice parameters of the GGA methods (GGA-PBE, GGA-RPBE, and GGA-PW91) are close to the experimental values ($a = 2.836$ Å, $b = 5.867$ Å, and $c = 4.955$ Å).

TABLE 1: Equilibrium lattice parameters (in Å), bulk modulus (in GPa), and total energy E (in eV) of uranium crystal.

Method	a	b	c	Bulk modulus	E
GGA-PBE[a]	2.854	5.869	4.955	114.524	−5615.273
GGA-RPBE[a]	2.826	5.962	4.898	132.124	−5616.184
GGA-PW91[a]	2.833	5.653	5.041	127.813	−5618.287
GGA-PAW (DFT)[b]	2.781	5.731	4.941	124.156	−5615.932
GGA- PAW (DFT + U)[b]	2.817	5.805	5.005	130.243	−5616.027
GGA- PAW (DFT + U + SOC)[b]	2.871	5.982	5.014	124.736	−5616.723
Expt [22]	2.836	5.866	4.936	115	—
FP [23]	2.845	5.818	4.996	130	—
PP [24]	2.809	5.447	4.964	—	—

[a] means results from CASTEP calculations. [b] means results from VASP calculations.

TABLE 2: Experimental and calculated elastic constants of uranium (M bar).

Method	C11	C22	C33	C44	C55	C66	C12	C13	C23
GGA-PBE[a]	2.05	1.96	2.94	1.26	1.03	0.86	0.51	0.15	1.02
GGA-RPBE[a]	1.91	1.86	3.33	1.33	1.11	1.04	0.55	0.40	0.98
GGA-PW91[a]	2.39	2.05	3.18	1.34	1.23	0.35	0.53	0.36	1.41
GGA-PAW (DFT)[b]	2.60	2.12	2.85	1.00	1.41	1.23	0.74	0.17	1.35
GGA- PAW (DFT + U)[b]	2.63	1.81	2.71	0.98	1.40	1.23	0.67	0.16	1.15
GGA-PAW (DFT + U + SOC)[b]	2.61	1.97	2.80	1.02	1.41	1.27	0.66	0.16	1.23
GGA-PAW [11]	2.87	2.41	3.16	1.40	1.05	0.96	0.43	0.17	1.10
Expt [23]	2.15	1.99	2.67	1.24	0.73	0.74	0.46	0.22	1.08

[a] Results from CASTEP calculations. [b] From VASP.

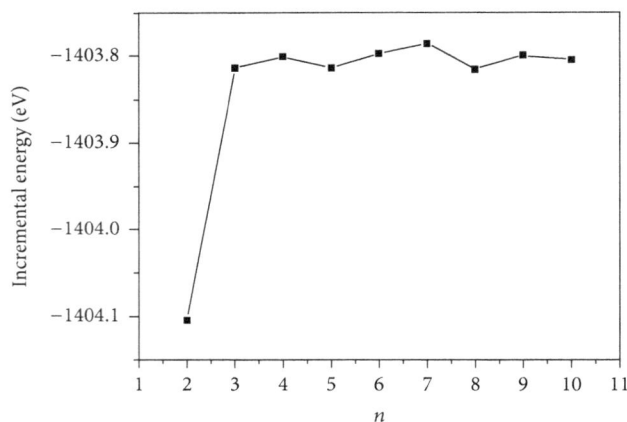

FIGURE 3: Variation of incremental energies of U(001) slabs as a function slab thickness.

It also shows that the results of the GGA-PBE method (a = 2.854 Å, b = 5.869 Å, and c = 5.035 Å, CASTEP) agree well with the experimental values, PP calculation (a = 2.809 Å, b = 5.447 Å, and c = 4.964 Å) and FP calculation (a = 2.845 Å, b = 5.818 Å, and c = 4.996 Å) and are more accurate than the VASP calculations of DFT (a = 2.781 Å, b = 5.731 Å, and c = 4.941 Å), DFT + U (a = 2.817 Å, b = 5.805 Å, and c = 5.005 Å), and DFT + U + SOC (a = 2.871 Å, b = 5.982 Å, and c = 5.014 Å). The similar investigations also can be seen in *Söderlind's* research [12]. After the original cell of the uranium being fully relaxed, the calculated total energy of the original cell is

−5615.273 eV (GGA-PBE) which is the lowest total energy of all the three GGA methods. Therefore, the GGA-PBE method is a feasible method. It is also further indicated that the properties of ground state alpha uranium obtained with the GGA-PBE method of CASTEP (ultrasoft pseudopotentials) are somewhat more accurate than the results obtained with the VASP (all electron pseudopotentials).

After optimizing and comparing the lattice constants via the GGA methods, the elastic and bulk moduli were also calculated using the GGA methods (PAW (VASP), PBE, RPBE, and PW91 (CASTEP)). The calculation results are compared with the experimental data and other works [11, 22–24] (Tables 1 and 2). The determination of elastic constants is important because these constants contain detailed information about the chemical bond and provide a very sensitive test of the DFT methods. As presented in Table 1, the values of bulk modulus of our calculation are very close to experimental value (115 GPa). Especially for the GGA-PBE method (CASTEP), the bulk modulus is 114.524 GPa and deviates only 1% from the experimental value. The same trend also can be seen in the elastic constants results of alpha uranium (Table 2). The elastic constants values by GGA-PBE method are close to the experimental value and agree well with those reported in literature. It is certificated that the GGA-PBE method (CASTEP) is an accurate method in predicting the ground state properties of alpha uranium bulk. In the following work, we decided to use only the GGA-PBE method (CASTEP) to calculate the surface properties of alpha uranium.

TABLE 3: Surface properties of U(001)[a].

N	$E_{tot}(n)$ (eV)	ΔE_n (eV)	Γ (J/m^2)	Φ (eV)	$\Delta d_{12}/d_0$ (%)	$\Delta d_{23}/d_0$ (%)
1	−1401.633		2.073	3.208		
2	−2805.738	−1404.105	1.788	3.122	−10.8	
3	−4209.553	−1403.815	1.769	2.682	−5.0	−4.9
4	−5613.355	−1403.802	1.767	3.486	−4.2	−2.7
5	−7017.170	−1403.815	1.754	3.303	−3.6	0.0
6	−8420.969	−1403.799	1.756	3.236	−3.5	0.0
7	−9824.756	−1403.787	1.756	3.209	−3.5	0.0
8	−11228.573	−1403.817	1.756	3.437	−3.5	0.0
9	−12632.374	−1403.801	1.756	3.461	−3.5	0.0
10	−14036.180	−1403.806	1.751	3.457	−3.5	0.0

[a]$\Delta d_{12}/d_0$ and $\Delta d_{23}/d_0$ represent the relaxation of the first and second interlayer spacing of the U(001) slab with respect to the bulk value d_0. $E_{tot}(n)$ is the total energy, ΔE_n is the slab incremental energy, Γ is the surface energy, and Φ is the work function.

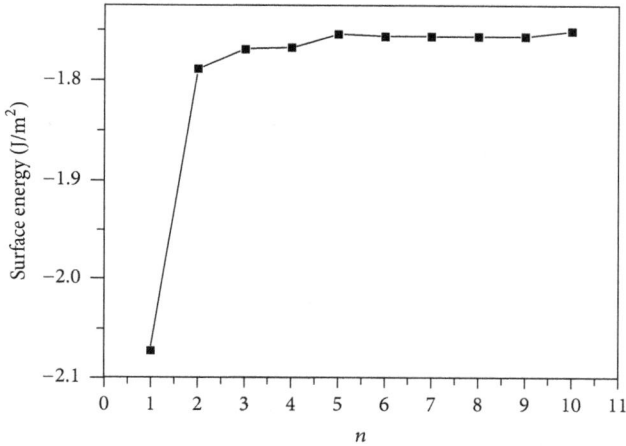

FIGURE 4: Variation of surface energies of U(001) slabs as a function slab thickness.

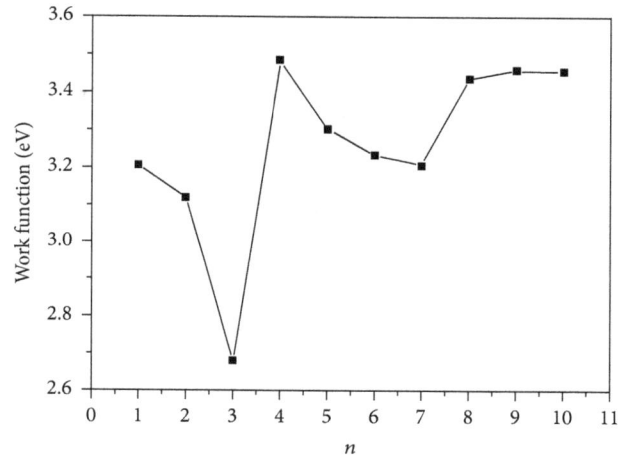

FIGURE 5: Variation of work functions of U(001) slabs as a function slab thickness.

3.2. Surface Properties of U(001). We selected the GGA-PBE computational method (CASTEP) to predict the surface properties of alpha U(001) surface. We also selected surface slab models by comparing the predicted surface properties of alpha U(001) surface with the experimental data and finally ensured that the number of slab layers is large enough to make an accurate description of surface properties. Table 3 lists the geometrical relaxation of the first and second interlayer spacing of the U(001) slab with respect to the bulk value d_0, the total energy $E_{tot}(n)$, slab incremental energy ΔE_n, surface energy Γ, and work function Φ. In order to facilitate the discussions, we plotted the aforementioned quantities versus n (Figures 3–5).

As can be seen in Figure 3, ΔE_n values oscillate largely until $n = 3$. This phenomenon can be attributed to the Quantum Size Effect (QSE), which is the dependence of thin film properties on its characteristic geometric dimensions. For the U(001) surface, QSE is small as the fluctuations in ΔE_n are small when $n = 3$ and beyond. It is indicated that the first three layers contribute dominantly to the surface properties of uranium. As shown in Table 3, in the same surface configurations, d_{12} has much larger shrinkage than d_{23} due to the

interaction of uranium atoms, which strengthens the interaction between the surface and subsurface, and finally narrows the distances between all the uranium layers. The decreases of surface distance ($\Delta d_{12}/d_0$, $\Delta d_{23}/d_0$) become much smaller as the surface layers increase. It is also indicated that the surface energies are greatly affected by the three outermost surface uranium atoms.

The surface energy is one of the most important quantities for characterizing the stability of a surface. As shown in Table 3 and Figure 4, the surface energies of U(001) surface have a great dependence on the number of slab layers. At $n = 6$, the variations in successive surface energies are quite small. Therefore, a slab of at least 6-layer thickness is appropriate for an accurate determination of the U(001) surface energy. The surface energies of our calculation are in the range of 1.751 to 1.756 J m^{-2} for the slab of at least 6-layer thickness, which are closer to the experimental data than the previous theoretical works [18].

The work function, Φ, is the smallest energy required to remove an electron deep inside the bulk crystal through the surface and far away from the surface on the microscopic

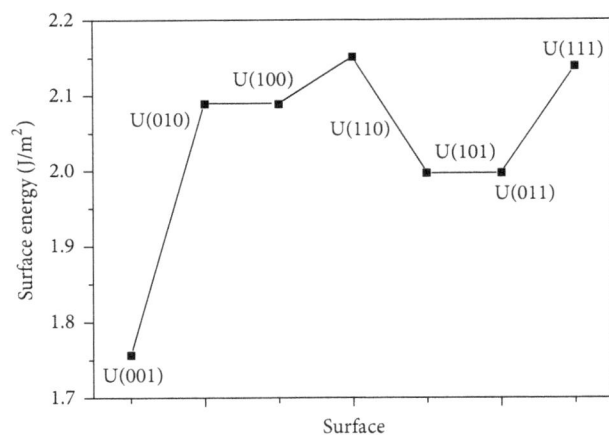

FIGURE 6: Variation of surface energies of seven basic uranium surfaces.

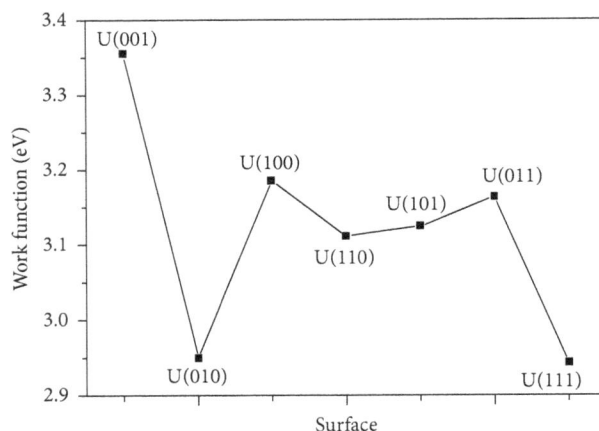

FIGURE 7: Variation of work functions of seven basic uranium surfaces.

TABLE 4: Surface energy (Γ) and work function (Φ).

Surface	Γ	Φ
U(001)	1.756	3.354
U(010)	2.089	2.951
U(100)	2.089	3.185
U(110)	2.151	3.111
U(101)	1.996	3.124
U(011)	1.996	3.163
U(111)	2.138	2.944

Γ in $J\,m^{-2}$ and Φ in eV.

scale at temperature of 0 K. The work functions for U(001) are listed in Table 3 and corresponding plot is presented in Figure 5. Clearly, a large QSE can be seen until $n = 8$, followed by a rapid convergence of work function. A close examination of the actual values in Table 3 revealed that, for $n = 8$, the value of work function is converged to the value about 3.450 eV, which is very close to the experimental value [25] of polycrystalline uranium (3.47 eV) and more accurate than the other research (3.58–3.82 eV). Thus, in the subsequent calculations of the other six basic uranium surfaces (U(010), U(100), U(110), U(011), U(101), and U(111)), we use the slab models of 9 layers to simulate all the properties of the seven low index surfaces.

3.3. Surface Properties of Seven Basic Uranium Surfaces.

The surface energies were obtained by (2) and listed in Table 4 as well as shown in Figure 6. The experimental surface energy of uranium is $1.8281\,J\,m^{-2}$ [18]. However, the experimental value corresponds to an isotropic crystal or an average value of different surfaces. Judged by this fact, our result is in good agreement with experiment. As shown in Figure 6, the descended order of surface energy for these seven surfaces is U(110) > U(111) > U(010) = U(100) > U(011) = U(101) > U(001). The surface energy of U(001) is $1.756\,J\,m^{-2}$, the lowest of all the seven surfaces. The U(001) surface was predicted to be the most stable one. On the contrary, the U(110) surface is the most unstable.

As shown in Table 4 and Figures 6 and 7, the values of the surface energies and work functions of these seven low index alpha uranium surfaces are roughly inversely proportional. A low work function value implies that electrons can easily escape from the bulk region of uranium surfaces, and react with impurities at the surfaces. The work function value of U(001) is 3.354 eV, the highest of all the seven surfaces. The U(001) surface was also predicted to be the most stable one. On the contrary, the U(111) surface is the surface which is the most likely to react with the impurities.

As can be seen in Figure 8, the total density of state (DOS) for uranium consists of $5f$ and $6d$ orbitals. Especially around the Fermi level, the $5f$ orbital's partial density of state is dominant; the s and the p orbitals contribute only with a small part of the total density of state. As shown in the density of state (DOS) for different surfaces (Figure 8), all the peaks of the density of states around the Fermi level shift down obviously due to the electronic orbit split and overlap together, compared to the DOS of α-uranium crystal. The widths of DOS around the Fermi level for all the seven surfaces are narrower in comparison with that of α-uranium bulk. It is indicated that the $6d$ orbital's partial density of state strengthens the hybrid between the $5f$-$6d$ orbitals for all the seven basic surfaces. The $5f$ orbital electrons are hybrid with the $6d$ electrons. The $5f$ orbital's electrons change from delocalization to locality, and the energy bands of $6d$ and $5f$ become narrower. Owing to the spin-orbit coupling the s, p, d, and f orbitals, the energy bands around the Fermi level split. Therefore, it can be demonstrated that the spin-orbit coupling has a large impact on the surface energies of uranium. With more peaks split and deformed, the surface energies become larger.

Compared to the bulk, the U(001) has the smallest hybrid between the $5f$-$6d$ orbitals, the least peak splitting, and the smallest deformation in the density of state. Therefore, the U(001) is the most stable surface with the lowest surface energy. On the contrary, the other six basic low index surfaces deform larger, which causes them more unstable and larger surface energies.

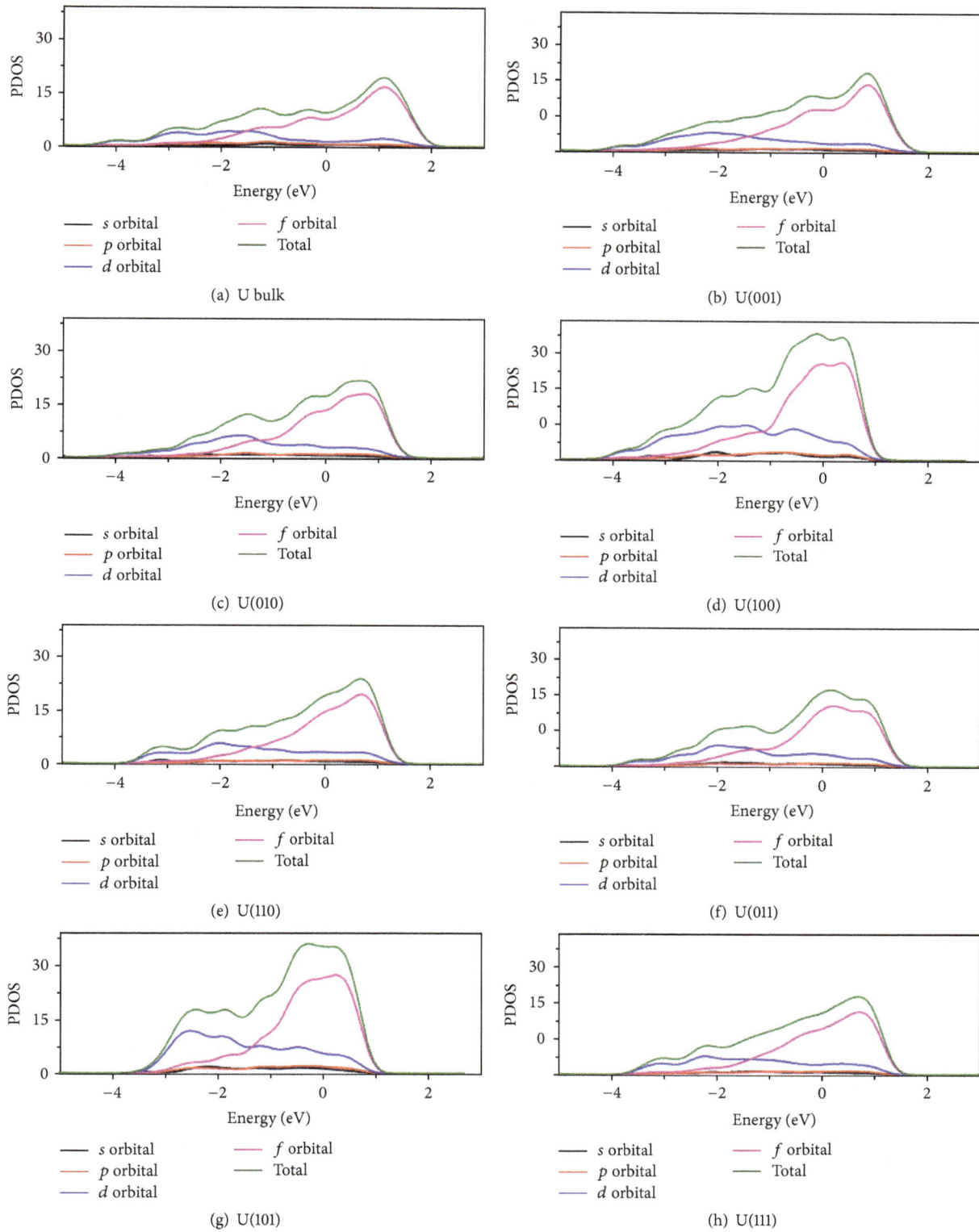

FIGURE 8: PDOSs of seven basic uranium surfaces.

4. Conclusions

The GGA methods are able to predict the ground state properties (lattice constants, elastic moduli, and bulk modulus) of alpha uranium bulk. The calculation results are in reasonable agreement with the experimental results and the available theoretical works [4, 11, 12, 18, 22–24]. The GGA-PBE (CASTEP) is the most accurate method among the GGA methods used in our research to predict the properties of alpha uranium.

The surface properties were calculated by GGA-PBE method for U(001) with n ($n = 1$–10) layers. By comparing the surface energies and work functions with the experimental data, we ensured that a slab with nine layers is thick enough to make an accurate description of surface properties. Judging by the relaxation of the first and second interlayer spacing of the U(001) slab, the first three layers contribute dominantly to the surface energy.

The surface energy and work function are related to the surface density of state. The analysis of density of states shows that the $5f$ orbital electrons are hybrid with the $6d$ orbital electrons. Consequently, the $5f$ orbital electrons change from delocalization to localization and energy bands become narrower and move down. Owing to the spin-orbit coupling, the splitting of the p, d, and f bands also has a large influence on the surface energies of uranium. With more peak splitting and deforming in the density of state, the surface energies become larger.

Competing Interests

The authors declare that they have no competing interests.

Acknowledgments

This work was financially supported by the funding of the Science and Technology on Combustion and Explosion Laboratory (Grant no. 9140C3501021101).

References

[1] J. W. Griffith, *The Uranium Industry: Its History, Technology and Prospects*, Mineral Resources Division, Ottawa, Canada, 1967.

[2] M. L. Rossi, M. D. Agostino, and J. W. Nostrand, *Activation Research in the Aerospace Industry*, Bethpage, New York, NY, USA, 1961.

[3] P. C. Burns, "The crystal chemistry of uranium," *Reviews in Mineralogy and Geochemistry*, vol. 38, no. 4, pp. 23–90, 1999.

[4] C. S. Barrett, M. H. Mueller, and R. L. Hitterman, "Crystal structure variations in alpha uranium at low temperatures," *Physical Review*, vol. 129, no. 2, pp. 625–629, 1963.

[5] D. E. Smirnova, S. V. Starikov, and V. V. Stegailov, "New interatomic potential for computation of mechanical and thermodynamic properties of uranium in a wide range of pressures and temperatures," *The Physics of Metals and Metallography*, vol. 113, no. 2, pp. 107–116, 2012.

[6] R. D. Rogers, C. B. Bauer, and A. H. Bond, "Crown ethers as actinide extractants in acidic aqueous biphasic systems: partitioning behavior in solution and crystallographic analyses of the solid state," *Journal of Alloys and Compounds*, vol. 213-214, pp. 305–312, 1994.

[7] G. H. Lander, E. S. Fisher, and S. D. Bader, "The solid-state properties of uranium A historical perspective and review," *Advances in Physics*, vol. 43, no. 1, pp. 1–111, 1994.

[8] J. L. Nie, H. Y. Xiao, X. T. Zu, and F. Gao, "Hydrogen adsorption, dissociation and diffusion on the α-U(001) surface," *Journal of Physics: Condensed Matter*, vol. 20, no. 44, Article ID 445001, 2008.

[9] X. Zeng, S. Huang, and X. Ju, "Ab initio study on the reaction of uranium with oxygen," *Journal of Radioanalytical and Nuclear Chemistry*, vol. 298, no. 1, pp. 481–484, 2013.

[10] B. Liang, R. D. Hunt, G. P. Kushto, L. Andrews, J. Li, and B. E. Bursten, "Reactions of laser-ablated uranium atoms with H2O in excess argon: a matrix infrared and relativistic DFT investigation of uranium oxyhydrides," *Inorganic Chemistry*, vol. 44, no. 7, pp. 2159–2168, 2005.

[11] P. Söderlind, "First-principles phase stability, bonding, and electronic structure of actinide metals," *Journal of Electron Spectroscopy and Related Phenomena*, vol. 194, pp. 2–7, 2014.

[12] P. Söderlind, A. Landa, and P. E. A. Turchi, "Comment on 'correlation and relativistic effects in U metal and U-Zr alloy: validation of ab *initio* approaches'," *Physical Review B*, vol. 90, Article ID 157101, 2014.

[13] M. L. Neidig, D. L. Clark, and R. L. Martin, "Covalency in f-element complexes," *Coordination Chemistry Reviews*, vol. 257, no. 2, pp. 394–406, 2013.

[14] X. Zhang, H. Zhang, J. Wang, C. Felser, and S.-C. Zhang, "Actinide topological insulator materials with strong interaction," *Science*, vol. 335, no. 6075, pp. 1464–1466, 2012.

[15] Y. Z. Zhang, X. L. Wang, and W. J. Guan, *Atom. Eng. Sci. Tech*, vol. 82, p. 451, 2003.

[16] J. Gao, S. Wu, and X. Yang, *Powder Metallurgy Technology*, vol. 28, no. 2, p. 140, 2010.

[17] J. Lu, H. Zhang, W. Deyun, and H. Yuxia, "Experimental computation process of the surface energy of leaves by acquiring drop image information," *Journal of Nanoelectronics and Optoelectronics*, vol. 7, no. 2, pp. 173–176, 2012.

[18] C. D. Taylor, *Physical Review B*, vol. 77, no. 9, p. 4119, 2009.

[19] M. E. Hoover, R. Atta-Fynn, and A. K. Ray, "Surface properties of uranium dioxide from first principles," *Journal of Nuclear Materials*, vol. 452, no. 1–3, pp. 479–485, 2014.

[20] G. Kresse and J. Furthmüller, "Efficient iterative schemes for ab initio total-energy calculations using a plane-wave basis set," *Physical Review B—Condensed Matter and Materials Physics*, vol. 54, no. 16, p. 11169, 1996.

[21] S. J. Clark, M. D. Segall, C. J. Pickard et al., "First principles methods using CASTEP," *Zeitschrift für Kristallographie—Crystalline Materials*, vol. 220, no. 5-6, pp. 567–570, 2009.

[22] W. A. Curtin and R. E. Miller, "Atomistic/continuum coupling in computational materials science," *Modelling and Simulation in Materials Science and Engineering*, vol. 11, no. 3, pp. R33–R68, 2003.

[23] P. Söderlind, "First-principles elastic and structural properties of uranium metal," *Physical Review B*, vol. 66, no. 8, Article ID 085113, 2002.

[24] A. A. Griffith, *Philosophical Transactions of the Royal Society of London A*, vol. 211, p. 163, 1920.

[25] E. G. Rauh and R. J. Thorn, "Thermionic properties of uranium," *The Journal of Chemical Physics*, vol. 31, no. 6, pp. 1481–1485, 1959.

19

Description of Weak Halogen Bonding Using Various Levels of Symmetry-Adapted Perturbation Theory Combined with Effective Core Potentials

Piotr Matczak

Department of Theoretical and Structural Chemistry, Faculty of Chemistry, University of Łódź, Pomorska 163/165, 90-236 Lodz, Poland

Correspondence should be addressed to Piotr Matczak; p.a.matczak@gmail.com

Academic Editor: Franck Rabilloud

The present work starts with providing a description of the halogen bonding (XB) interaction between the halogen atom of MH_3X (where M = C–Pb and X = I, At) and the N atom of HCN. This interaction leads to the formation of stable yet very weakly bound $MH_3X \cdots NCH$ complexes for which the interaction energy (E_{int}) between MH_3X and HCN is calculated using various symmetry-adapted perturbation theory (SAPT) methods combined with the def2-QZVPP basis set and midbond functions. This basis set assigns effective core potentials (ECPs) not only to the I or At atom directly participating in the XB interaction with HCN but also to the M atom when substituted with Sn or Pb. Twelve SAPT methods (or levels) are taken into consideration. According to the SAPT analysis of E_{int}, the XB interaction in the complexes shows mixed electrostatic-dispersion nature. Next, the accuracy of SAPT E_{int} is evaluated by comparing with CCSD(T) reference data. This comparison reveals that high-order SAPT2+(3) method and the much less computationally demanding SAPT(DFT) method perform very well in describing E_{int} of the complexes. However, the accuracy of these methods decreases dramatically if they are combined with the so-called Hartree-Fock correction.

1. Introduction

Weak intermolecular interactions play an important role in establishing the structure and stability of a broad range of chemical systems, from simple molecular complexes to macromolecular assemblies [1–3]. Computational methods based on electronic structure calculations are useful in determining the magnitude of weak intermolecular interactions as well as in understanding their nature [4–6]. However, the ability to describe weak intermolecular interactions accurately and efficiently differs significantly between various families of computational electronic structure methods. Among these families, the symmetry-adapted perturbation theory (SAPT) [7–11] is capable of providing weak interaction energies accurate even up to 2–4% for benchmark systems of rare-gas dimers [9] but the computational cost of the corresponding calculations in principle scales like N^7 with the system size. This makes such calculations very time-consuming, especially if they are carried out for large chemical systems. The high computational cost of SAPT calculations is associated, to a great extent, with the inclusion of intramolecular electron correlation effects. It is particularly true for the traditional formulation of SAPT which uses many-body electron correlation wave functions for molecules forming a complex [7]. An alternative approach to handling the intramolecular electron correlation within SAPT is to utilize the density functional theory (DFT) for the molecular wave functions [12–17]. The resulting SAPT(DFT) formulation scales better than the traditional formulation, and, therefore, the former can be applied to systems with hundreds of atoms [18]. In order to adjust the computational demands of SAPT calculations to the size of a given system, various methods (or levels) have been introduced for the traditional SAPT formulation [10, 19]. In general, these levels differ in their scaling behavior through the omission of some energy correction terms from the SAPT expansion of intermolecular interaction energy.

Weak intermolecular interactions sometimes occur between molecular sites, of which one contains an atom of element with high nuclear charge. Halogen bonding (XB)

[20] involving iodine constitutes a good example of such interactions (historically speaking, the XB formation in $H_3N \cdots I_2$ was probably the very first observation of XB interaction [21]). From the computational point of view, describing weak intermolecular interactions involving heavier elements is a real challenge for electronic structure methods for two main reasons. Firstly, the large number of electrons in atoms of heavier elements makes electronic structure calculations very costly. Secondly, the incorporation of relativistic effects in the calculations is essential to obtain accurate energies of weak interactions involving heavier elements. These difficulties can be largely overcome by the use of effective core potentials (ECPs) for atoms of heavier elements (usually with the nuclear charge $Z > 36$) [22]. Within the ECP approximation the explicit treatment of core electrons of heavier atoms is replaced by introducing potential operators fitted to the results of relativistic all-electron calculations for such atoms. Therefore, the ECP approximation is able to implicitly account for leading relativistic effects at low computational cost. The use of ECPs has also been implemented for SAPT calculations of interaction energies [23].

In this work a systematic examination of the accuracy of a combination of SAPT and ECP for the description of weak XB is provided relative to the level of SAPT. The accuracy of a variety of SAPT methods is established across a series of 10 heterodimers containing MH_3X (where M = C–Pb and X = I, At) complexed with HCN. These complexes exemplify the $X \cdots N$ type of XB in which the halogen atom possesses a very high nuclear charge (53 and 85 for iodine and astatine, resp.). Thus, the ECP treatment of I- and At-atom core electrons is employed here for the SAPT description of the XB interaction in the $MH_3X \cdots NCH$ complexes.

2. Methods

The geometries of the $MH_3X \cdots NCH$ complexes have been optimized at the MP2/def2-QZVPP level of theory [24–27]. These geometries are characterized by the absence of imaginary vibrational frequencies, and, therefore, they correspond to minima on the potential energy surfaces of the complexes. These geometries will also be kept fixed in all subsequent calculations of the interaction energy (E_{int}) between the MH_3X and HCN parts of the complexes. The MP2 method has been combined with the resolution-of-the-identity approximation [28]. The def2-QZVPP basis set makes use of Stuttgart relativistic ECPs [26] to describe the core electrons of Sn, Pb, I, and At. These calculations have been carried out with the TURBOMOLE 6.6 program [29].

The values of E_{int} between MH_3X and HCN in the complexes have been calculated using such SAPT methods as SAPT0, SAPT2, SAPT2+, SAPT2+(3), SAPT2+3, and SAPT(DFT) [10, 19]. The first five methods are based on the traditional (or regular) wave function-based formulation of SAPT. They truncate the SAPT expansion of E_{int} either at the second or at the third order in the perturbation operator of intermolecular interaction, and they differ in the degree of the completeness of included intramolecular

electron correlation effects. Details of the truncation of SAPT expansion for individual SAPT methods are shown in the following definitions:

$$
\begin{aligned}
E_{int}^{SAPT0} &= E_{elst}^{(10)} + E_{exch}^{(10)} + E_{ind,resp}^{(20)} + E_{exch-ind,resp}^{(20)} \\
&\quad + E_{disp}^{(20)} + E_{exch-disp}^{(20)} \\
E_{int}^{SAPT2} &= E_{int}^{SAPT0} + E_{elst,resp}^{(12)} + E_{exch}^{(11)} + E_{exch}^{(12)} + {}^{t}E_{ind}^{(22)} \\
&\quad + {}^{t}E_{exch-ind}^{(22)} \\
E_{int}^{SAPT2+} &= E_{int}^{SAPT2} + \varepsilon_{exch}^{(1)}(CCSD) - E_{exch}^{(11)} - E_{exch}^{(12)} \\
&\quad + E_{disp}^{(21)} + E_{disp}^{(22)} \\
E_{int}^{SAPT2+(3)} &= E_{int}^{SAPT2+} + E_{elst,resp}^{(13)} + E_{disp}^{(30)} \\
E_{int}^{SAPT2+3} &= E_{int}^{SAPT2+(3)} + E_{ind,resp}^{(30)} + E_{exch-ind,resp}^{(30)} \\
&\quad + E_{exch-disp}^{(30)} + E_{ind-disp}^{(30)} + E_{exch-ind-disp}^{(30)} \\
E_{int}^{SAPT(DFT)} &= E_{elst}^{(1)}(KS) + E_{exch}^{(1)}(KS) + E_{ind}^{(2)}(CKS) \\
&\quad + E_{exch-ind}^{(2)}(CKS) + E_{disp}^{(2)}(CKS) \\
&\quad + E_{exch-disp}^{(2)}(CKS).
\end{aligned}
\tag{1}
$$

The explanation of SAPT energy correction terms on the right sides of (1) can be found in a number of reviews covering the fundamentals of SAPT [7–11]. Suffice it to say here that SAPT0, SAPT2, and SAPT2+ treat the inter-molecular interaction perturbation up to second order, while SAPT2+(3) and SAPT2+3 additionally include third-order terms in the perturbation operator of intermolecular interaction. Regarding the intramolecular electron correlation, it is completely neglected by SAPT0 and included up to second order for electrostatic, exchange, and induction contributions to E_{int} in SAPT2 and additionally for the dispersion contribution by SAPT2+, SAPT2+(3), and SAPT2+3. The PBE0 density functional [30] with the Fermi-Amaldi asymptotic correction [31] and the Tozer-Handy splicing scheme [32] have been used in SAPT(DFT). To compute the asymptotic correction, either available experimental or calculated ionization potentials have been employed (see Supplementary Section 1 in Supplementary Material available online at https://doi.org/10.1155/2017/9031494). Both the wave function-based SAPT methods and SAPT(DFT) can be supplemented with the so-called Hartree-Fock (HF) correction ($\delta^{(2)}E_{int}^{HF}$ or $\delta^{(3)}E_{int}^{HF}$, depending on the SAPT method to which the correction is added). For SAPT0, SAPT2, SAPT2+, SAPT2+(3), and SAPT(DFT) such a correction is determined as

$$
\delta^{(2)}E_{int}^{HF} = E_{int}^{HF} - E_{elst}^{(10)} - E_{exch}^{(10)} - E_{ind,resp}^{(20)} - E_{exch-ind,resp}^{(20)}
\tag{2}
$$

whereas a different form is employed for the SAPT2+3 method

$$
\delta^{(3)}E_{int}^{HF} = \delta^{(2)}E_{int}^{HF} - E_{ind,resp}^{(30)} - E_{exch-ind,resp}^{(30)}.
\tag{3}
$$

The term $E_{\text{int}}^{\text{HF}}$ in (2) denotes the supermolecular Hartree-Fock interaction energy in the complex for which the SAPT interaction energy is calculated. The HF correction is a way to capture some higher-order terms not explicitly evaluated by SAPT. All SAPT results have been produced using the dimer-centered plus basis set (DC$^+$BS) scheme, with the def2-QZVPP basis set extended with a set of midbond (mb) functions [33]. The resulting basis set will be denoted by def2-QZVPP+mb further in the text. The set of mb functions is of the $3s3p2d2f$ form, with the exponents 0.9, 0.3, and 0.1 for s and p and 0.6 and 0.2 for d and f. These functions are centered midway between the X and N atoms of the complexes. All SAPT calculations have been done using the SAPT 2012.2 [34] program interfaced with DALTON 2.0 [35]. The SAPT(DFT) calculations have additionally been carried out with the SAPT(DFT) implementation available in the MOLPRO 2012.1 program [36], but these calculations do not end with successful reproduction of the results obtained with SAPT 2012.2.

Apart from the SAPT description of the XB interaction in MH$_3$X$\cdots$NCH, additional supermolecular estimations of E_{int} are provided in this work for comparison purposes. A number of popular wave function theory methods, mainly those belonging to the family of Møller-Plesset methods [24], will be taken into account. MP2 [37], SCS-MP2 [38], SCS(MI)-MP2 [39], MP2C [40], MP2.5 [41], MP3 [37], MP4(SDQ) [37], MP4 [37], and CCSD [42] are selected. The resolution-of-the-identity approximation is used wherever possible. The aforementioned methods are combined with the def2-QZVPP+mb basis set in order to calculate the magnitude of the XB interaction in the MH$_3$X$\cdots$NCH complexes. The resulting E_{int} values are corrected for the basis set superposition error by utilizing a counterpoise correction [43]. All the supermolecular calculations of E_{int} have been performed using TURBOMOLE 6.6.

The accuracy of SAPT methods is evaluated by comparing their E_{int} values calculated for MH$_3$X$\cdots$NCH with the corresponding reference data. The reference values of E_{int} have been obtained from the CCSD(T) method [42] combined with the def2-QZVPP+mb basis set. The CCSD(T)/def2-QZVPP+mb values of E_{int} are corrected for the basis set superposition error [43]. Additionally, the CCSD(T) E_{int} values extrapolated to the complete basis set (CBS) limit [44–47] will also serve as reference values. The CCSD(T)/CBS level of theory is commonly perceived as the "gold standard" for predicting reliable energies of weak intermolecular interactions [48]. This level of theory was previously used to provide benchmark E_{int} values in a set of 40 complexes that were stabilized by a variety of weak intermolecular interactions in which halogens participated [49]. Details of the CCSD(T)/CBS calculations performed for MH$_3$X$\cdots$NCH are given in Supplementary Section 3.

3. Results and Discussion

3.1. Fundamental Properties of the Complexes. The geometry optimization of the complexes with the assumed X$\cdots$N XB interaction leads to structures in which the HCN molecule

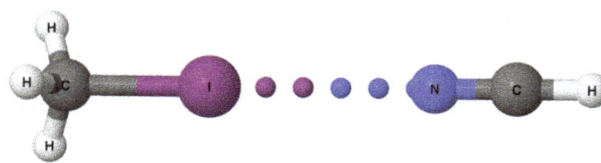

FIGURE 1: Structure of the investigated MH$_3$X$\cdots$NCH complexes. The optimized geometry of CH$_3$I$\cdots$NCH is shown as an example. The XB interaction is marked by a dotted line.

is oriented precisely along the M–X bond axis (see Figure 1). The values of the distance between the X and N atoms ($d_{\text{X}\cdots\text{N}}$) for the complexes in their optimized geometries are listed in Table 1. These values are rather large ($d_{\text{X}\cdots\text{N}} > 3$ Å) but in the case of the complexes with X = I they are still shorter than the sum of the I- and N-atom van der Waals radii. This sum is equal to 3.70 Å if the van der Waals radii proposed by Alvarez [50] are taken. The fulfillment of the geometrical criterion formulated as $d_{\text{I}\cdots\text{N}} < 3.70$ Å suggests the existence of XB in the MH$_3$I$\cdots$NCH complexes. Such a geometrical criterion cannot be applied to the complexes with X = At because the van der Waals radius of astatine is unknown [50].

The structure of the MH$_3$X$\cdots$NCH complexes results from the highly directional ability of the X atom, covalently bonded to the M atom, to attract the N-atom lone electron pair of HCN, acting as a nucleophilic species in this case. A widely accepted mechanism of XB formation postulates that XB originates from the interaction between two centers possessing opposite electrostatic potentials [51]. For the MH$_3$X$\cdots$NCH complexes the center exhibiting a positive electrostatic potential is located on the X atom in the region on the extension of the M–X bond (i.e., the σ-hole), whereas the negative electrostatic potential characterizes the region around the lone electron pair of the N atom. The maximal value of electrostatic potential on the molecular surface delineated by the electron density contour of 0.001 au ($V_{s,\text{max}}$) in the σ-hole region on the X atom is a convenient tool for quantifying the σ-hole [52]. The calculated values of $V_{s,\text{max}}$ are presented in Table 1. An increase in $V_{s,\text{max}}$ values is observed if the X atom changes from I to At, and thus it becomes heavier and more polarizable [52]. Moreover, $V_{s,\text{max}}$ values increase with the growing electron-attracting character of the MH$_3$ group [53]. The electron-attracting character can be represented by the so-called group electronegativity [54] which is calculated for the MH$_3$ groups (for details, see Supplementary Section 4). The electronegativity of the MH$_3$ groups increases gradually while going up group 14 from M = Pb to M = C, and, therefore, the most positive $V_{s,\text{max}}$ values are achieved for CH$_3$X in either of the sequences of the MH$_3$X$\cdots$NCH complexes with the X atom kept fixed.

The CCSD(T)/CBS E_{int} values shown in Table 1 are negative for all MH$_3$X$\cdots$NCH complexes. This implies their energetic stabilization and this also constitutes an energetic criterion for XB occurrence in the complexes. The values of E_{int} fall in a narrow range from −0.55 to −2.36 kcal/mol. The XB interaction in MH$_3$X$\cdots$NCH can be classified as weak in view of the fact that the strongest XB interactions reported in

TABLE 1: Distance between the X and N atoms ($d_{X\cdots N}$; in Å), maximal value of electrostatic potential in the σ-hole region on the X atom ($V_{s,max}$; in kcal/mol), CCSD(T)/CBS and SAPT(DFT) interaction energies between MH$_3$X and HCN (E_{int}; in kcal/mol), and SAPT(DFT) principal interaction energy components (E_{elst}, E_{ind}, E_{disp}, and E_{exch}; in kcal/mol) for the investigated MH$_3$X$\cdots$NCH complexes. The percentage of each attractive SAPT(DFT) component relative to the total attraction is given in parentheses.

Complex	$d_{X\cdots N}$	$V_{s,max}$	E_{int} CCSD(T)/CBS	SAPT(DFT)	E_{elst}	E_{ind}	E_{disp}	E_{exch}
CH$_3$I$\cdots$NCH	3.261	12.2	−1.46	−1.23	−2.23 (43)	−0.52 (10)	−2.45 (47)	3.97
SiH$_3$I$\cdots$NCH	3.451	8.7	−1.14	−1.08	−1.41 (39)	−0.35 (10)	−1.83 (51)	2.51
GeH$_3$I$\cdots$NCH	3.426	7.3	−1.01	−0.94	−1.34 (36)	−0.38 (11)	−1.95 (53)	2.73
SnH$_3$I$\cdots$NCH	3.470	4.4	−0.79	−0.74	−1.02 (31)	−0.36 (11)	−1.86 (58)	2.50
PbH$_3$I$\cdots$NCH	3.481	1.6	−0.55	−0.49	−0.73 (25)	−0.37 (12)	−1.87 (63)	2.47
CH$_3$At$\cdots$NCH	3.171	18.4	−2.36	−2.02	−3.95 (49)	−0.84 (11)	−3.19 (40)	5.96
SiH$_3$At$\cdots$NCH	3.357	13.6	−1.80	−1.72	−2.58 (47)	−0.55 (10)	−2.41 (43)	3.83
GeH$_3$At$\cdots$NCH	3.322	12.9	−1.74	−1.59	−2.68 (46)	−0.60 (10)	−2.59 (44)	4.28
SnH$_3$At$\cdots$NCH	3.353	10.8	−1.52	−1.38	−2.37 (44)	−0.57 (10)	−2.51 (46)	4.08
PbH$_3$At$\cdots$NCH	3.343	8.1	−1.33	−1.15	−2.25 (41)	−0.59 (11)	−2.61 (48)	4.30

literature exhibit interaction energies exceeding −40 kcal/mol [55]. The CH$_3$I molecule forms weaker XB with HCN than with NH$_3$ [56] or CH$_2$O [57]. The weakness of the XB interaction in the MH$_3$X$\cdots$NCH complexes, as indicated by their E_{int} values, is connected with the large $d_{X\cdots N}$ distances. If pairs of the complexes with the M atom fixed are considered, MH$_3$At always forms stronger XB than MH$_3$I. This is reflected in the geometrical relation $d_{At\cdots N} < d_{I\cdots N}$ for each such pair. The XB interaction becomes weaker and weaker when the M atom traverses down group 14 for two sequences of the complexes with the X atom kept fixed and varying M. The clear trend in E_{int} is only roughly followed by a trend in $d_{X\cdots N}$ for these sequences. These trends do not correlate with each other ideally but the monotonous weakening of the XB interaction is often accompanied by the gradual elongation of $d_{X\cdots N}$. It is noteworthy that $V_{s,max}$ correlates well with E_{int} for all MH$_3$X$\cdots$NCH complexes: the greater $V_{s,max}$ is developed in the σ-hole region, the stronger the XB occurs.

In order to establish the nature of the XB interaction in MH$_3$X$\cdots$NCH, the energy correction terms from the SAPT(DFT) expansion of E_{int} have been grouped into four principal components corresponding to electrostatics, induction, dispersion, and exchange. Details of the grouping scheme can be found in Supplementary Section 5. The values of the electrostatic (E_{elst}), induction (E_{ind}), dispersion (E_{disp}), and exchange (E_{exch}) components of E_{int} obtained from SAPT(DFT) are listed in Table 1. The most important aspect of these results is the fact that E_{elst} and E_{disp} play the central

role in the attraction between MH$_3$X and HCN. These two components are responsible for at least 88% of the total attraction occurring in the complexes. Thus, the XB interaction in MH$_3$X$\cdots$NCH shows mixed electrostatic-dispersion nature. This finding is in agreement with the analogical conclusion drawn in an earlier work taking various types of XB into account [58]. It is interesting to examine trends in the attractive components for the sequences of the complexes with the X atom kept fixed. For such sequences the percentage of electrostatics decreases (and simultaneously the percentage of dispersion increases) while descending group 14 from M = C to M = Pb. This leads to the conclusion that the role of dispersion in the stabilization of the complexes is most significant for the weakest XB interaction. E_{ind} component does not change monotonously if M descends group 14. Moreover, the percentage share of E_{ind} in the total attraction between MH$_3$X and HCN remains practically constant for all 10 complexes.

3.2. Assessment of Various SAPT Methods.
In order to evaluate the performance of various SAPT methods in describing the XB interaction in the MH$_3$X$\cdots$NCH complexes, E_{int} values predicted by these methods are compared with CCSD(T) reference results in a statistical manner. This statistical comparison is based on three metrics of errors in the SAPT E_{int} values. The root mean-squared error (RMSE), the percentage relative root mean-squared error (rRMSE (%)),

and the mean signed error (MSE) are calculated using the following formulas:

$$\text{RMSE} = \sqrt{\frac{1}{M}\sum_{i=1}^{M}\left(E_{\text{int}}^{\text{SAPT}}(i) - E_{\text{int}}^{\text{CCSD(T)}}(i)\right)^2}$$

$$\text{rRMSE}\,(\%) = 100$$

$$\cdot\sqrt{\frac{1}{M}\sum_{i=1}^{M}\left(\frac{E_{\text{int}}^{\text{SAPT}}(i) - E_{\text{int}}^{\text{CCSD(T)}}(i)}{E_{\text{int}}^{\text{CCSD(T)}}(i)}\right)^2} \qquad (4)$$

$$\text{MSE} = \frac{1}{M}\sum_{i=1}^{M}\left(E_{\text{int}}^{\text{SAPT}}(i) - E_{\text{int}}^{\text{CCSD(T)}}(i)\right),$$

where M is the number of the complexes studied ($M = 10$). RMSE is used to characterize the overall performance of each SAPT method in predicting E_{int}. Because E_{int} values are small themselves and the same applies to the RMSE values, then rRMSE (%) should more conveniently and reliably illustrate the accuracy of the SAPT E_{int} values relative to the corresponding CCSD(T) results. Finally, MSE provides information about systematic errors occurring in E_{int} values obtained from each SAPT method. If MSE is negative (positive) for a given SAPT method, then this method tends to overestimate (underestimate) the strength of the XB interaction in the $MH_3X\cdots NCH$ complexes.

The evaluation of SAPT accuracy has been performed for 12 SAPT methods. Table 2 lists these methods, together with the three mean errors determined for their E_{int} values. Based on the RMSE values one can think that SAPT generally performs very well because RMSE is less than 1 kcal/mol for all methods. This is not necessarily true if rRMSE (%) is analyzed as well. rRMSE (%) is high for several SAPT methods, which indicates that the RMSE values are in fact quite large when compared to both the CCSD(T)/def2-QZVPP+mb and CCSD(T)/CBS values. It is justified to rank SAPT methods' performances relative to the rRMSE (%) values. From the magnitude of rRMSE (%) one can conclude that the SAPT2+(3) method provides the best approximation of the CCSD(T) reference results. As with SAPT2+(3), the SAPT(DFT) method reproduces perfectly the CCSD(T)/def2-QZVPP+mb E_{int} energies. A slightly larger rRMSE (%) value is observed for SAPT2+3 but this value still does not exceed 10%. The values of MSE for the three methods indicate that SAPT2+(3) tends to slightly overestimate the strength of the XB interaction in $MH_3X\cdots NCH$, while SAPT(DFT) and SAPT2+3 show the opposite tendency (the mean underestimation of E_{int} reaches a maximum of 0.14 kcal/mol for SAPT(DFT)). Unsurprisingly, two regular high-order SAPT methods SAPT2+(3) and SAPT2+3 yield E_{int} with very good accuracy. It is rather puzzling, however, that SAPT2+3 performs slightly worse than SAPT2+(3) while the reverse ordering of these two methods could actually be expected. This means that the inclusion of all third-order interaction energy correction terms does not lead to any improvement in E_{int} for the $MH_3X\cdots NCH$ complexes. The SAPT2+(3) accuracy superior to that of SAPT2+3 is mostly

TABLE 2: Errors (RMSE, rRMSE (%), and MSE; in kcal/mol or %) in E_{int} obtained from various SAPT methods. The errors without parentheses have been calculated relative to the CCSD(T)/def2-QZVPP+mb results, while those put in parentheses are based on the CCSD(T)/CBS reference values of E_{int}.

Method	RMSE	rRMSE (%)	MSE
SAPT0	0.28 (0.23)	21 (16)	−0.26 (−0.21)
SAPT0 + $\delta^{(2)}E_{\text{int}}^{\text{HF}}$	0.67 (0.62)	51 (44)	−0.64 (−0.59)
SAPT2	0.39 (0.34)	36 (29)	−0.38 (−0.33)
SAPT2 + $\delta^{(2)}E_{\text{int}}^{\text{HF}}$	0.78 (0.73)	66 (58)	−0.76 (−0.71)
SAPT2+	0.30 (0.25)	27 (21)	−0.30 (−0.25)
SAPT2+ + $\delta^{(2)}E_{\text{int}}^{\text{HF}}$	0.70 (0.65)	58 (50)	−0.67 (−0.62)
SAPT2+(3)	0.06 (0.04)	7 (3)	−0.05 (−0.01)
SAPT2+(3) + $\delta^{(2)}E_{\text{int}}^{\text{HF}}$	0.44 (0.39)	37 (30)	−0.43 (−0.38)
SAPT2+3	0.13 (0.16)	9 (9)	0.03 (0.08)
SAPT2+3 + $\delta^{(3)}E_{\text{int}}^{\text{HF}}$	0.51 (0.46)	41 (34)	−0.49 (−0.44)
SAPT(DFT)	0.12 (0.16)	7 (10)	0.08 (0.14)
SAPT(DFT) + $\delta^{(2)}E_{\text{int}}^{\text{HF}}$	0.30 (0.25)	26 (20)	−0.29 (−0.24)

due to the fact that the third-order induction and exchange-induction terms are poorly recovered by SAPT combined with ECPs [23] (see also Supplementary Section 2).

From the results in Table 2 the effect of the HF correction on E_{int} can be established. This correction collects mainly higher-order induction and exchange-induction terms but it may also contain some spurious, unphysical terms. It is evident that the HF correction adversely affects the accuracies of all SAPT methods in describing the weak XB interaction in $MH_3X\cdots NCH$. Six SAPT methods utilizing this correction (the so-called hybrid approach) show many times larger RMSE values than the same methods without $\delta^{(2)}E_{\text{int}}^{\text{HF}}$ or $\delta^{(3)}E_{\text{int}}^{\text{HF}}$. As illustrated by the negative values of MSE, the majority of SAPT methods tend to predict an overbinding between MH_3X and HCN. Adding the HF correction results in further overestimation of the XB strength in the complexes. A similar overestimation of $I\cdots O$ XB strength was also detected in a previous SAPT study of the $CH_3I\cdots OCH_2$ complex [57].

The deterioration of SAPT accuracy due to the incorporation of $\delta^{(2)}E_{\text{int}}^{\text{HF}}$ or $\delta^{(3)}E_{\text{int}}^{\text{HF}}$ into E_{int} is likely caused by a spurious, unphysical part of the HF correction. It is known that spurious, unphysical effects contribute considerably to the HF correction if a molecular complex is bound primarily by dispersion [59]. This seems to be the case for the

TABLE 3: Errors (RMSE, rRMSE (%), and MSE; in kcal/mol or %) in E_{int} obtained from various wave function theory methods within the supermolecular approach. The errors without parentheses have been calculated relative to the CCSD(T)/def2-QZVPP+mb results, while those put in parentheses are based on the CCSD(T)/CBS reference values of E_{int}.

Method	RMSE	rRMSE (%)	MSE
HF	2.19 (2.24)	200 (193)	2.16 (2.22)
MP2	0.46 (0.40)	41 (34)	−0.45 (−0.39)
SCS-MP2	0.11 (0.16)	10 (14)	0.10 (0.15)
SCS(MI)-MP2	0.08 (0.05)	7 (3)	−0.06 (−0.01)
MP2C	0.06 (0.08)	3 (5)	0.01 (0.06)
MP2.5	0.06 (0.11)	4 (8)	0.04 (0.10)
MP3	0.55 (0.60)	48 (50)	0.54 (0.59)
MP4(SDQ)	0.23 (0.28)	23 (26)	0.22 (0.27)
MP4	0.34 (0.29)	31 (25)	−0.33 (−0.28)
CCSD	0.40 (0.45)	36 (38)	0.39 (0.44)

MH$_3$X$\cdots$NCH complexes for which the dispersion component of their E_{int} is always significant or even dominant (the percentage of $E_{disp} \geq 40\%$; see Table 1).

The simplest SAPT method, that is, SAPT0, outperforms both SAPT2 and SAPT2+, although the latter methods incorporate more energy correction terms, and, therefore, they are computationally more demanding. This superior accuracy of SAPT0 seems to be due to a fairly good compensation of energy correction terms appearing in the SAPT0 expansion of E_{int}. It has been demonstrated in another recent work [19] that SAPT0 yields in many cases not much worse interaction energies compared to more advanced SAPT methods which account for intramolecular electron correlation effects. The combination of SAPT0, SAPT2, or SAPT2+ with $\delta^{(2)}E_{int}^{HF}$ fails badly in describing the XB interaction in MH$_3$X$\cdots$NCH; rRMSE (%) of the resulting hybrid levels exceeds 50% relative to the CCSD(T)/def2-QZVPP+mb reference results.

It is instructive to compare the performance of the SAPT methods with the performance of a series of popular wave function theory methods that compute E_{int} within the supermolecular approach. Statistical metrics measuring the accuracy of supermolecular E_{int} values obtained from 10 wave function theory methods against the CCSD(T) reference data are presented in Table 3. Of the 10 methods, SCS(MI)-MP2, MP2C, and MP2.5 perform best. They afford very small RMSE and rRMSE (%) and these errors are even slightly smaller than those made by SAPT2+(3) and SAPT2+3. What is particularly important is that the SCS(MI)-MP2 calculations entail much lower computational cost (MP2 formally

scales as N^5 with the system size). The good performance of SCS(MI)-MP2 has actually been expected because this method has been specially designed for weak intermolecular interactions. Similarly, the MP2C and MP2.5 methods were also developed for providing accurate interaction energies in various molecular complexes. For the MH$_3$X$\cdots$NCH complexes, these modifications of the original MP2 method do show a great improvement in the reproduction of E_{int} over their ancestor. The performances of MP4 and CCSD are rather disappointing and these advanced methods yield less accurate values of E_{int} than the regular high-order SAPT and SAPT(DFT) methods. According to the HF method, the interaction between MH$_3$X and HCN destabilizes the MH$_3$X$\cdots$NCH complexes, which disqualifies this method as a computational tool for describing weak XB interactions reliably.

4. Conclusions

A description of the X$\cdots$N XB interaction in a series of 10 MH$_3$X$\cdots$NCH complexes (where M = C–Pb and X = I, At) has been obtained from various SAPT methods combined with the def2-QZVPP basis set and midbond functions. This basis set assigns ECPs not only to the I or At atom directly participating in the XB interaction with HCN but also to the M atom when substituted with Sn or Pb. The accuracy of SAPT E_{int} has been evaluated by comparing with CCSD(T) reference data. The main findings made in this work can be summarized as follows:

(1) The CCSD(T)/CBS calculations of E_{int} reveal that the complexes are stabilized by a weak X$\cdots$N XB interaction whose values fall in the range between −0.55 and −2.36 kcal/mol. Astatine forms stronger XB interaction with HCN than iodine, by no more than 1 kcal/mol in each pair of the complexes with a given M. A relationship between E_{int} and the kind of group 14 element substituting the M atom has been identified for two sequences of the complexes with the X atom kept fixed. The strength of the XB interaction decreases gradually when the M atom varies down group 14 from C to Pb. This is rationalized in terms of $V_{s,max}$ and MH$_3$-group electronegativity.

(2) Grouping the terms from the SAPT(DFT) expansion of E_{int} into four principal components indicates the mixed electrostatic-dispersion nature of the XB interaction in MH$_3$X$\cdots$NCH. The XB weakening caused by the substitution of the M and X atoms clearly affects the percentage shares of E_{elst} and E_{disp}. This weakening is associated with the decrease of E_{elst} percentage share and the simultaneous increase of E_{disp} percentage share.

(3) Regular high-order SAPT methods such as SAPT2+(3) and SAPT2+3 provide XB description very close to that obtained at the CCSD(T)/def2-QZVPP+mb and CCSD(T)/CBS levels of theory. The accuracy of E_{int} obtained from the computationally favorable SAPT(DFT) method is comparable to that

found for SAPT2+(3) interaction energies. The HF correction present in the hybrid approach leads to a significant deterioration in the accuracy of the SAPT description of the XB interaction in the $MH_3X\cdots NCH$ complexes.

Competing Interests

The author declares no competing interests.

Acknowledgments

This work was partially supported by PL-Grid Infrastructure.

References

[1] G. A. Jeffrey and W. Saenger, *Hydrogen Bonding in Biological Structures*, Springer, Berlin, Germany, 1991.

[2] I. Haiduc and F. T. Edelmann, *Supramolecular Organometallic Chemistry*, Wiley-VCH, Weinheim, Germany, 1999.

[3] A. Karshikoff, *Non-Covalent Interactions in Proteins*, Imperial College Press, Singapore, 2006.

[4] S. Tsuzuki, "Interactions with aromatic rings," *Structure and Bonding*, vol. 115, pp. 149–193, 2005.

[5] I. G. Kaplan, *Intermolecular Interactions: Physical Picture, Computational Methods and Model Potentials*, John Wiley & Sons, Chichester, UK, 2006.

[6] P. Hobza and K. Müller-Dethlefs, *Non-Covalent Interactions. Theory and Experiment*, RSC Publishing, Cambridge, UK, 2010.

[7] B. Jeziorski, R. Moszynski, and K. Szalewicz, "Perturbation theory approach to intermolecular potential energy surfaces of van der Waals complexes," *Chemical Reviews*, vol. 94, no. 7, pp. 1887–1930, 1994.

[8] K. Szalewicz, K. Patkowski, and B. Jeziorski, "Intermolecular interactions via perturbation theory: from diatoms to biomolecules," *Structure and Bonding*, vol. 116, pp. 43–117, 2005.

[9] K. Szalewicz, "Symmetry-adapted perturbation theory of intermolecular forces," *Wiley Interdisciplinary Reviews: Computational Molecular Science*, vol. 2, no. 2, pp. 254–272, 2012.

[10] E. G. Hohenstein and C. D. Sherrill, "Wavefunction methods for noncovalent interactions," *Wiley Interdisciplinary Reviews: Computational Molecular Science*, vol. 2, no. 2, pp. 304–326, 2012.

[11] G. Jansen, "Symmetry-adapted perturbation theory based on density functional theory for noncovalent interactions," *Wiley Interdisciplinary Reviews: Computational Molecular Science*, vol. 4, no. 2, pp. 127–144, 2014.

[12] H. L. Williams and C. F. Chabalowski, "Using Kohn-Sham orbitals in symmetry-adapted perturbation theory to investigate intermolecular interactions," *The Journal of Physical Chemistry A*, vol. 105, no. 3, pp. 646–659, 2001.

[13] A. J. Misquitta and K. Szalewicz, "Intermolecular forces from asymptotically corrected density functional description of monomers," *Chemical Physics Letters*, vol. 357, no. 3-4, pp. 301–306, 2002.

[14] A. Heßelmann and G. Jansen, "First-order intermolecular interaction energies from Kohn-Sham orbitals," *Chemical Physics Letters*, vol. 357, no. 5-6, pp. 464–470, 2002.

[15] A. Heßelmann and G. Jansen, "Intermolecular induction and exchange-induction energies from coupled-perturbed Kohn-Sham density functional theory," *Chemical Physics Letters*, vol. 362, no. 3-4, pp. 319–325, 2002.

[16] A. J. Misquitta, B. Jeziorski, and K. Szalewicz, "Dispersion energy from density-functional theory description of monomers," *Physical Review Letters*, vol. 91, no. 3, Article ID 033201, 4 pages, 2003.

[17] A. Heßelmann and G. Jansen, "Intermolecular dispersion energies from time-dependent density functional theory," *Chemical Physics Letters*, vol. 367, no. 5-6, pp. 778–784, 2003.

[18] R. Podeszwa, W. Cencek, and K. Szalewicz, "Efficient calculations of dispersion energies for nanoscale systems from coupled density response functions," *Journal of Chemical Theory and Computation*, vol. 8, no. 6, pp. 1963–1969, 2012.

[19] T. M. Parker, L. A. Burns, R. M. Parrish, A. G. Ryno, and C. D. Sherrill, "Levels of symmetry adapted perturbation theory (SAPT). I. Efficiency and performance for interaction energies," *The Journal of Chemical Physics*, vol. 140, no. 9, Article ID 094106, 16 pages, 2014.

[20] P. Metrangolo and G. Resnati, "Halogen bonding: a paradigm in supramolecular chemistry," *Chemistry—A European Journal*, vol. 7, no. 12, pp. 2511–2519, 2001.

[21] F. Guthrie, "On the iodide of iodammonium," *Journal of the Chemical Society*, vol. 16, pp. 239–244, 1863.

[22] P. Schwerdtfeger, *Relativistic Electronic Structure Theory. Part 1. Fundamentals*, Elsevier, Amsterdam, The Netherlands, 2002.

[23] K. Patkowski and K. Szalewicz, "Frozen core and effective core potentials in symmetry-adapted perturbation theory," *The Journal of Chemical Physics*, vol. 127, no. 16, Article ID 164103, 17 pages, 2007.

[24] C. Møller and M. S. Plesset, "Note on an approximation treatment for many-electron systems," *Physical Review*, vol. 46, no. 7, pp. 618–622, 1934.

[25] F. Weigend and R. Ahlrichs, "Balanced basis sets of split valence, triple zeta valence and quadruple zeta valence quality for H to Rn: design and assessment of accuracy," *Physical Chemistry Chemical Physics*, vol. 7, no. 18, pp. 3297–3305, 2005.

[26] B. Metz, H. Stoll, and M. Dolg, "Small-core multiconfiguration-Dirac-Hartree-Fock-adjusted pseudopotentials for post-d main group elements: application to PbH and PbO," *The Journal of Chemical Physics*, vol. 113, no. 7, pp. 2563–2569, 2000.

[27] C. Hättig, A. Hellweg, and A. Köhn, "Distributed memory parallel implementation of energies and gradients for second-order Møller-Plesset perturbation theory with the resolution-of-the-identity approximation," *Physical Chemistry Chemical Physics*, vol. 8, no. 10, pp. 1159–1169, 2006.

[28] F. Weigend and M. Häser, "RI-MP2: first derivatives and global consistency," *Theoretical Chemistry Accounts*, vol. 97, no. 1–4, pp. 331–340, 1997.

[29] R. Ahlrichs, M. K. Armbruster, R. A. Bachorz et al., TURBOMOLE 6.6, a development of University of Karlsruhe and Forschungszentrum Karlsruhe GmbH, 1989–2007, TURBOMOLE GmbH, since 2007, 2014, http://www.turbomole.com.

[30] C. Adamo and V. Barone, "Toward reliable density functional methods without adjustable parameters: the PBE0 model," *The Journal of Chemical Physics*, vol. 110, no. 13, pp. 6158–6170, 1999.

[31] E. Fermi and E. Amaldi, "Le orbite ∞s degli elementi," *Memorie dell'Accademia d'Italia*, vol. 6, pp. 119–149, 1934.

[32] D. J. Tozer and N. C. Handy, "Improving virtual Kohn-Sham orbitals and eigenvalues: application to excitation energies and

static polarizabilities," *The Journal of Chemical Physics*, vol. 109, no. 23, pp. 10180–10189, 1998.

[33] H. L. Williams, E. M. Mas, K. Szalewicz, and B. Jeziorski, "On the effectiveness of monomer–, dimer–, and bond–centered basis functions in calculations of intermolecular interaction energies," *The Journal of Chemical Physics*, vol. 103, no. 17, pp. 7374–7391, 1995.

[34] R. Bukowski, W. Cencek, P. Jankowski et al., "SAPT2012: an ab initio program for symmetry-adapted perturbation theory calculations of intermolecular interaction energies. Sequential and parallel versions," University of Delaware, Newark, Del, USA; University of Warsaw, Warsaw, Poland, 2013.

[35] C. Angeli, K. L. Bak, V. Bakken et al., "DALTON 2.0, a molecular electronic structure program," 2005, http://www.kjemi.uio.no/software/dalton/dalton.html.

[36] H.-J. Werner, P. J. Knowles, G. Knizia et al., "MOLPRO 2012.1, a package of ab initio programs," University College Cardiff Consultants Limited, Cardiff, UK, 2012, http://www.molpro.net.

[37] D. Cremer, "Møller-Plesset perturbation theory," in *Encyclopedia of Computational Chemistry*, P. V. R. Schleyer, N. L. Allinger, T. Clark et al., Eds., pp. 1706–1735, John Wiley & Sons, Chichester, UK, 1998.

[38] S. Grimme, "Improved second-order Møller-Plesset perturbation theory by separate scaling of parallel- and antiparallel-spin pair correlation energies," *The Journal of Chemical Physics*, vol. 118, no. 20, pp. 9095–9102, 2003.

[39] R. A. Distasio Jr. and M. Head-Gordon, "Optimized spin-component scaled second-order Møller-Plesset perturbation theory for intermolecular interaction energies," *Molecular Physics*, vol. 105, no. 8, pp. 1073–1083, 2007.

[40] A. Heßelmann, "Improved supermolecular second order Møller-Plesset intermolecular interaction energies using time-dependent density functional response theory," *The Journal of Chemical Physics*, vol. 128, no. 14, Article ID 144112, 9 pages, 2008.

[41] M. Pitoňák, P. Neogrády, J. Černý, S. Grimme, and P. Hobza, "Scaled MP3 non-covalent interaction energies agree closely with accurate CCSD(T) benchmark data," *ChemPhysChem*, vol. 10, no. 1, pp. 282–289, 2009.

[42] J. Gauss, "Coupled-cluster theory," in *Encyclopedia of Computational Chemistry*, P. V. R. Schleyer, N. L. Allinger, T. Clark et al., Eds., pp. 615–636, John Wiley & Sons, Chichester, UK, 1998.

[43] S. F. Boys and F. Bernardi, "The calculation of small molecular interactions by the differences of separate total energies. Some procedures with reduced errors," *Molecular Physics*, vol. 19, no. 4, pp. 553–566, 1970.

[44] A. Karton and J. M. L. Martin, "Comment on: 'Estimating the Hartree-Fock limit from finite basis set calculations' [Jensen F (2005) Theoretical Chemistry Accounts 113: 267]," *Theoretical Chemistry Accounts*, vol. 115, no. 4, pp. 330–333, 2005.

[45] A. Halkier, T. Helgaker, P. Jørgensen et al., "Basis-set convergence in correlated calculations on Ne, N_2, and H_2O," *Chemical Physics Letters*, vol. 286, no. 3-4, pp. 243–252, 1998.

[46] D. G. Truhlar, "Basis-set extrapolation," *Chemical Physics Letters*, vol. 294, no. 1–3, pp. 45–48, 1998.

[47] F. Neese and E. F. Valeev, "Revisiting the atomic natural orbital approach for basis sets: robust systematic basis sets for explicitly correlated and conventional correlated ab initio methods?" *Journal of Chemical Theory and Computation*, vol. 7, no. 1, pp. 33–43, 2011.

[48] K. E. Riley, M. Pitoňák, P. Jurečka, and P. Hobza, "Stabilization and structure calculations for noncovalent interactions in extended molecular systems based on wave function and density functional theories," *Chemical Reviews*, vol. 110, no. 9, pp. 5023–5063, 2010.

[49] J. Řezáč, K. E. Riley, and P. Hobza, "Benchmark calculations of noncovalent interactions of halogenated molecules," *Journal of Chemical Theory and Computation*, vol. 8, no. 11, pp. 4285–4292, 2012.

[50] S. Alvarez, "A cartography of the van der Waals territories," *Dalton Transactions*, vol. 42, no. 24, pp. 8617–8636, 2013.

[51] T. Clark, M. Hennemann, J. S. Murray, and P. Politzer, "Halogen bonding: the σ-hole," *Journal of Molecular Modeling*, vol. 13, no. 2, pp. 291–296, 2007.

[52] P. Politzer, J. S. Murray, and M. C. Concha, "Halogen bonding and the design of new materials: organic bromides, chlorides and perhaps even fluorides as donors," *Journal of Molecular Modeling*, vol. 13, no. 6-7, pp. 643–650, 2007.

[53] A. Bundhun, P. Ramasami, J. S. Murray, and P. Politzer, "Trends in σ-hole strengths and interactions of F_3MX molecules (M = C, Si, Ge and X = F, Cl, Br, I)," *Journal of Molecular Modeling*, vol. 19, no. 7, pp. 2739–2746, 2013.

[54] F. De Proft, W. Langenaeker, and P. Geerlings, "Ab initio determination of substituent constants in a density functional theory formalism: calculation of intrinsic group electronegativity, hardness, and softness," *The Journal of Physical Chemistry*, vol. 97, no. 9, pp. 1826–1831, 1993.

[55] P. Metrangolo, H. Neukirch, T. Pilati, and G. Resnati, "Halogen bonding based recognition processes: a world parallel to hydrogen bonding," *Accounts of Chemical Research*, vol. 38, no. 5, pp. 386–395, 2005.

[56] J.-W. Zou, Y.-J. Jiang, M. Guo et al., "Ab initio study of the complexes of halogen-containing molecules RX (X=Cl, Br, and I) and NH_3: towards understanding the nature of halogen bonding and the electron-accepting propensities of covalently bonded halogen atoms," *Chemistry—A European Journal*, vol. 11, no. 2, pp. 740–751, 2005.

[57] K. E. Riley and P. Hobza, "Investigations into the nature of halogen bonding including symmetry adapted perturbation theory analyses," *Journal of Chemical Theory and Computation*, vol. 4, no. 2, pp. 232–242, 2008.

[58] K. E. Riley and P. Hobza, "The relative roles of electrostatics and dispersion in the stabilization of halogen bonds," *Physical Chemistry Chemical Physics*, vol. 15, no. 41, pp. 17742–17751, 2013.

[59] K. Patkowski, K. Szalewicz, and B. Jeziorski, "Third-order interactions in symmetry-adapted perturbation theory," *The Journal of Chemical Physics*, vol. 125, no. 15, Article ID 154107, 20 pages, 2006.

20

Influence of Al_2O_3, CaO/SiO_2, and B_2O_3 on Viscous Behavior of High Alumina and Medium Titania Blast Furnace Slag

Lingtao Bian[1] and Yanhong Gao[2]

[1]*Chongqing College of Electronic Engineering, Chongqing 401331, China*
[2]*School of Metallurgy and Materials Engineering, Chongqing University of Science and Technology, Chongqing 401331, China*

Correspondence should be addressed to Yanhong Gao; gyh3636@hotmail.com

Academic Editor: Sedat Yurdakal

The effect of Al_2O_3, CaO/SiO_2, and B_2O_3 on the viscosity of high alumina and medium titania blast furnace slag was analyzed. An increase in CaO/SiO_2 ratio from 1.14 to 1.44 resulted in higher slag viscosity and break point temperature. They also increased with increasing Al_2O_3 content but decreased with adding B_2O_3 and Al_2O_3 simultaneously at a fixed CaO/SiO_2 ratio of 1.14, which suggested that the effect of B_2O_3 on viscosity and break point temperature is predominant compared to Al_2O_3. Apparent activation energies of $CaO–SiO_2–MgO–Al_2O_3–TiO_2–B_2O_3$ slag were found to be between 74 and 169 kJ/mol.

1. Introduction

As high-quality iron ore resources in the world decrease, vanadium–titanium magnetite ores from china or iron ores with high Al_2O_3 from Australia and India have become an alternative choice in the blast furnace (BF) smelting process [1–3]. As a result, more alumina (exceeding 15% in many enterprises) or titania (more than 10%) occur in the BF slag, resulting in higher viscosity, worse fluidity, and poor operational stability [4–10]. So it is essential to studying the viscous behavior of BF slag with more alumina (above 15%) or titania (above 10%) for optimizing the iron-making operation. It has been found that the little amount of B_2O_3 remarkably improved the properties of slag [11–15]. Influence of B_2O_3 on high titanium (above 20%) BF slag [11–13, 16], medium titanium (10%–20%) BF slag [14], and high alumina (above 15%) BF slag [17, 18] has been studied in previous reports. However, it is hard to find investigations on the viscous behavior of high alumina (above 15%) and medium titania (15%–20%) BF slag (HAMT BF slag) and the influence of B_2O_3, Al_2O_3, and basicity on viscous behavior. This research work appears just in this background.

In this study, the viscous behavior of the $CaO–SiO_2–(13\%–19\%)$ $Al_2O_3–MgO–(17\%–20\%)$ TiO_2 slags was measured to clarify the effect of Al_2O_3 and C/S. In addition, B_2O_3 was added to the slag in order to improve its fluidity. A series of slags containing different Al_2O_3, B_2O_3 content, and C/S were designed and the viscosity, break point temperature (the critical temperature at which the measured viscosity changes abruptly during the cooling cycle), and apparent activation energy were measured and analyzed.

2. Experimental

All samples were prepared by adding analytical-grade reagents CaO, SiO_2, Al_2O_3, and B_2O_3 to basic slag obtained from BF and analyzed by chemical processing method. Three slag series were designed with binary basicity (C/S) range of 1.14–1.44, Al_2O_3 content of 13%–19%, and B_2O_3 content of 1%–4%. The chemical compositions of the designed slag are listed in Table 1. Slag samples were put inside a molybdenum crucible and melted in a high temperature furnace.

The slag viscosity was measured by rotating cylinder method and recorded during the cooling cycle and the experiment was not ended until a steep increase in the viscosity value. The experimental setup and experimental procedure in detail can be found in our earlier studies [19, 20].

TABLE 1: Chemical compositions of slag samples (wt.%).

Number	CaO	SiO$_2$	MgO	Al$_2$O$_3$	TiO$_2$	FeO	V$_2$O$_5$	MnO	B$_2$O$_3$	C/S
Basic slag	28.33	24.85	8.64	14.24	19.48	1.11	0.20	0.65	—	1.14
R1	29.76	24.04	9.27	13.77	18.84	1.07	0.19	0.63	—	1.24
R2	31.46	23.45	9.06	13.44	18.38	1.05	0.19	0.62	—	1.34
R3	33.02	22.91	8.87	13.13	17.95	1.02	0.18	0.60	—	1.44
A1	27.45	24.08	9.28	15.98	18.87	1.08	0.19	0.63	—	1.14
A2	27.12	23.79	9.18	17.00	18.64	1.06	0.19	0.63	—	1.14
A3	26.78	23.50	9.08	18.01	18.41	1.05	0.19	0.62	—	1.14
B1	27.12	23.79	9.18	16.00	18.64	1.06	0.19	0.63	1.00	1.14
B2	26.45	23.20	8.98	17.02	18.18	1.04	0.19	0.61	2.00	1.14
B3	25.80	22.64	8.78	17.97	17.74	1.01	0.18	0.60	3.00	1.14
B4	25.13	22.05	8.58	19.00	17.28	0.98	0.18	0.58	4.00	1.14

3. Results and Discussion

3.1. Effect of CaO/SiO$_2$ on Viscous Behavior. The effect of C/S on viscosity at different temperatures is shown in Figure 1. As clearly observed, the viscosity strongly depends on the temperature for each sample. The viscosities of all slag samples are under 0.5 Pa·s at above 1460°C. As temperature reduces to 1440°C, the viscosity of sample R3 (C/S = 1.44) goes up to 0.92 Pa·s and the viscosities of sample R2 (C/S = 1.34) and sample R3 both exceed 1.0 Pa·s with temperature continuous decreasing to 1420°C. When temperature further drops to 1400°C, the viscosity of sample R1 (C/S = 1.24) increases sharply to 3.38 Pa·s and that of sample R2 or R3 is even higher. For basic slag sample (C/S = 1.14), its viscosity does not reach 0.91 Pa·s until 1360°C because of low basicity. It illustrates that these slags are short slags. When the temperature is lower than break point temperature, the viscosity increases sharply in a narrow temperature range. Meanwhile, the fluidity and the stability of slags become worse. Besides, with an increase of the CaO/SiO$_2$, the change of viscosity is much sharper, and the viscosity becomes more sensitive to the temperature change, resulting from the precipitation of phases with the high melting point [21, 22].

Figure 2 indicates that break point temperature raises with increasing C/S and shows relatively gentle change at the stage where C/S varies from 1.24 to 1.34. As we all know, although CaO can modify the melt structure effectively by providing additional free oxygen ions (O^{2-}) as a typical basic oxide, more addition of CaO exerts a negative effect on the viscosity and the break point temperature because of its high melting temperature.

3.2. Effect of Al$_2$O$_3$ on Viscous Behavior. Figure 3 shows the effect of Al$_2$O$_3$ on the viscosity of CaO–SiO$_2$–MgO–Al$_2$O$_3$–(18%-19%) TiO$_2$ slag at various temperature and a fixed C/S of 1.14. It is noted that the viscosity increases with increasing Al$_2$O$_3$ content from 14.24% to 18.01%. When temperature is 1360°C, the viscosity goes up rapidly, especially as Al$_2$O$_3$ content is more than 16%. Viscosity of sample A1 is more than 1.5 Pa·s and the value of sample A3 goes up to

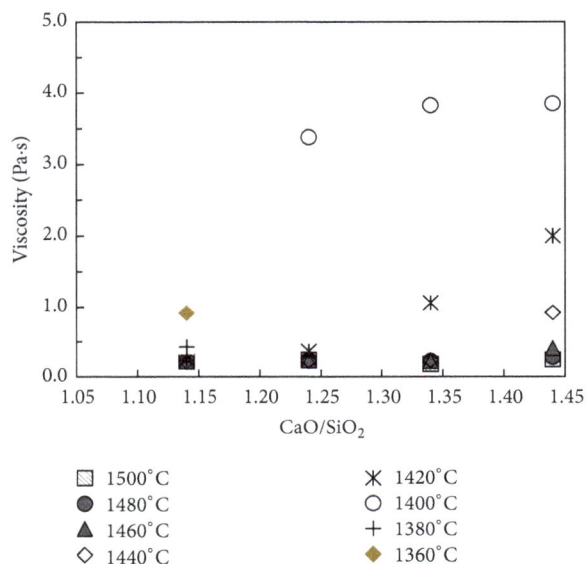

FIGURE 1: Viscosity isotherms for slag series R and basic slag.

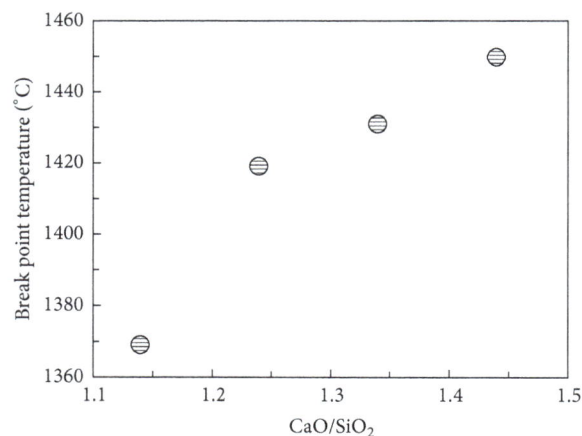

FIGURE 2: Break point temperature for slag series R and basic slag.

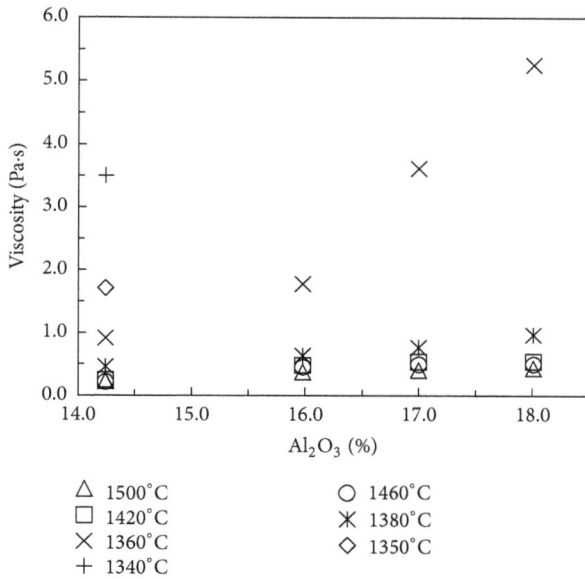

FIGURE 3: Viscosity isotherms for slag series A and basic slag.

Legend for Figure 3:
△ 1500°C   ○ 1460°C
□ 1420°C   ✳ 1380°C
✕ 1360°C   ◇ 1350°C
+ 1340°C

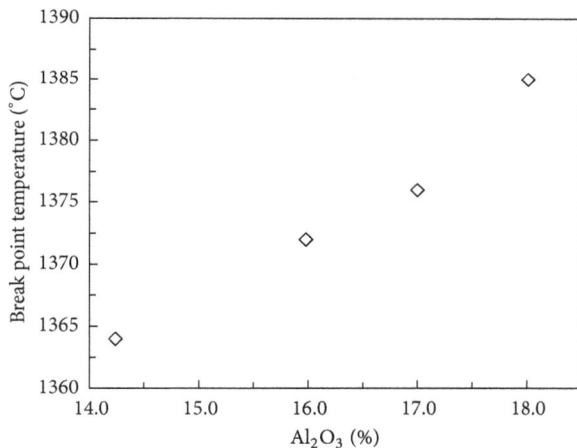

FIGURE 4: Break point temperature for slag series A and basic slag.

5.5 Pa·s. When temperature decreases to 1340°C, viscosity of basic slag reaches up to 3.5 Pa·s. It is inferred that these slags also take on the characteristics of short slag and more Al_2O_3 responds to larger viscosity. With an increase of the Al_2O_3 content, viscosity varies more quickly in a narrow temperature range and the fluidity deteriorates rapidly, attributing to the precipitation of phases with the high melting point [21, 22]. The break point temperature shows a similar tendency, as marked in Figure 4. When Al_2O_3 content in slag exceeds 17%, it dramatically increases. It is reported that Al_2O_3 behaves as an amphoteric oxide and initially Al^{3+} cations can replace Si^{4+} to form $[AlO_4]^{5-}$ tetrahedral units [23–25]. However, after further addition of Al_2O_3, it behaves as a network modifier and exists in the $[AlO_6]^{9-}$ octahedral configuration [3]. Thus complex structure corresponds to higher break point temperature and larger viscosity. In this investigation,

Al_2O_3 behaves as a network former and increases the slag viscosity, agreeing with previous study [3].

3.3. Combined Effect of B_2O_3 and Al_2O_3 on Viscous Behavior. The viscosity-temperature curves of the slag samples are shown in Figure 5. Viscosity is closely related to temperature for each sample and B_2O_3 addition decreases the viscosity and improves the fluidity of slags regardless of Al_2O_3 content in slag. The viscosity of CaO–13% Al_2O_3–SiO_2–MgO–18% TiO_2–B_2O_3 slags decreases with increasing B_2O_3 addition in our previous investigation [14]. Similar results are obtained by Sun et al. [11], Fox et al. [26], and Wang et al. [27] despite different compositions. Besides, the break point temperature of slag was decreased by the addition of B_2O_3, as shown in Figure 6. The more B_2O_3 in slag, the faster break point temperature drops.

According to some studies on boron bearing multicomponent slag [28, 29], B_2O_3 exists mainly in the form of $[BO_3]^-$ triangle and $[BO_4]^-$ tetrahedral units; B_2O_3 additions could decrease the $[AlO_4]^-$ tetrahedral structural units and transformed the 3D network structures such as pentaborate and tetraborate into 2D network structures of boroxol and boroxyl rings by breaking the bridged oxygen atoms (O^0) to produce nonbridged oxygen atoms (O^-) leading to a decrease in the slag viscosity.

The isoviscosity curves of boron containing HAMT BF slag are constructed in Figure 7 and its iso-break point temperature curves are plotted in Figure 8. As can be seen, viscosities and break point temperatures decrease gradually with the addition of B_2O_3 and Al_2O_3, which suggests that the B_2O_3 effect is predominant compared to the Al_2O_3 additions. The isoviscosity curves become closer and closer as the B_2O_3 content decreases and the Al_2O_3 content increases, which indicates that the thermal stability of slag starts to deteriorate. However, increase of Al_2O_3 must be accompanied by decrease of B_2O_3 to maintain constant viscosity (in Figure 7) in several domains where B_2O_3 is greater than 3.5 or Al_2O_3 is more than 18.5%. Furthermore, these domains where increase of Al_2O_3 should be accompanied by decrease of B_2O_3 for constant break point temperature are more widespread in Figure 8. Significantly, in present study, break point temperatures are all under 1350°C after adding B_2O_3, which can meet the requirement of BF operation.

3.4. Effect of B_2O_3 on the Apparent Activation Energy for Viscous Flow. Temperature dependence of viscosity (η) is given by the Arrhenius equation (see (1)), from which the apparent activation energy can be derived.

$$\ln \eta = \ln A + \frac{Ea}{RT}, \quad (1)$$

where η is viscosity of the slag; A is constant; Ea is the apparent activation energy; R is molar gas constant; T is absolute temperature. Variations in Ea can reveal changes in the frictional resistance of viscous flow and suggest a change in the structure of the molten slag. The variation of apparent activation energy is constructed in Figure 9 based on (1). It decreases with an increase of B_2O_3 content

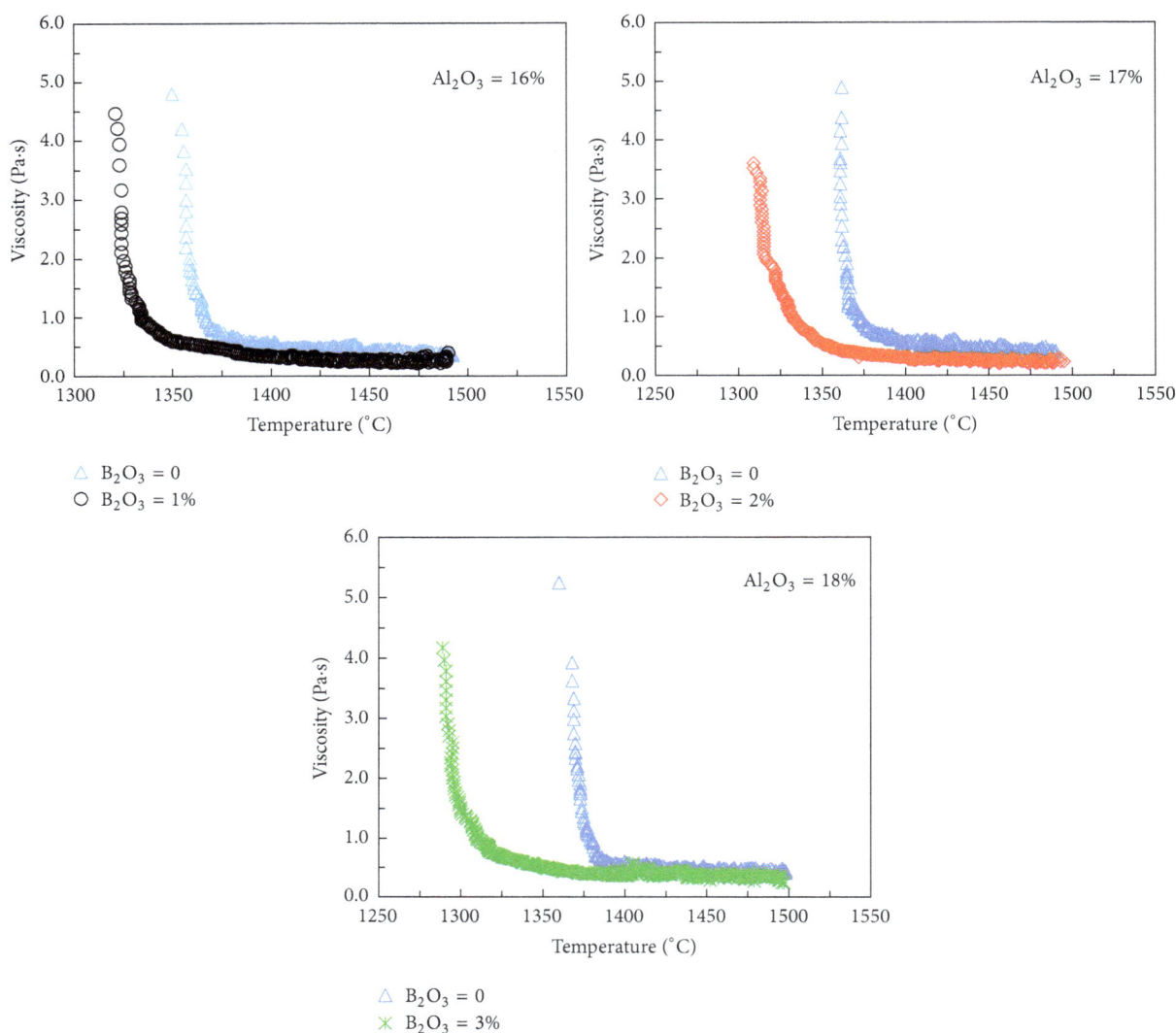

FIGURE 5: Viscosity changes of the slag with temperature at different B_2O_3 content.

and Al_2O_3 content, which indicates that the resistance for viscous flow becomes smaller. So the slag structure becomes simpler and less complex. This is attributed to a weakening of the bond energy by increasing of B–O bonds in the network structure of BO_4 and BO_3 units [30]. In other words, although increasing Al_2O_3 can make the network of slag melts complex, it was modified by adding B_2O_3.

4. Conclusions

In the present study, viscous behavior of HAMT BF slag was analyzed when Al_2O_3 content and B_2O_3 content in slag varied. The important results were summarized as follows:

(1) The thermal stability of the slag is better at higher temperature. Viscosity begins to deteriorate rapidly when temperature is below 1420°C and C/S varies from 1.14 to 1.44. For HAMT BF slag with fixed C/S of 1.14, the temperature is 1360°C.

(2) Break point temperature increases faster with Al_2O_3 content in high aluminum and medium titanium slag; Al_2O_3 content especially exceeds 17%. Therefore, Al_2O_3 content in slag should be under 17% during iron-making process.

(3) B_2O_3 addition can improve fluidity of HAMT BF slag and decline significantly break point temperature regardless of Al_2O_3 content.

(4) Apparent activation energy decreases with an increase of B_2O_3 content and Al_2O_3 content.

Conflicts of Interest

The authors declare that there are no conflicts of interest regarding the publication of this paper.

Authors' Contributions

The authors contributed equally to this work.

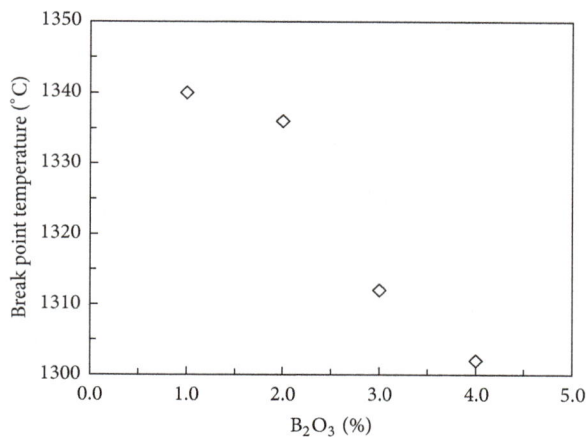

FIGURE 6: Break point temperature of the slag with B_2O_3 content.

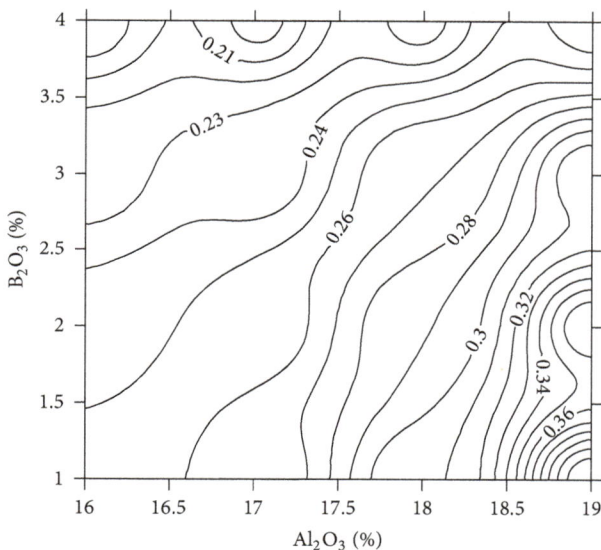

FIGURE 7: Effects of Al_2O_3 and B_2O_3 content on viscosity.

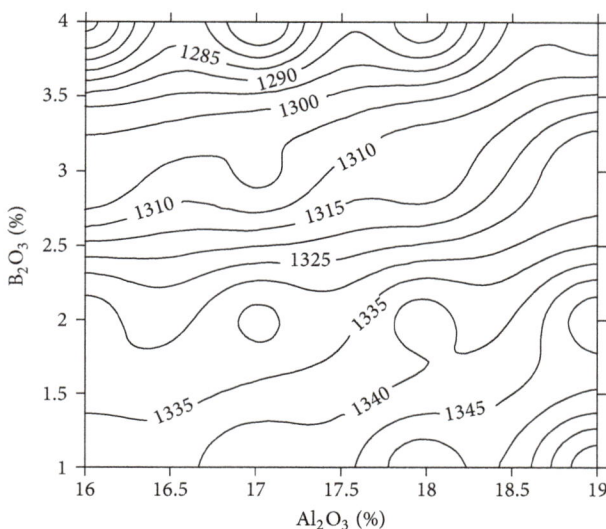

FIGURE 8: Effects of Al_2O_3 and B_2O_3 content on break point temperature.

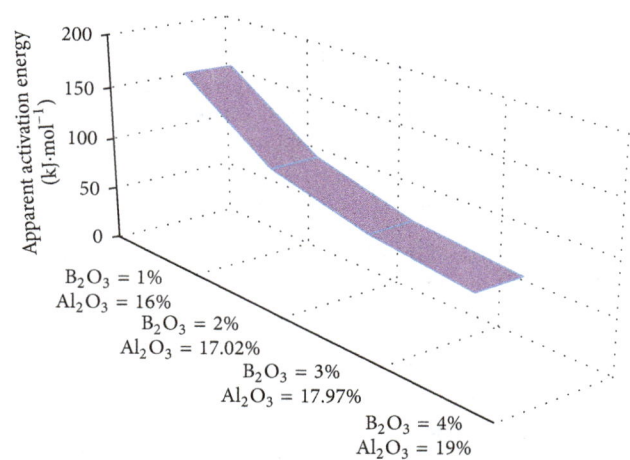

FIGURE 9: Relationship of apparent activation energy and B_2O_3 and Al_2O_3 content in slag.

Acknowledgments

This research was funded by the National Natural Science Foundation of China (no. 51374267), Chongqing Research Program of Basic Research and Frontier Technology (no. cstc2017jcyjAX0236 and no. cstc2016jcyjA0142), and the Scientific and Technological Research Program of Chongqing Municipal Education Commission (no. KJ1713326).

References

[1] S. Y. Chen and M. S. Chu, "Metalizing reduction and magnetic separation of vanadium titano-magnetite based on hot briquetting," *International Journal of Minerals, Metallurgy, and Materials*, vol. 255, 2014.

[2] X. H. Liu, G. S. Gai, Y. F. Yang, Z. T. Sui, L. Li, and J. X. Fu, "Kinetics of the leaching of TiO_2 from Ti-bearing blast furnace slag," *Journal of China University of Mining and Technology*, vol. 18, no. 2, pp. 275–278, 2008.

[3] Z. Yan, X. Lv, J. Zhang, Y. Qin, and C. Bai, "Influence of MgO, Al_2O_3 and CaO/SiO_2 on the viscosity of blast furnace type slag with high Al_2O_3 and 5 wt-% TiO_2," *Canadian Metallurgical Quarterly*, vol. 55, article 186, 2016.

[4] H. Kim, H. Matsuura, F. Tsukihashi, W. L. Wang, D. J. Min, and I. Sohn, "Effect of Al_2O_3 and CaO/SiO_2 on the viscosity of calcium-silicate-based slags containing 10 mass Pct MgO," *Metallurgical and Materials Transactions B*, vol. 44, no. 1, pp. 5–12, 2013.

[5] C. G. Bai, Z. M. Yan, S. P. Li et al., Thermo-physical-chemical properties of blast furnace slag bearing high TiO_2, Proc. of the 10th Int. Conf. on Molten Slags, Fluxes and Salts, Seattle, Washington State, USA, 405, 2016.

[6] L. Yao, S. Ren, X. Wang et al., "Effect of Al_2O_3, MgO, and CaO/SiO_2 on viscosity of high alumina blast furnace slag," *Steel Research International*, vol. 87, no. 2, pp. 241–249, 2016.

[7] K. Sunahara, K. Nakano, M. Hoshi, T. Inada, S. Komatsu, and T. Yamamoto, "Effect of high Al_2O_3 slag on the blast furnace operations," *ISIJ International*, vol. 48, no. 4, pp. 420–429, 2008.

[8] L. Forsbacka, L. Holappa, T. Iida, Y. Kita, and Y. Toda, "Experimental study of viscosities of selected CaO-MgO-Al_2O_3-SiO_2

slags and application of the Iida model," *Scandinavian Journal of Metallurgy*, vol. 32, no. 5, pp. 273–280, 2003.

[9] J. R. Kim, Y. S. Lee, D. J. Min, S. M. Jung, and S. H. Yi, "Influence of MgO and Al_2O_3 contents on viscosity of blast furnace type slags containing FeO," *ISIJ International*, vol. 44, no. 8, pp. 1291–1297, 2004.

[10] Z. Wang, Y. Sun, S. Sridhar, M. Zhang, M. Guo, and Z. Zhang, "Effect of Al_2O_3 on the viscosity and structure of $CaO-SiO_2$-$MgO-Al_2O_3$-FetO slags," *Metallurgical and Materials Transactions B*, vol. 46, no. 2, pp. 537–541, 2015.

[11] Y. Sun, J. Liao, K. Zheng, X. Wang, and Z. Zhang, "Effect of B_2O_3 on the structure and viscous behavior of Ti-bearing blast furnace slags," *JOM: The Journal of The Minerals, Metals & Materials Society (TMS)*, vol. 66, no. 10, pp. 2168–2175, 2014.

[12] S. Ren, J. L. Zhang, L. S. Wu et al., "Influence of B_2O_3 on viscosity of high Ti-bearing blast furnace slag," *ISIJ International*, vol. 52, no. 6, pp. 984–991, 2012.

[13] Y. H. Gao, L. T. Bian, and Z. Y. Liang, "Influence of B_2O_3 and TiO_2 on viscosity of titanium-bearing blast furnace slag," *Steel Research International*, vol. 86, no. 4, pp. 386–390, 2015.

[14] L. T. Bian and Y. H. Gao, "Influence of B_2O_3 and basicity on viscosity and structure of medium titanium bearing blast furnace slag," *Journal of Chemistry*, vol. 1, 2016.

[15] X. Yu, G. H. Wen, P. Tang, and H. Wang, "Effect of B_2O_3 on the physico-chemical properties of mold slag used for high-Al steel," *Journal of Chongqing University*, vol. 34, no. 1, pp. 66–71, 2011.

[16] Y. H. Lin, Y. C. Wen, W. G. Fu, and J. T. Rao, "Effect and mechanism of B_2O_3 on apparent viscosity of slag during smelting of vanadium-titanium magnetite in the blast furnace," *Metallurgical Research & Technology*, vol. 113, article 506, 2016.

[17] R. Z. Xu, J. L. Zhang, Z. Y. Wang, and K. X. Jiao, "Influence of Cr_2O_3 and B_2O_3 on viscosity and structure of high alumina slag," *Steel Research International*, vol. 88, no. 4, Article ID 1600241, pp. 1–4, 2017.

[18] Q. Chenglin, Z. Jianliang, S. Jiugang, L. Weijia, Z. Zhixing, and Z. Xuesong, "Study of boronizing mechanism of high-alumina slag," *Steel Research International*, vol. 82, no. 11, pp. 1319–1324, 2011.

[19] Y. H. Gao, Z. Y. Liang, Q. C. Liu, and L. T. Bian, "Effect of TiO_2 on the slag properties for $CaO-SiO_2$-$MgO-Al_2O_3$-TiO_2 system," *Asian Journal of Chemistry*, vol. 24, no. 11, pp. 5337–5340, 2012.

[20] Y. H. Gao, Z. Y. Liang, and L. T. Bian, "Influence of TiO_2 and comprehensive alkalinity on the viscous characteristics of blast furnace type slag," *Applied Mechanics and Materials*, vol. 291–294, pp. 2617–2620, 2013.

[21] Y. Zhang, J. Tang, M.-S. Chu, Y. Liu, S.-Y. Chen, and X.-X. Xue, "Optimization of BF slag for high Cr_2O_3 vanadium-titanium magnetite," *Journal of Iron and Steel Research, International*, vol. 21, no. 2, pp. 144–150, 2014.

[22] C. Feng, M. Chu, J. Tang, Y. Tang, and Z. Liu, "Effect of CaO/SiO_2 and Al_2O_3 on viscous behaviors of the titanium-bearing blast furnace slag," *Steel Research International*, vol. 87, no. 10, pp. 1274–1283, 2016.

[23] K. C. Mills, "The influence of structure on the physico-chemical properties of slags," *ISIJ International*, vol. 33, no. 1, pp. 148–155, 1993.

[24] J. H. Park, D. J. Min, and H. S. Song, "Amphoteric behavior of alumina in viscous flow and structure of $CaO-SiO_2$ (-MgO)-Al_2O_3 slags," *Metallurgical and Materials Transactions B*, vol. 35, no. 2, pp. 269–275, 2004.

[25] B. Mysen and P. Richet, *Silicate Glasses and Melts: Properties and Structure*, vol. 10 of *Developments in Geochemistry*, Elsevier, Washington, DC, USA, 2005.

[26] A. B. Fox, K. C. Mills, D. Lever et al., "Development of fluoride-free fluxes for billet casting," *ISIJ International*, vol. 45, no. 7, pp. 1051–1058, 2005.

[27] Z. Wang, Q. Shu, and K. Chou, "Viscosity of fluoride-free mold fluxes containing B_2O_3 and TiO_2," *Steel Research International*, vol. 84, no. 8, pp. 766–776, 2013.

[28] G. H. Kim and I. Sohn, "Role of B_2O_3 on the viscosity and structure in the $CaO-Al_2O_3$-Na_2O-based system," *Metallurgical and Materials Transactions B*, vol. 45, no. 1, pp. 86–95, 2014.

[29] X. H. Huang, J. L. Liao, K. Zheng, H. H. Hu, F. M. Wang, and Z. T. Zhang, "Effect of B_2O_3 addition on viscosity of mould slag containing low silica content," *Ironmaking and Steelmaking*, vol. 41, no. 1, pp. 67–74, 2014.

[30] Y. Kim and K. Morita, "Relationship between molten oxide structure and thermal conductivity in the $CaO-SiO_2$-B_2O_3 system," *ISIJ International*, vol. 54, no. 9, pp. 2077–2083, 2014.

Permissions

All chapters in this book were first published in JC, by Hindawi Publishing Corporation; hereby published with permission under the Creative Commons Attribution License or equivalent. Every chapter published in this book has been scrutinized by our experts. Their significance has been extensively debated. The topics covered herein carry significant findings which will fuel the growth of the discipline. They may even be implemented as practical applications or may be referred to as a beginning point for another development.

The contributors of this book come from diverse backgrounds, making this book a truly international effort. This book will bring forth new frontiers with its revolutionizing research information and detailed analysis of the nascent developments around the world.

We would like to thank all the contributing authors for lending their expertise to make the book truly unique. They have played a crucial role in the development of this book. Without their invaluable contributions this book wouldn't have been possible. They have made vital efforts to compile up to date information on the varied aspects of this subject to make this book a valuable addition to the collection of many professionals and students.

This book was conceptualized with the vision of imparting up-to-date information and advanced data in this field. To ensure the same, a matchless editorial board was set up. Every individual on the board went through rigorous rounds of assessment to prove their worth. After which they invested a large part of their time researching and compiling the most relevant data for our readers.

The editorial board has been involved in producing this book since its inception. They have spent rigorous hours researching and exploring the diverse topics which have resulted in the successful publishing of this book. They have passed on their knowledge of decades through this book. To expedite this challenging task, the publisher supported the team at every step. A small team of assistant editors was also appointed to further simplify the editing procedure and attain best results for the readers.

Apart from the editorial board, the designing team has also invested a significant amount of their time in understanding the subject and creating the most relevant covers. They scrutinized every image to scout for the most suitable representation of the subject and create an appropriate cover for the book.

The publishing team has been an ardent support to the editorial, designing and production team. Their endless efforts to recruit the best for this project, has resulted in the accomplishment of this book. They are a veteran in the field of academics and their pool of knowledge is as vast as their experience in printing. Their expertise and guidance has proved useful at every step. Their uncompromising quality standards have made this book an exceptional effort. Their encouragement from time to time has been an inspiration for everyone.

The publisher and the editorial board hope that this book will prove to be a valuable piece of knowledge for researchers, students, practitioners and scholars across the globe.

List of Contributors

Ji An Joung, Mi Na Park, Ji Young You, Bong Joon Song and Joon Ho Choi
Department of Food Science and Biotechnology, Wonkwang University, Iksan 54538, Republic of Korea

Jun Wang, Huijuan Yang, Hongzhi Shi, Mengyue Zhang and Tong Jin
Henan Agricultural University, National Tobacco Cultivation & Physiology & Biochemistry Research Center, Zhengzhou 450002, China

Jun Zhou and Ruoshi Bai
Beijing Cigarette Factory, Shanghai Tobacco Group Co. Ltd., Beijing 100024, China

Leili Mohammadi, Edris Bazrafshan, Alireza Ansari-Moghaddam and Davoud Balarak
Health Promotion Research Center, Zahedan University of Medical Sciences, Zahedan, Iran

Meissam Noroozifar
Analytical Research Laboratory, Department of Chemistry, University of Sistan and Baluchestan, Zahedan, Iran

Farahnaz Barahuie
Engineering Faculty, Sistan and Baluchestan University, Zahedan, Iran

Xiaohu Li, Jianqiang Wang and Hao Wang
Key Laboratory of Submarine Geosciences, Second Institute of Oceanography, State Oceanic Administration, Hangzhou 310012, China

J. Castañeda-Díaz, T. Pavón-Silva, E. Gutiérrez-Segura and A. Colín-Cruz
Facultad de Química, Universidad Autònoma del Estado de México, Paseo Colòn y Tollocan s/n, 50000 Toluca, MEX, Mexico

Ju Lin, Yinian Zhu, Huili Liu and Zhangnan Jiang
College of Environmental Science and Engineering, Guilin University of Technology, Guilin 541004, China

Lihao Zhang and Zongqiang Zhu
Guangxi Key Laboratory of Environmental Pollution Control Theory and Technology, Guilin University of Technology, Guilin 541004, China

K. M. Sachin, Sameer A. Karpe and Man Singh
School of Chemical Sciences, Central University of Gujarat, Gandhinagar, India

Ajaya Bhattarai
School of Chemical Sciences, Central University of Gujarat, Gandhinagar, India
Department of Chemistry, MMAMC, Tribhuvan University, Biratnagar 56613, Nepal

Shu-Ling Huang, Hsin-Fu Yu and Yung-Sheng Lin
Department of Chemical Engineering, National United University, Miaoli 36003, Taiwan

Wenguang Du, Song Yang, Feng Pan, Ju Shangguan and Huiling Fan
Key Laboratory for Coal Science and Technology of Ministry of Education and Shanxi Province, Institute for Chemical Engineering of Coal, Taiyuan University of Technology, Taiyuan 030024, China

Jie Lu and Shoujun Liu
College of Chemistry and Chemical Engineering, Taiyuan University of Technology, Taiyuan 030024, China

Phan Ha Nu Diem
Dong Nai University, 4 Le Quy Don, Bien Hoa City, Vietnam
College of Sciences, Hue University, 77 Nguyen Hue, Hue City, Vietnam

Tran Thai Hoa
College of Sciences, Hue University, 77 Nguyen Hue, Hue City, Vietnam

Doan Thi Thu Thao and Hoang Thi Dong Quy
University of Science, Vietnam National University in Ho Chi Minh City, 227 Nguyen Van Cu, Ho Chi Minh City, Vietnam

Dang Van Phu, Nguyen Ngoc Duy and Nguyen Quoc Hien
Research and Development Center for Radiation Technology, Vietnam Atomic Energy Institute, 202A Street 11, Linh XuanWard, Thu Duc District, Ho Chi Minh City, Vietnam

Maja Marasović and Mladen Miloš
Faculty of Chemistry and Technology, University of Split, Ruđera Boškovića 35, 21000 Split, Croatia

Tea Marasović
Faculty of Electrical Engineering, Mechanical Engineering and Naval Architecture, University of Split, Ruđera Boškovića 32, 21000 Split, Croatia

Liliana Giraldo
Departamento de Química, Facultad de Ciencias, Universidad Nacional de Colombia, Sede Bogotà, Colombia

Juan Carlos Moreno-Piraján
Departamento de Química, Facultad de Ciencias, Universidad de Los Andes, Grupo de Investigaciòn en Sòlidos Porosos y Calorimetría, Bogotà, Colombia

Ali Murat Soydan
Institute of Energy Technologies, Gebze Technical University, Cayirova, Gebze, 41400 Kocaeli, Turkey

Recep Akdeniz
Department of Materials Science and Engineering, Gebze Technical University, Cayirova, Gebze, 41400 Kocaeli, Turkey

Jian Shi, Jiayin Hu, Long Li, Xiaoping Yu and Tianlong Deng
Tianjin Key Laboratory of Marine Resources and Chemistry, College of Chemical Engineering and Materials Science, Tianjin University of Science and Technology, Tianjin 300457, China

Yafei Guo
Tianjin Key Laboratory of Marine Resources and Chemistry, College of Chemical Engineering and Materials Science, Tianjin University of Science and Technology, Tianjin 300457, China
College of Chemistry and Materials Science, Northwest University, Xi'an 710127, China

U. Wanderlingh, C. Branca, C. Crupi, S. Rifici and G. D'Angelo
Dipartimento di Scienze Matematiche e Informatiche, Scienze Fisiche e Scienze della Terra, Università di Messina, Via F. Stagno d'Alcontres 31, 98166Messina, Italy

V. Conti Nibali
Institute for Physical Chemistry II, Ruhr-University Bochum, Bochum, Germany

G. La Rosa
Physik-Department, Technische Universit`at München, Munich, Germany

J. Ollivier
Institut Laue-Langevin, 6 rue J. Horowitz, BP 156, 38042 Grenoble, France

Jelena D. Lović
ICTM, Institute of Electrochemistry, University of Belgrade, Njegoševa 12, Belgrade, Serbia

Mingxu Yi, Jun Huang and Lifeng Wang
School of Aeronautic Science and Engineering, Beihang University, Beijing 100191, China

Shan-Qisong Huang and Xue-Hai Ju
Key Laboratory of Soft Chemistry and Functional Materials of MOE, School of Chemical Engineering, Nanjing University of Science and Technology, Nanjing 210094, China

Piotr Matczak
Department of Theoretical and Structural Chemistry, Faculty of Chemistry, University of Łódź, Pomorska 163/165, 90-236 Lodz, Poland

Lingtao Bian
Chongqing College of Electronic Engineering, Chongqing 401331, China

Yanhong Gao
School of Metallurgy and Materials Engineering, Chongqing University of Science and Technology, Chongqing 401331, China

Index

www.ingramcontent.com/pod-product-compliance
Lightning Source LLC
Chambersburg PA
CBHW050456200326
41458CB00014B/5200